Do not desensitize computer disk

44 0485473 X

University of Hertfordshire

Learning and Information Services

Hatfield Campus Learning Resources Centre
College Lane Hatfield Herts AL10 9AB
Renewals: Tel 01707 284673 Mon-Fri 12 noon-8pm only

This book is in heavy demand and is due back strictly by the last date stamped below. A fine will be charged for the late return of items.

ONE WEEK LOAN

WITHDRAWN

TOPICS IN DIGITAL SIGNAL PROCESSING

C. S. Burrus and T. W. Parks: *DFT/FFT AND CONVOLUTION ALGORITHMS: THEORY AND IMPLEMENTATION*

John R. Treichler, C. Richard Johnson, Jr., and Michael G. Larimore: *THEORY AND DESIGN OF ADAPTIVE FILTERS*

T. W. Parks and C. S. Burrus: *DIGITAL FILTER DESIGN*

Rulph Chassaing and Darrell W. Horning: *DIGITAL SIGNAL PROCESSING WITH THE TMS320C25*

Rulph Chassaing: *DIGITAL SIGNAL PROCESSING WITH C AND THE TMS320C30*

Digital Signal Processing with C and the TMS320C30

Rulph Chassaing
School of Engineering
Roger Williams University

A WILEY-INTERSCIENCE PUBLICATION
JOHN WILEY & SONS, INC.
New York   Chichester   Brisbane   Toronto   Singapore

A NOTE TO THE READER:
This book has been electronically reproduced from digital information stored at John Wiley & Sons, Inc. We are pleased that the use of this new technology will enable us to keep works of enduring scholarly value in print as long as there is a reasonable demand for them. The content of this book is identical to previous printings.

The publisher assumes no responsibility for errors, omissions, or damages, including without limitation damages caused by the use of the programs or from the use of the information contained therein.

In recognition of the importance of preserving what has been written, it is a policy of John Wiley & Sons, Inc., to have books of enduring value published in the United States printed on acid-free paper, and we exert our best efforts to that end.

Copyright © 1992 by John Wiley & Sons, Inc.

All rights reserved. Published simultaneously in Canada.

Reproduction or translation of any part of this work beyond that permitted by Section 107 or 108 of the 1976 United States Copyright Act without the permission of the copyright owner is unlawful. Requests for permission or further information should be addressed to the Permissions Department, John Wiley & Sons, Inc.

Library of Congress Cataloging in Publication Data:
Chassaing, Rulph.
    Digital signal processing with C and the TMS320C30/Rulph Chassaing.
        p. cm.—(Topics in digital signal processing)
    "A Wiley-Interscience Publication."
    Includes index.
    1. Signal processing—Digital techniques—Data processing. 2. C (Computer program language) 3. Texas Instruments TMS320 series microprocessors. I. Title. II. Series.
TK5102.5.C472 1992
621.382'2—dc20   92-10635
ISBN 0-471-55780-3   CIP
ISBN 0-471-57777-4 (pbk.)

Printed in the United States of America

Preface

Special-purpose microprocessors, such as the TMS320 family of processors, have been used in digital signal processing (DSP) since the early 1980s, when the devices were first introduced. These devices, made possible through advances in integrated circuits, are found in a wide range of applications such as in speech processing, telecommunications, instrumentation, and so on. Having added a new element to the environment of DSP, these devices have also found their way into the university classroom, where they provide an economical way to introduce real-time signal processing to the student. For the most part, the DSP algorithms on these processors have been implemented in assembly language in both the classroom and engineering design environments. Some time must be spent in learning the architecture and the instruction set of the specific processor involved. In many situations, algorithms written in higher-level languages are not immediately portable to the real-time arena or to the specific digital signal processor.

With the introduction of Texas Instruments' (TI) third-generation TMS320C30 processor, floating-point instructions and a new architecture that supports features that facilitate the development of high-level language compilers appeared. Considerable gains were made in the implementation of the C compiler for the TMS320C30. C is rapidly becoming the high-level language of choice for microprocessor-based systems. The C optimizing compiler takes advantage of the special features of the TMS320C30 such as parallel instructions and delayed branches. Generally, the price paid for going to a high-level language is a reduction in speed and a similar increase in the size of the executable file. Although assembly language produces fast code, problems with documentation and maintenance may exist. A compromise solution is to write in assembly language time-critical routines that can be called from C.

This book was developed at Roger Williams University from the Digital Signal Processing and Senior Project courses and from several workshops given to faculty and engineers. It is intended primarily for senior undergraduate and first-year graduate students in electrical and computer engineering and as a tutorial for the practicing engineer. This book is written with the conviction that the principles of DSP can best be learned through interaction in a laboratory setting, where the student can appreciate the concepts of DSP through real-time implementation of experiments and projects. The background assumed is an electrical engineering system course and a knowledge of assembly language and a high-level language such as C. Most chapters begin with a theoretical discussion, followed by representative examples. A total of 85 examples and 11 projects are discussed.

This text can be used in the following ways:

1. For a senior undergraduate or first-year graduate project course, using Chapters 1 through 6 (Chapter 3 partially) and selected materials from Chapters 7 and 8.
2. For a DSP lab course using Chapters 1 through 8 (Chapters 3, and parts of Chapters 6 through 8). The beginning of the semester can be devoted to short experiments with the programming examples and the remainder of the semester used for a final project.
3. For the practicing engineer as a tutorial.

Programming examples using both C and TMS320C30 code are included throughout the text. This book will be very useful to the reader who is familiar with both digital signal processing and C programming, but who is not necessarily an expert in both. Although the reader who elects to study the programming examples in either C or TMS320C30 will benefit from this book, the ideal reader is one with an appreciation for both C and TMS320C30 code.

Chapter 1 introduces the software and hardware tools associated with the TMS320C30. These tools, such as the assembler, linker, simulator, C optimizing compiler, and the evaluation module (EVM) with on-board analog interface, are demonstrated through short programming examples using both C and TMS320C30 code. Chapter 2 covers the architecture and the instruction set of the TMS320C30. Special instructions that are useful in DSP are discussed. Examples with the simulator using both C and TMS320C30 code are included. Chapter 3 discusses several input and output (I/O) alternatives: the TI two-input analog interface chip (AIC), which can be interfaced through the serial port connector on the EVM; the TI analog interface board (AIB); and the Burr–Brown two-channel, 18-bit, 200-kHz sampling rate analog evaluation fixture. Simulation and real-time programming examples, in both C and TMS320C30 code, are included.

Chapter 4 introduces the Z-transform and the solution of difference equations and discusses finite impulse response (FIR) filters and the effect of window functions on these filters. Chapter 5 covers infinite impulse response (IIR) filters. Simulation and real-time programming examples to implement FIR and IIR filters, in both C and TMS320C30 code, are included.

Chapter 6 covers the development of transform methods, the decimation-in-frequency and the decimation-in-time of the fast Fourier transform (FFT). Both radix-2 and radix-4 FFTs are included, concluding with the introduction and development of the fast Hartley transform (FHT). Simulation and real-time programming examples of the FFT are provided. Chapter 7 introduces adaptive filtering and demonstrates the usefulness of the adaptive approach through simulation and real-time applications of the least mean square (LMS) algorithm. Chapter 8 discusses a number of DSP applications.

A disk included with this book contains all the source code programs discussed in the text.

I would like to thank all the students who have made the project course on applications in digital signal processing very rewarding. In particular, I am indebted to two recent graduates with whom I have worked very closely: Bill Bitler, who spent so much time and effort assisting me throughout the writing of this book, and Peter Martin for his contributions over the last few years. The suggestions made by Dr. Winfred Anakwa from Bradley University, Antoine Ataya from Roger Williams University, Dr. Panos Papamichalis from Texas Instruments, Inc., Dr. Albert Richardson from California State University at Chico, and Dr. Janet Rutledge from Northwestern University have been useful. Discussions with Dr. D. Horning from the University of New Haven, with whom I coauthored the text *Digital Signal Processing with the TMS320C25*, have been very invaluable. I am also grateful for the support of Roger Williams University School of Engineering and Research Foundation. The support of A. Goldgar and Dr. B. Gordon from Texas Instruments' University program is appreciated. A special thanks to Sheryl Rachmil and Nicole Bousquet for typing the manuscript.

RULPH CHASSAING

Bristol, Rhode Island
July 1992

Contents

1 Digital Signal Processing Development System   **1**

   1.1 Introduction   1
   1.2 Testing the Software Tools   3
   1.3 Testing the Hardware Tools   10
         References   15

2 The TMS320C30 Digital Signal Processor   **17**

   2.1 Introduction   17
   2.2 TMS320C30 Architecture and Memory Organization   18
   2.3 Memory Addressing Modes   22
   2.4 TMS320C30 Instruction Set   23
   2.5 Data and Floating-Point Formats   28
   2.6 Programming Examples Using both C and TMS320C30 Code with the Simulator   30
         References   48

3 Altenative Input/Output and Extended Development System   **50**

   3.1 Introduction   51
   3.2 The Analog Interface Chip   52
   3.3 Interrupts   56
   3.4 Programming Examples   57
   3.5 PC-Host–TMS320C30 Communication   82

x   Contents

3.6	Burr–Brown Two-Channel Analog Evaluation Fixture	87
3.7	Texas Instruments' Analog Interface Board	88
3.8	Extended Development System XDS1000 Emulator	88
	References	89

4 Finite Impulse Response Filters — 90

4.1	Introduction to the Z-Transform	91
4.2	Discrete Signals	96
4.3	Finite Impulse Response Filters	97
4.4	FIR Lattice Structure	99
4.5	FIR Implementation Using Fourier Series	103
4.6	Window Functions	106
4.7	Filter Design Packages	109
4.8	Programming Examples Using C and TMS320C30 Code	110
4.9	Filter Development Package (FDP) and Digital Filter Design Package (DFDP)	139
	References	150

5 Infinite Impulse Response Filters — 153

5.1	Introduction	153
5.2	IIR Filter Structures	155
5.3	Bilinear Transformation	166
5.4	Utility Programs for BLT and Magnitude and Phase Responses	170
5.5	Programming Examples Using C and TMS320C30 Code	174
	References	195

6 Fast Fourier Transform — 197

6.1	Introduction	197
6.2	Development of the FFT Algorithm: Radix-2	198
6.3	Decimation-in-Frequency FFT Algorithm	199
6.4	Decimation-in-Time FFT Algorithm	205
6.5	Bit Reversal for Unscrambling	209
6.6	Development of the FFT Algorithm: Radix-4 DIF	211
6.7	Inverse Fast Fourier Transform	215
6.8	Fast Hartley Transform	215
6.9	FFT Programming Examples Using C and TMS320C30 Code	222
	References	233

Contents   xi

7 Adaptive Filters   **236**

     7.1 Introduction   236
     7.2 Adaptive Structures   238
     7.3 Linear Adaptive Combiner and the LMS Algorithm   239
     7.4 Programming Examples Using C and TMS320C30 Code   242
     References   267

8 Real-Time Digital Signal Processing Applications with C and the TMS320C30: Student Projects   **269**

     8.1 Parametric Equalizer   269
     8.2 Adaptive Notch Filter Using TMS320C30 Code   274
     8.3 Adaptive Filter for Noise Cancellation Using C Code   279
     8.4 Swept Frequency Response   287
     8.5 Multirate Filter   291
     8.6 Introduction to Image Processing: Video Line Rate Analysis   301
     8.7 PID Controller   310
     8.8 Wireguided Submersible   318
     8.9 Frequency Shift Using Modulation   320
     8.10 Four-Channel Multiplexer for Fast Data Acquisition   321
     8.11 Neural Network for Signal Recognition   324
     References   328

A Introduction to C Programming   **331**

B Instruction Set, Registers, and Memory Maps   **340**

     B.1 TMS320C30 Instruction Set   340
     B.2 Register Formats and Memory Maps   340

C Digital Signal Processing Tools   **353**

     C.1 Introduction   353
     C.2 Time and Frequency Domain Utilities   354
     C.3 FIR Filter Design and Code Generation   359
     C.4 IIR Filter Design and Code Generation   366

D Burr–Brown Analog Evaluation Fixture   **373**

     D.1 Testing Channel A Using Polling   375
     D.2 Testing Both Channels Using Polling   375
     D.3 Testing Channel A Using Interrupt   375

	D.4	Testing Channels A and B Using Interrupt	376
	D.5	Burr–Brown Communication Routines	379
E	**Extended Development System**		**380**
	E.1	The XDS1000 Emulator	380
F	**Programs for Multirate Filter and Video Line Rate Analysis Projects**		**390**
	F.1	Multirate Filter Program	390
	F.2	Video Module Parts List	390
	F.3	Supporting Programs for Video Line Rate Analysis	408

Index **409**

List of Examples

Example

1.1 Matrix multiplication using C code
1.2 Matrix multiplication using TMS320C30 code
1.3 Real-time loop program with the EVM and AIC using C code
2.1 Addition of five values using TMS320C30 code
2.2 Multiplication of two arrays using TMS320C30 code
2.3 Circular buffer and data transfer using TMS320C30 code
2.4 Background programming for digital filtering using TMS320C30 code
2.5 Matrix multiplication using TMS320C30 code
2.6 Input and output using C code
2.7 Matrix multiplication using C code
2.8 Addition with both C and C-called function using TMS320C30 code
2.9 Matrix multiplication with both C and C-called function using TMS320C30 code
3.1 Output rate controlled by an interrupt, using TMS320C30 Code
3.2 Pseudorandom noise generator with interrupt, using TMS320C30 code
3.3 Alternative pseudorandom noise generator program using TMS320C30 code
3.4 Sine generation with four points and with interrupt, using TMS320C30 code
3.5 Sine generation with four points and with interrupt, using C code
3.6 Real-time sine generation with four points, using C code
3.7 Real-time loop program without interrupt, using C code
3.8 Alternative loop program suitable for either port 0 or 1, using C code
3.9 Loop program using TMS320C30 code

List of Examples

- 3.10 Loop program to test both AIC inputs, using TMS320C30 code
- 3.11 Loop program with amplitude control, using C code
- 3.12 Resetting and running the TMS320C30 with PC host
- 3.13 DMA Communication between the PC host and the TMS320C30
- 4.1 ZT of exponential sequence
- 4.2 ZT of sinusoidal sequence
- 4.3 FIR lattice structure
- 4.4 FIR lowpass filter
- 4.5 FIR lowpass filter using TMS320C30 code
- 4.6 FIR lowpass filter with two circular buffers, using TMS320C30 code
- 4.7 FIR bandpass filter with 45 coefficients, using TMS320C30 code
- 4.8 FIR bandpass filter using C and C-called assembly function
- 4.9 FIR bandpass filter using C code with the modulo operator
- 4.10 FIR bandpass filter with data move using C code
- 4.11 FIR bandpass filter without modulo operation, using C code
- 4.12 FIR bandpass filter with samples shifted, using C code
- 4.13 Real-time FIR bandpass filter using C code
- 4.14 Real-time FIR bandpass filter using mixed C and TMS320C30 code
- 4.15 Real-time FIR bandpass filter using TMS320C30 code
- 4.16 Real-time FIR bandpass filter using TMS320C30 code with macro
- 4.17 Real-time FIR bandpass filter with noise generation, using TMS320C30 code
- 4.18 Real-time FIR filter using a noncommercial filter development package
- 5.1 All-pole lattice structure
- 5.2 Lattice structure with poles and zeros
- 5.3 First-order Butterworth highpass filter
- 5.4 Second-order Butterworth bandstop filter
- 5.5 Second-order bandstop filter using the computer program BLT.BAS
- 5.6 Sine generation using C code
- 5.7 Real-time sine generation using C code
- 5.8 Sine generation using TMS320C30 code
- 5.9 Cosine generation using TMS320C30 code
- 5.10 Sixth-order IIR bandpass filter using C code
- 5.11 Real-time implementation of sixth-order IIR bandpass filter using C code
- 5.12 Sixth-order IIR bandpass filter using TMS320C30 code
- 5.13 Real-time IIR bandpass filter using TMS320C30 code
- 6.1 Eight-point FFT using decimation-in-frequency
- 6.2 16-Point FFT using decimation-in-frequency
- 6.3 16-Point radix-4 FFT using decimation-in-frequency
- 6.4 Eight-point fast Hartley transform
- 6.5 16-Point fast Hartley transform
- 6.6 Complex eight-point FFT using C code
- 6.7 Eight-point FFT with real-valued input, using mixed C and TMS320C30 code with the simulator

6.8	Real-time 512-point real-valued input FFT, using mixed C and TMS320C30 code
7.1	Adaptation using C code without the TI simulator
7.2	Adaptive filter for noise cancellation, using C code
7.3	Adaptive notch filter with two weights, using TMS320C30 code
7.4	Adaptive predictor using TMS320C30 code
7.5	Adaptive filter for noise cancellation, using TMS320C30 code
7.6	Real-time adaptive filter for noise cancellation, using TMS320C30 code
A.1	Monitor display
A.2	Use of variables and arithmetic operations
A.3	Concept of arrays and loops
A.4	Volatile variable
A.5	Recursive IIR filter
C.1	Plot of a sine sequence
C.2	Edge enhancement using C code
C.3	FIR bandpass filter with 45 coefficients
C.4	Sixth-order IIR bandpass filter design
C.5	Sixth-order IIR direct form II transpose filter using TMS320C30 code
C.6	Real-time IIR bandpass filter using TMS320C30 code
C.7	Second-order IIR lowpass filter
E.1	Generation of a square wave with the XDS1000/AIB, using TMS320C30 code
E.2	Loop program with the AIC on the AIB, using TMS320C30 code
E.3	Real-time FFT with the XDS1000/AIB, using TMS320C30 code

1
Digital Signal Processing Development System

CONCEPTS AND PROCEDURES
- *Use of the* TMS320C30-*based evaluation module* (EVM)
- *Use of the software tools such as the simulator and* C *compiler*
- *Execution of* C *and* TMS320C30 *code using both simulation and real-time processing*

Chapter 1 introduces the development tools for digital signal processing (DSP) *using both simulation and real-time processing. These tools include the evaluation module* (EVM) *based on the third-generation* TMS320C30 *floating-point digital signal processor and the analog interface chip* (AIC) *on board the* EVM. *Three short examples using* C *and* TMS320C30 *assembly code illustrate these development tools. Alternative development tools and input/output capabilities are discussed in Chapter 3.*

1.1 INTRODUCTION

Digital signal processors, such as the TMS320C30, made possible through advances in integrated circuits have added a new element to the environment of digital signal processing (DSP). DSP processors are just special-purpose fast microprocessors with specialized instruction sets appropriate for signal processing. In 1982, Texas Instruments, Inc., introduced the first-generation (N-MOS) fixed-point TMS32010 digital signal processor. This 16-bit processor boasted on-chip memory and multiplier, with an instruction cycle time of 200 ns. The second-generation (C-MOS) TMS320C25 followed with more on-chip memory and an instruction cycle time of 100 ns [1]. The third-generation floating-point TMS320C30 is a (full) 32-bit processor, with more on-chip memory and an instruction cycle time of 60 ns, or 16.6 million instructions per second (MIPS). Many of the instructions can be performed in parallel, producing a maximum instruction cycle time of 30 ns (33 MIPS). A number of special instructions are available to facilitate common signal processing operations such as filtering and spectral analysis. The architecture of the TMS320C30 was designed to take advantage of available supporting tools such as the optimizing C compiler.

DSP processors have found their way into a number of applications, including communications and controls, graphics, and speech and image processing. They are being used in talking toys, music synthesizers, spectrum analyzers, adaptive echo suppressors, and so on. Today's DSP processors have emerged as a response to the ever-increasing number of applications. A number of articles have been written that address the importance of digital signal processors for a variety of applications [2–5].

Both this book and digital signal processors are concerned primarily with real-time signal processing. In non-real-time processing, timing is not a constraint. However, in a real-time environment, the processing must keep pace with some external event, usually an analog input. Various technologies have been used for real-time processing, from fiber optics for very high frequency processing to DSP processors very suitable for the audio-frequency range. The more common applications using these processors have been primarily for frequencies from 0 to 20 kHz. Most of the work presented will involve the design of a program to implement a digital signal processing application, using software and hardware tools as an aid in this task.

To perform the DSP experiments, the following tools are needed:

1. An IBM AT (the EVM can be used with an IBM XT, ISA bus) compatible, with a coprocessor essential for the use of commercially available filter design packages
2. Texas Instruments' evaluation module (EVM), based on the third-generation floating-point digital signal processor TMS320C30 [6–8]. Equivalent DSP development systems are available from companies such as Hyperception, Inc., and Atlanta Signal Processors, Inc.
3. Assembler, linker, and C compiler with an optimizer option, all versions 4.4 (some of the files were compiled, assembled, and linked using versions 4.1), and a simulator version 1.3 [8–11]
4. Oscilloscope, function generator, signal analyzer (optional)

Input/output (I/O) alternatives, using a Burr–Brown high sampling rate analog interface and a Texas Instruments two-input analog interface chip (AIC) module, are discussed in Chapter 3 and Appendix D. Texas Instruments' powerful, but rather expensive DSP development system, the XDS1000, is discussed in Chapter 3 and Appendix E. The XDS1000 emulator can be connected to either the Texas Instruments analog interface board (AIB) or the AIC module through a target system containing the TMS320C30.

1.2 TESTING THE SOFTWARE TOOLS

In order to learn how to use the software tools, two program examples are introduced, one in C and the other in TMS320C30 assembly code. At this point, do not worry too much about the program code, because the emphasis is to illustrate the use of the software tools. To develop an executable program, the following steps are taken:

1. Use a word processor to create a source program in C or TMS320C30 assembly code.
2. Compile this program (if the source program is in C) to create an assembly program. This step is not necessary if the source program is written directly in TMS320C30 assembly code.
3. Assemble this source program to create an object program.
4. Link the object program and create an executable program that can be downloaded into either the simulator or the EVM for real-time processing. Linking also can be accomplished to include separate program modules and library support functions.

Example 1.1 Matrix Multiplication Using C Code. The reader, without programming knowledge of C (but with knowledge of a higher-level language), can still benefit from this example. Appendix A contains a brief tutorial on the C programming language [12–14]. Figure 1.1 shows the listing of the matrix multiplication program in C, `MATRIXC.C`. All the programs listed in the text are also on the accompanying disk. In order to verify the result of this matrix example through the simulator, it is first necessary to compile, assemble, and link this program. It is assumed that all the tools are within the same directory in a hard drive, with the file path properly defined.

Compiling

To compile, type the following commands (a more efficient way is presented later in this chapter):

1. `AC30 MATRIXC.C`   to invoke the parser and create `MATRIXC.IF`
2. `OPT30 MATRIXC.IF`   to compile with optimization and create `MATRIXC.OPT`
3. `CG30 MATRIXC.OPT`   to invoke the code generator and create the TMS320C30 assembly code `MATRIXC.ASM`

```
/*MATRIXC.C-MATRIX MULTIPLICATION (3x3)(3x1)=(3x1)*/
main()
{
  volatile int *IO_OUTPUT = (volatile int *) 0x804001;
  float A[3][3] = {{1,2,3},
                   {4,5,6},
                   {7,8,9}};
  float B[3] = {1,2,3};
  float result;
  int i, j;
  for (i = 0; i < 3; i++)
   {
     result = 0;
     for (j = 0; j < 3; j++)
       {
         result += A[i][j] * B[j];
       }
     *IO_OUTPUT = (int)result;   /*result=14,32,50*/
   }
}
```

FIGURE 1.1. Matrix multiplication program using C code (MATRIXC.C)

The compiled TMS320C30 source code MATRIXC.ASM can now be assembled.

Assembling

To assemble, type

ASM30 MATRIXC.ASM

to create an object file MATRIXC.OBJ that can then be linked (the extension .ASM is not required).

Linking

To link, type

LNK30 MATRIXC.OBJ -O MATRIXC.OUT

This creates an output file MATRIXC.OUT. This output file is a linked common object file format (COFF) that can be downloaded into the simulator (or the EVM). COFF was originally developed by AT&T for their work in Unix-based systems [15]. This format recently has been adopted by several makers of digital processors, including Texas Instruments (TI) and Motorola. TI adopted the COFF format, changing from their TI-tagged format, initially used with the first two generations of the TMS320 family of processors. The COFF format makes it easier for modular programming. Managing code segments becomes easier. In the command file for linking, one can specify where various sections of code are to be stored in memory. For example, data can be specified to be stored within a data section, or a section called bss (block started by symbol—a block of memory that is not initialized), and code for the instructions can be specified to be stored within a text section.

Generally, it is more appropriate to specify the output file within a user-defined command file. Create the following command file for the matrix example, and save it as MATRIXC.CMD:

```
/*MATRIXC.CMD         COMMAND FILE                */
-c                    /*LINK USING C CONVENTIONS */
matrixc.obj           /*OBJECT CODE              */
-l rts30.lib          /*RUN-TIME LIBRARY SUPPORT */
-o matrixc.out        /*LINKED OUTPUT FILE       */
```

The -c linking convention, in the command file, is required by the compiler. Another option (-cr) can enhance boot time by reducing the memory space utilized by the initialization process, with global variables initialized at load-time instead of at run-time as in the case of the -c convention. However, this option requires a special loader. The object file matrixc.obj, and the COFF output file matrixc.out are specified within this command file. Note that comments within the program are included between /* and */. The -l option specifies to the linker that rts30.lib is an object library, which contains run-time support functions, written in C. This is necessary for linking C code. This linking process, without specifying memory addresses, or sections, assumes that memory begins at address 0 and that 2^{24} words are available to allocate object code. We will see later how we can have more control over where our code and data reside in memory.

Compiling, Assembling, and Linking in One Step

An alternative and preferable procedure to the preceding steps for compiling, assembling, and linking, which will be used throughout the text, is to compile, assemble, and link using

```
CL30 -sq -o2 MATRIXC.C -z MATRIXC.CMD
```

The s option is to interlist C code with the assembly code (useful for debugging), the q option is to suppress messages (quiet option), the o is to invoke the optimization option level 2 (default) within the compiler, and the z is to invoke the linker option.

It is desirable to use `CL30 MATRIX` when compiling and assembling a C program for the first time in order to verify that there are no errors. Note that the `.C` extension in the file name `MATRIXC.C` is not required. Then, to link, use `LNK30 MATRIXC.CMD`. Invoking the `CL30` command produces an extensive menu of various options for compiling, assembling, and linking. The executable file has a `.OUT` extension. Verify that the linked COFF output file `MATRIXC.OUT` has been created.

Loading an Executable File into the Simulator

To simulate, type the following:

`SIM30`	to access the simulator
`C`	to configure
`M`	the memory map
`A`	to add
`804000h`	the starting address in hex
`8043FFh`	the ending address in hex
`0`	for 0 wait states
`S`	to save and reuse this configuration of the memory map file
`MATRIXC.CFM`	the memory configuration file

Repeat the save command `S` in order to save this memory configuration file as `M.CFM` as well, because this file will be utilized throughout the text when using the simulator with the other examples. Press `F1` to toggle to the port map configuration. Type the following:

`A`	to add
`I`	for input or output
`O`	for output
`MATRIXC.DAT`	the simulator output file name (after running the program)
`4001h`	the output port address (the simulator will not accept 804001h); note the hex representation with the letter h.
`S`	to save the port configuration file
`MATRIXC.CFP`	the port configuration file name
`ESC`	to access the simulator main menu
`L`	to load
`O`	an object file to be downloaded
`MATRIXC.OUT`	the linked COFF executable output file
`D`	to display
`M`	memory address
`804000h`	starting address to be displayed

1.2 Testing the Software Tools

```
Simulator: Display eXecute Reg Mem Bpoint Eval Config Trace Load Save ->
                                                HALTED
==========================code===========    =====extended precision registers====
   00002F  XXXXXXXX  _c_int00:                R0 00 00000000  =  1.00000000e+00
   00002F  08750000  LDI 0,ST                  R1 00 00000000  =  1.00000000e+00
   000030  50700000  LDIU 0,DP                 R2 00 00000000  =  1.00000000e+00
   000031  0834002D  LDI @02dH,SP              R3 00 00000000  =  1.00000000e+00
   000032  080B0014  LDI SP,AR3                R4 00 00000000  =  1.00000000e+00
   000033  50700000  LDIU 0,DP                 R5 00 00000000  =  1.00000000e+00
   000034  0828002E  LDI @02eH,AR0             R6 00 00000000  =  1.00000000e+00
   000035  04E8FFFF  CMPI -1,AR0               R7 00 00000000  =  1.00000000e+00
   000036  6A05000D  BZ 044H                  =============cpu status============
   000037  08402001  LDI *AR0++(1),R0          PC      0000002F    CLK   00000000
==========================display=========   ST      00000000    DP    00000000
   804000  00000000 00000000 00000000 00000000 SP      00000000    BK    00000000
   804004  00000000 00000000 00000000 00000000 AR0     00000000    AR1   00000000
   804008  00000000 00000000 00000000 00000000 AR2     00000000    AR3   00000000
   80400C  00000000 00000000 00000000 00000000 AR4     00000000    AR5   00000000
   804010  00000000 00000000 00000000 00000000 AR6     00000000    AR7   00000000
   804014  00000000 00000000 00000000 00000000 IR0     00000000    IR1   00000000
   804018  00000000 00000000 00000000 00000000 IE      00000000    IF    00000000
   80401C  00000000 00000000 00000000 00000000 IOF     00000000    RS    00000000
   804020  00000000 00000000 00000000 00000000 RE      00000000    RC    00000000
F1Inspect                 F8Command F9Update F10Help
```

FIGURE 1.2. Simulator screen for matrix multiplication example

The simulator can also be accessed and the executable COFF file downloaded using

SIM30   MATRIXC.OUT

Figure 1.2 shows a display of the simulator screen. Note the various registers of the TMS320C30 being displayed. Press F1 to toggle to the different windows. With the cursor (using F1) at the window displaying the code, use the down-arrow key to place it adjacent to memory address 2F (line 2F). Press B to set a breakpoint there. Press X to execute and G to run until the set breakpoint. To step through this process, press

X	to execute
I	for instruction
2	with a repeat count of 2, to step through two instructions at a time (a repeat count of 1 is to single-step).

Press the ⟨SPACE BAR⟩ to continue stepping through the code. Note the four levels of pipelining. While an instruction is being executed (x), the next instruction is being read (r) and two subsequent instructions are being decoded (d) and fetched (f). Press (and hold on) the ⟨SPACE BAR⟩ while monitoring memory address 804001h, which contains the output result. Verify that the resulting values are *e*, then 20, then 32, in hex. These values are also written into the file MATRIXC.DAT. View the file MATRIXC.DAT for the output result.

```
;MATRIX.ASM-MATRIX MULTIPLICATION IN ASSEMBLY CODE
            .GLOBAL BEGIN
            .DATA                 ;ASSEMBLE INTO DATA SECTION
A           .FLOAT  1,2,3,4,5,6,7,8,9 ;VALUES FOR MATRIX A
B           .FLOAT  1,2,3         ;VALUES FOR MATRIX B
A_ADDR      .WORD   A             ;ADDR OF MATRIX A
B_ADDR      .WORD   B             ;ADDR OF MATRIX B
IO_OUT      .WORD   804001H       ;ADDR OF OUTPUT PORT
            .TEXT                 ;ASSEMBLE INTO TEXT SECTION
BEGIN       LDI     @A_ADDR,AR0   ;AR0=START ADDR OF A
            LDI     @B_ADDR,AR1   ;AR1=START ADDR OF B
            LDI     @IO_OUT,AR2   ;AR2=OUTPUT PORT ADDR
            LDI     3,R4          ;R4 IS LOOPI COUNTER
LOOPI       LDF     0,R0          ;INITIALIZE R0
            LDI     2,AR4         ;AR4 IS LOOPJ COUNTER
LOOPJ       MPYF    *AR0++,*AR1++,R1  ;A[I,J]*B[J]=R1
            ADDF    R1,R0         ;R0 ACCUMULATES RESULT
            DB      AR4,LOOPJ     ;DEC AR4 & BRANCH TIL AR4<0
            FIX     R0            ;FLOAT TO INTEGER
            STI     R0,*AR2       ;OUTPUT RESULT TO IO_OUT
            LDI     @B_ADDR,AR1   ;RELOAD START ADDR OF B
            SUBI    1,R4          ;DECREMENT R4
            BNZ     LOOPI         ;BRANCH WHILE  R4 <> 0
WAIT        BR      WAIT          ;WAIT INDEFINITELY
```

FIGURE 1.3. Matrix multiplication program using TMS320C30 code (MATRIX.ASM)

To run the program, without stepping through, press X to execute, then G. Wait until the data (the three values) is written to the disk before pressing ESC to halt the program and Q to quit. However, before executing the matrix program again, delete the file MATRIXC.DAT, since the resulting values (e, 20, 32) will be *appended* to the first set of values obtained in the previous step.

The execution modes available with the simulator include execution until a condition is met or while a condition exists, as well as execution for a set loop count. Trace execution is also available. You may wish to trace or monitor a specific register. For example, save all the values of R0 into a file, during the execution of a program.

Example 1.2 *Matrix Multiplication Using TMS320C30 Code.* The following matrix multiplication example produces the same result as the previous example written in C. Figure 1.3 lists the program MATRIX.ASM, which is the assembly "counterpart" of the C version. Although a C-coded program is more portable and maintainable than assembly code, it does not execute as fast as assembly code. For many time-critical DSP applications, it may be necessary to use assembly code. Preferably, time-critical functions or routines can be written in TMS320C30 assembly code and called from a C program.

The assembler expects the source file MATRIX.ASM to be in a certain format, which follows:

LABEL   INSTRUCTION OR ASSEMBLER DIRECTIVE   OPERAND   COMMENT

For example, the following line of code,

BEGIN   SUBI 1,R0   ; DECREMENT R0

consists of a label BEGIN, the subtract integer instruction SUBI, and the operand 1,R0. This line of code is commented with DECREMENT R0 (subtract 1 from the value contained in R0).

One or more blank spaces must separate these various fields. TAB keys can be used in lieu of blank spaces for ease of aligning the different fields. Labels such as A, B, etc., must be in column 1. A statement must begin with a label, a blank, a semicolon, or a *. Comments are optional. Comments after the *operand* must begin with a semicolon, and comments in column 1 can begin either with a semicolon or a *. An assembler directive [10], such as .GLOBAL BEGIN, is a message for the assembler and not a machine instruction. Other assembler directives include .FLOAT and .WORD. The matrix program will be discussed again in Chapter 2, so do not worry much about the code for now. Proceed with the following:

1. Assemble this program using

ASM30 MATRIX.ASM

to create the object code MATRIX.OBJ. The extension .ASM can be omitted.

2. Create and save the following command file, and name it MATRIX.CMD:

```
/*MATRIX.CMD        COMMAND FILE                          */
-E BEGIN            /*SPECIFIES ENTRY POINT FOR OUTPUT   */
MATRIX.OBJ          /*OBJECT FILE                         */
-O MATRIX.OUT       /*LINKED COFF OUTPUT FILE             */
```

3. Link, using

LNK30

Command files: MATRIX.CMD
Object files: ⟨ENTER⟩
Output file: ⟨ENTER⟩
Options: ⟨ENTER⟩

Enter the command file name, `MATRIX.CMD` (which you just created), for the first prompt. Press the ⟨ENTER⟩ key, when prompted to enter the object and output files, because they are already specified within the command file. No options are required for this example, although a useful option is the map file. The map file can be helpful during debugging a program. Such a file contains addresses of variables such as `BEGIN`, which is defined globally in the matrix program. Linking can also be accomplished using

`LNK30 MATRIX.CMD`

4. Access the simulator, as before. The `-E` option in the command file specifies the entry point. When the code is downloaded, it is displayed in the code window, starting with the first instruction within the text section. Type `C` and `M` to configure the memory, and type `L` to load `M.CFM`, the memory configuration file obtained and saved in the previous example. Press `F1` to configure the port file, and press (enter the commands separately) `A I O` to add an output file. Type `MATRIX.DAT` and `4001h` to specify the simulator output file name at address `804001h`. Press (enter the commands separately) `ESC L O MATRIX.OUT X G` to download and run the linked COFF executable output file. Press `ESC` and `Q` to quit the simulation after the output data (the three values) is written to the disk, and view the file `MATRIX.DAT`, which contains the output result, listed in hex. It is instructive not to skip the first example. You may wish to access the simulator and download the executable COFF file `MATRIX.OUT` in one step using `SIM30 MATRIX.OUT`.

1.3 TESTING THE HARDWARE TOOLS

The evaluation module (EVM) is an eight-bit half-card that plugs into an IBM AT compatible slot. The EVM includes the third-generation TMS320C30 floating-point digital signal processor, 16K of static RAM, and the TLC32044 analog interface chip (AIC) for input and output. The TLC32044, a member of the TLC32040 [1, 16] family of voice-band analog interface circuits, contains 14-bit ADC and DAC, switched capacitor input and output filters on a single C-MOS chip. Although the TLC32044 AIC has two inputs and one output, the AIC on-board the EVM has one of its inputs connected to ground. Chapter 3 covers alternative input and output (I/O) devices, including a two-input AIC module connected through the serial port connector available on the EVM, and a Burr–Brown high-sampling-rate analog fixture.

Example 1.3 *Real-Time Loop Program with the EVM and AIC Using C Code.* This example tests the EVM and the on-board AIC. At this point, again do not worry too much about what the program is doing. Instead, concentrate on the use of the tools. Connect a sinusoid source, with an amplitude of less

1.3 Testing the Hardware Tools

than 3 V peak-to-peak and a frequency between 1 and 4 kHz, to the input connector on the EVM. Connect the input also to an oscilloscope. Connect the EVM output to a second channel on the oscilloscope. The loop program LOOPALL.C, listed in Figure 1.4, produces a delayed sinusoidal output waveform of the same frequency as the input and will be analyzed in Chapter 3 in more detail. The command file for the loop program is listed in Figure

```
/*LOOPALL.C-LOOP AND COMUNICATIONS ROUTINES FOR SERIAL 0 PORT AIC    */
#define TWAIT while (!(PBASE[0x40] & 0x2)) /*wait till XMIT buffer clear*/
int AICSEC[4]= {0x1428,0x1,0x4A96,0x67};    /*config data for SP0 AIC  */
volatile int *PBASE = (volatile int *) 0x808000; /*peripherals base addr*/

void AICSET()                       /*function to initialize AIC      */
{
 volatile int loop;                 /*declare local variables         */
 PBASE[0x28] = 0x00000001;          /*set timer period                */
 PBASE[0x20] = 0x000002C1;          /*set timer control register      */
 asm("    LDI   00000002h,IOF");    /*set IOF low to reset AIC        */
 for (loop = 0; loop < 50; loop++); /*keep IOF low for a while        */
 PBASE[0x42] = 0x00000111;          /*set xmit port control           */
 PBASE[0x43] = 0x00000111;          /*set receive port control        */
 PBASE[0x40] = 0x0E970300;          /*set serial port global reg      */
 PBASE[0x48] = 0x00000000;          /*clear xmit register             */
 asm("    OR    00000006h,IOF");    /*set IOF high to enable AIC      */
 for (loop = 0; loop < 4; loop++)   /*loop to configure AIC           */
   {
    TWAIT;                          /*wait till XMIT buffer clear     */
    PBASE[0x48] = 0x3;              /*enable secondary comm           */
    TWAIT;                          /*wait till XMIT buffer clear     */
    PBASE[0x48] = AICSEC[loop];     /*secondary command for SP0       */
   }
}

int UPDATE_SAMPLE(int output)       /*function to update sample       */
{
  int input;                        /*declare local variables         */
  TWAIT;                            /*wait till XMIT buffer clear     */
  PBASE[0x48] = output << 2;        /*left shift and output sample    */
  input = PBASE[0x4C] << 16 >> 18;  /*input sample and sign extend    */
  return(input);                    /*return new sample               */
}

main()
{
 int data_in, data_out;             /*initialize variables            */
 AICSET();                          /*call function to congig AIC     */
 while (1)                          /*create endless loop             */
  {
   data_in = UPDATE_SAMPLE(data_out); /*call function to update sample*/
   data_out = data_in;              /*loop input to output            */
  }
}
```

FIGURE 1.4. Loop program using C code (LOOPALL.C)

```
/*LOOPALL.CMD-COMMAND FILE FOR LINKING        */
-c                    /*USING C CONVENTION         */
vecsir.obj            /*INTERR/RESET DEF           */
loopall.obj           /*MAIN PROGRAM               */
-O loopall.out        /*LINKED COFF OUTPUT FILE */
-l rts30.lib          /*RUN-TIME LIBRARY SUPPORT*/
MEMORY
{
 VECS:  org = 0          len = 0x40    /*INTERRUPT VECTORS*/
 SRAM:  org = 0x40       len = 0x3FC0  /*USER STATIC RAM   */
 RAM:   org = 0x809800   len = 0x800   /*INTERNAL RAM      */
}
SECTIONS
{
 .text:   {} > SRAM       /*CODE                       */
 .cinit:  {} > RAM        /*INITIALIZATION TABLES      */
 .stack:  {} > RAM        /*SYSTEM STACK               */
 .bss:    {} > RAM        /*BSS SECTION                */
 vecs:    {} > VECS       /*VECTOR SECTION             */
}
```

FIGURE 1.5. Command file for loop program (LOOPALL.CMD)

1.5. The procedure that follows to run this loop program on the EVM assumes that the appropriate support files are already loaded on a hard drive. However, before running the loop program, type

`EVMTEST`

to run a utility "loop" program, provided by TI with the EVM package. This program tests the communication between the EVM and the PC host. The batch file `EVMTEST` downloads and executes a TMS320C30 program as well as a PC host program. Run this test program first before proceeding with the loop program (same input/output connections). Note that the delayed output depends on the input frequency.

Now, compile, assemble, and link the loop program using

`CL30 -kg -O2 LOOPALL.C -z LOOPALL.CMD`

The k compiler option is to keep the .asm file LOOPALL.ASM, and the g option is to enable symbolic debugging. This option produces a larger output file LOOPALL.OUT resulting in slower timing (execution). However, it is useful when debugging a program using the C-source debugger [17]. Within the C-source debugger, it is possible to list both the C and TMS320C30 code.

After successful debugging, the source file can then be recompiled without the g option (removing the symbolic directives).

To download and run the executable file LOOPALL.OUT, type

EVMLOAD LOOPALL.OUT

Verify that the output waveform is delayed from the input, but is of the same frequency as the input. The amplitude of the output is $\simeq 8$ times that of the input, due to an amplifier gain of 2 at the input side and a gain of 4 at the output side. The amplification circuitry on the EVM is used to drive an external speaker. This loop program includes the communication routines to interface the TMS320C30 to the on-board analog interface chip AIC (through serial port 0). In Chapter 3, we will discuss how to interface a two-input AIC through the serial port 1 connector on the EVM. We will also discuss the AIC communication routines as a separate program, which can be used with either port 0 or 1, with or without interrupt, and which can be called from a main program.

Accessing the C-Source Debugger

Available with the EVM package is a C-source debugger [17], useful for debugging C code. To access the C-source debugger, first load the loop program LOOPALL.OUT into the EVM. Type or press the following commands or keys (labels are case-sensitive):

1.	EVM30	LOOPALL.OUT	to download the loop program into the EVM
2.	FILE	LOOPALL.C	loads the C file into a separate window
3.	GO	main	goes to the main function. Observe the cursor in both windows pointing to main
4.	BA	UPDATE_SAMPLE	sets a breakpoint at the address specified with the label UPDATE_SAMPLE
5.	F5		runs/executes the program until the set breakpoint at the address specified by UPDATE_SAMPLE
6.	BR		removes the set breakpoint
7.	MEM	0x808040	displays a separate memory window starting at address 808040h
8.	MEM	SP	displays the memory in the stack pointer SP

9. wa   PBASE[0x40]   sets a "watch" window to show data at the specified memory address; for example, the command wa R0 monitors the value in R0

10. Go   UPDATE_SAMPLE   executes to UPDATE_SAMPLE. Observe the data change at address PBASE[0x40] in the watch window

11. F8   to single-step through the assembly code program

Figure 1.6 shows the debugger screen displaying the various windows. Use the function keys to select the available options. For example, type or press ALT-M in order to select the memory window, F to fill, 0x809800 (selected starting address), 2 (length), and 0x20 (data value). Type the command mem 0x809800 and verify that memory locations 809800h and 809801h contain the data value 20h. Note that the program LOOPALL.OUT could have been loaded using ALT-L (after accessing the debugger with EVM30). It is quite useful to use a "mouse," if you have one, for the various commands such as setting breakpoints and single-stepping, as well as for moving the locations of selected windows.

Within the Mode Window option, select the C mode option to single-step through the C code if the C code is compiled with symbolic directive using the -g option. With the mixed-mode (default) selection, single-stepping can

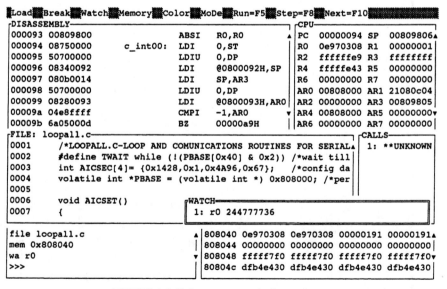

FIGURE 1.6. Debugger screen for loop program

be done through both the C instruction and the equivalent assembly instruction(s). In Chapter 4, we will discuss filter examples using mixed code (C and C-called assembly functions). Type the command `quit` to exit the debugger.

Amplitude Control of Loop Program

Run the batch file `LOOPCTRL.BAT` (on disk). Use the same input/output connections as with the loop program of the previous example. On execution, the program `LOOPCTRL.OUT` is downloaded. A C-called program `PCCTRL.C`, which provides communication between the PC host and the TMS320C30, is also executed. Note that the graphics device driver file `EGAVGA.BGI` (on the accompanying disk) should be compiled because it provides the graphics support to run the amplitude control loop program. Follow the prompts and verify that pressing `F5` (or `F6`) results in increasing (or decreasing) the output amplitude. This communication feature will be discussed in Chapter 3.

REFERENCES

[1] R. Chassaing and D. W. Horning, *Digital Signal Processing with the TMS320C25*, Wiley, New York, 1990.

[2] H. M. Ahmed and R. B. Kline, Recent advances in DSP systems, *IEEE Communications Magazine*, May 1991, pp. 32–45.

[3] A. Aliphas and J. A. Feldman, The versatility of digital signal processing, *IEEE Spectrum*, June 1987, pp. 40–45.

[4] B. C. Mather, Embedding DSP, *IEEE Spectrum*, November 1991, pp. 52–55.

[5] R. Goering, DSP tools: Hammering out new solutions, *High Performance Systems*, February 1990, pp. 20–59.

[6] *Digital Signal Processing Applications with the TMS320C30 Evaluation Module—Selected Application Notes*, Texas Instruments, Inc., Dallas, Tex., 1991.

[7] *TMS320C30 Evaluation Module Technical Reference*, Texas Instruments, Inc., Dallas, Tex., 1990.

[8] *TMS320C3X User's Guide*, Texas Instruments, Inc., Dallas, Tex., 1991.

[9] *TMS320 Floating-Point DSP Optimizing C Compiler User's Guide*, Texas Instruments, Inc., Dallas, Tex., 1991.

[10] *TMS320 Floating-Point DSP Assembly Language Tools User's Guide*, Texas Instruments, Inc., Dallas, Tex., 1991.

[11] *TMS320C30 Simulator User's Guide*, Texas Instruments, Inc., Dallas, Tex., 1989.

[12] B. W. Kernigan and D. M. Ritchie, *The C Programming Language*, Prentice-Hall, Englewood Cliffs, N.J., 1988.

[13] S. G. Kochan, *Programming in C*, Hayden Books, 1991.

[14] P. M. Embree and B. Kimble, *C Language Algorithms for Digital Signal Processing*, Prentice-Hall, Englewood Cliffs, N.J., 1990.

[15] G. R. Gircys, *Understanding and Using COFF*, O'Reilly & Assoc., Inc., Newton, Mass., 1988.

[16] *TMS320C2X User's Guide*, Texas Instruments, Inc., Dallas, Tex., 1989.

[17] *TMS320C30 C Source Debugger User's Guide*, Texas Instruments, Inc., Dallas, Tex., 1991.

[18] P. Papamichalis (editor), *Digital Signal Processing Applications with the TMS320 Family—Theory, Algorithms, and Implementations*, Vol. 3, Texas Instruments, Inc., Dallas, Tex., 1990.

2
The TMS320C30 Digital Signal Processor

CONCEPTS AND PROCEDURES
- *Architecture and instruction set of the* TMS320C30
- *Program structure and use of special instructions*
- *Programming examples using* TMS320C30 *assembly code,* C *code, and C-callable* TMS320C30 *assembly functions*

This chapter covers the architecture, memory organization, addressing modes, and the instruction set of the TMS320C30. Special instructions, such as the multiply instruction MPYF in conjunction with the repeat instruction RPT, and special features such as circular buffering for implementing real-time digital filters are illustrated. A special addressing mode for handling bit reversal within a fast Fourier transform (FFT) algorithm is introduced. Several examples in TMS320C30 assembly, C, and mixed modes are covered.

2.1 INTRODUCTION

Special purpose microprocessors, such as the TMS320 family of processors, have been used in digital signal processing since the early 1980s. The Intel 2920 appeared first, followed by NEC's uPD7720. In 1982, Texas Instruments, Inc., introduced the first-generation TMS32010 digital signal processor, followed by the second-generation TMS32020 in 1985, and the faster C-MOS version TMS320C25 in 1986.

The first-generation TMS32010 has 144 words (16-bit) of on-chip data RAM with a 200-ns instruction cycle time and features such as a 16-bit by 16-bit integer multiply in one instruction cycle. With most instructions requiring only a single cycle, it is capable of executing 5 million instructions per second (MIPS) [1–4]. Many members of the TMS32010 family of processors are now available, including a C-MOS version with a 160-ns instruction cycle time. The second-generation C-MOS version TMS320C25 is based on the same modified Harvard architecture as the first-generation TMS32010, with upward code compatibility. This architecture is such that data and instructions reside in separate memory spaces, allowing for concurrent accesses. Its features include 544 words (16-bit) of on-chip data RAM, 4K words of on-chip maskable program ROM, and separate program memory and data memory address spaces, each 64K words. It has an instruction cycle time of 100 ns, compared to the 200-ns instruction cycle time of the TMS32020 or TMS32010, enabling the TMS320C25 to execute 10 million instructions per second [5, 6].

Although the first two generations of digital signal processors are considered to be fixed-point 16-bit processors, the third-generation TMS320C30 supports both fixed- and floating-point processing and is a true 32-bit processor. With features such as a 32-bit by 32-bit floating-point multiply in one instruction cycle and special addressing modes for circular buffering and bit reversal, the TMS320C30 is well suited to implement many DSP applications, such as digital filtering and the fast Fourier transform [7–10].

The TMS320C31 is a different version of the TMS320C30 processor, with the same execution speed, but with only one primary bus (no expansion bus) and one serial port (as opposed to two).

Several examples in C and TMS320C30 assembly code are covered in this chapter, including examples with functions in TMS320C30 code that can be called from C. Chapter 4 includes discussions and comparisons of execution time of digital filters, written either in C, in assembly code, or in mixed modes [11]. In many applications where execution time is critical, it may be necessary to write specific functions in assembly code. Although C programs are more portable and can be maintained more easily than assembly code programs, they cannot achieve the same processing speed as assembly code programs.

Although the ideal reader is one with an appreciation for both C and TMS320C30 code, the reader interested in either C or TMS320C30 code will also benefit from the many examples and projects in this text.

2.2 TMS320C30 ARCHITECTURE AND MEMORY ORGANIZATION

The TMS320C30 is a 32-bit digital signal processor, capable of performing floating-point, integer, and logical operations. Its architecture allows four levels of pipelining. While an instruction is being executed, the next three instructions are being consequently fetched, decoded, and read. Many instructions can be executed in parallel, such as load with store, multiply with add, and so on.

2.2 TMS320C30 Architecture and Memory Organization

The TMS320C30 has 2K words (32-bit) of on-chip memory (4K words of ROM in the microcomputer mode) and a total addressable range of 16 million words (32-bit) of memory containing program, data, and input/output space. Separate program, data, and direct memory access (DMA) busses enable the TMS320C30 to perform concurrent data read and write, program fetches, and DMA operations. The TMS320C30 has a 60-ns instruction cycle time, with most instructions requiring only a single cycle. Hence, it can execute 16.66 million instructions per second. Furthermore, because many instructions can be performed in parallel, such as load with store, and multiply with add, the TMS320C30 effectively can execute up to 33.3 million instructions per second [12].

The functional block diagram of the TMS320C30 is shown in Figure 2.1. The TMS320C30 processor contains the following registers:

1. Eight 32-bit auxiliary registers (AR0–AR7) that can be used for general purpose and, in particular, for indirect addressing
2. Eight 40-bit registers (R0–R7) that can be used for extended-precision floating-point results (These registers can store 32-bit integer and 40-bit floating-point numbers.)
3. IR0 and IR1, for indexing the address
4. SP, the system stack pointer, which contains the address of the top of the stack
5. ST, for status of the CPU
6. IE and IF, for interrupt enable and flag, respectively
7. IOF, for I/O flag
8. BK, to specify the data block size within a circular buffer
9. DP, as a pointer to one of 256 pages of data being addressed, with each data page being 64K words long
10. RC, the repeat count to specify how many times a block of code is to be executed
11. RS and RE, which contain the start address and end address, respectively, of a block of code to be repeated
12. PC, the program counter, which contains the address of the next instruction to be fetched; can be modified to control the flow of the program

When the multiplier performs a multiplication of two 32-bit floating-point numbers, the result is a 40-bit floating-point number, which can be stored in a 40-bit extended-precision register, such as R0.

Two serial ports are available on the TMS320C30. Serial port 0 is used for the on-board AIC. Chapter 3 shows how serial port 1, available through the

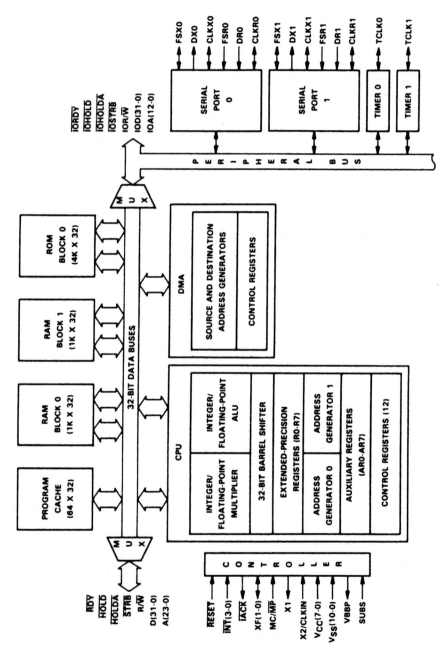

FIGURE 2.1. TMS320C30 functional block diagram (Reprinted by permission of Texas Instruments)

2.2 TMS320C30 Architecture and Memory Organization   21

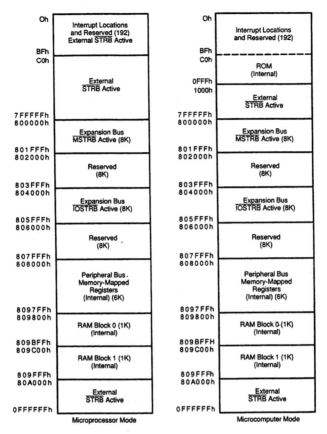

FIGURE 2.2. TMS320C30 memory organization (Reprinted by permission of Texas Instruments)

EVM, can be used to connect alternative I/O devices. The two serial ports are completely independent, each with a set of control registers. Two timer/event counters are also available.

Figure 2.2 shows the memory organization of the TMS320C30. Two on-chip RAM blocks of memory are available, RAM 0 and RAM 1, for a total of 2K words (32-bit). The memory map section shows the expansion bus, with starting address 804000h. This expansion bus can be used for input/output to communicate with the host via the PC bus. Starting with RESET at memory address 0, interrupts and trap vectors occupy the first 64 (40h) memory locations of the TMS320C30. The instruction set of the TMS320C30 along with the memory maps for the reset and interrupt vectors, the peripheral bus, the DMA registers, the peripheral timers, the serial ports, and the primary and expansion bus control registers are shown in Appendix B.

2.3 MEMORY ADDRESSING MODES

The TMS320C30 supports a number of different modes of addressing. The more frequently used modes follow.

1. *Register Addressing.* A register can be directly addressed with an instruction, such as

    ```
    FIX   R0
    ```

    where R0 is a register that may contain a floating-point value and is to be converted from a floating-point to an equivalent integer value (with the instruction FIX).

2. *Direct Addressing.* A data value at a specific memory location can be added to a register with an instruction such as

    ```
    ADDI  @809800h, R0
    ```

    with the data value at the memory address location 809800h added to R0. Note the @ symbol to represent direct addressing.

3. *Indirect Addressing.* Several modes of indirect addressing are available, including circular and bit-reversed addressing, using displacements and indexing; for example, using auxiliary register ARn (n = 0, ..., 7), to obtain the address of an operand to be fetched:

    (a) *ARn   The indirect addressing is represented with the * symbol; ARn contains the address to be used; for example, the auxiliary register AR0 may contain the *address* of a memory location where the desired data is stored.

    (b) *+ARn(d)   With ARn plus the displacement d (an eight-bit unsigned integer including 1) containing the address; in this case, ARn is not updated; the plus (+) can be replaced by a minus (−) for a predisplacement subtract.

    (c) *++ARn(d)   With the address in (ARn + d); after the operand is fetched, ARn is updated (modified); a double minus (−−) would be used for a predisplacement subtract.

    (d) *ARn++(d)   With the address being the content of ARn; after the operand is fetched, ARn is updated (postincremented) to (ARn + d); a double minus (−−) would update ARn to (ARn − d).

    (e) *ARn++(d)%   With the address being the content of ARn; after the operand is fetched, ARn is updated to (ARn + d) as previously, but in a circular mode (used, for example, to model delays in digital filtering); this important and special mode of addressing, using the modulo operator %, will be illustrated later through a programming example; a double minus (--) would update ARn to (ARn - d).

Similar types of indirect addressing as in (b)–(e) are also available using the index register IR0 or IR1 instead of the displacement d.

4. *Short- and Long-Immediate Addressing.* The operand using short-immediate addressing is a 16-bit immediate value. For example,

    SUBI   1,R0

would decrement R0 by 1. In long-immediate addressing, the operand is a 24-bit immediate value.

5. *Bit-Reversed Addressing.* A special type of indirect addressing is for bit reversal. The bit-reversal process is encountered when implementing the fast Fourier transform, for proper resequencing of data. For example, the address of the operand to be fetched, using the register *ARn + + (IR0)B is ARn. After the operand is fetched, ARn is updated to (ARn + IR0) in a reversed-carry propagation format. That is, the carry bit is propagated in the reverse direction. This mode of addressing will be illustrated in Chapter 6 in conjunction with the fast Fourier transform.

6. *Circular Addressing.* A circular buffer is necessary to implement the delays associated with convolution and correlation equations, using a circular mode of addressing [see (3e)]. The block size of the circular buffer in memory is specified within a special register BK. For example, using

    *AR1 + + %

causes AR1 to be incremented, to point at the next-higher memory location, until it reaches the bottom of the circular buffer. After AR1 reaches (points at) the bottom address of the circular buffer, it will point to the top of the buffer the next time it is incremented.

2.4 TMS320C30 INSTRUCTION SET

Although the instruction set of the second generation of processors TMS320C25/TMS32020 is upwardly compatible with the first-generation TMS32010, the third-generation TMS320C30 has an architecture and instruction set quite different from the first two generations of processors. Even though the TMS320C30 contains a richer and more powerful set of instructions as compared with the first two generations of processors, it is not any harder to program. Table B.1 in Appendix B contains a summary of the TMS320C30 instruction set.

Instruction Types

1. *Load and Store Instructions.* A word can be loaded from memory into a register or stored from a register into memory. Loading instructions can be performed using LDI for integer format or LDF for floating-point format. Note the use of I or F at the end of the instruction to represent integer or floating-point. For example, the instruction

   ```
   LDI    @IN_ADDR,AR0
   ```

   loads (using integer format) directly (using the @ symbol) the address (an integer value) specified with the label IN_ADDR into auxiliary register AR0. The instruction

   ```
   STF    R0,*AR1++
   ```

   stores the floating-point value from R0 into a memory location specified by AR1. AR1 contains the address in memory where R0 is stored. AR1 is then incremented by 1 to point at the next-higher memory address. A displacement value of 1 is implied.

   Instructions such as PUSH and POP are available to manipulate data on the system stack. For example,

   ```
   POPF R1
   ```

   restores the floating-point value in R1 from the stack.

2. *Math Instructions to Add, Subtract, or Multiply.* For example,

   ```
   MPYF3   *AR0++,*AR1++,R0
   ```

   is a three-operand multiply instruction, using floating-point format (the 3 at the end is implied and can be omitted). It multiplies the content in memory specified by AR0 by the content in memory specified by AR1 and stores the resulting floating-point value in R0. After the multiplication operation, both AR0 and AR1 are post-incremented by 1. Note again that AR0 and AR1 contain the memory *addresses* (not the data) where the data is stored. The instruction

```
ADDF    R0,R2,R1
```

adds the floating-point values in R0 and R2 and stores the result in R1. The instruction ADDF R0,R2 stores the result in R2 (the third operand defaults to the second operand).

3. **Logical Instructions.** For example, the instruction

```
XOR     R1,R1
```

performs an exclusive OR between R1 and R1, storing the result again in R1. This effectively clears R1.

4. **Input and Output Instructions**
   (a) *Input.* An input sample can be obtained from an analog-to-digital converter (ADC), using the following program segment:

```
LDI     @IN_ADDR,AR2
FLOAT   *AR2,R3
```

The input address (IN_ADDR) is directly loaded into AR2. The content of the input address (specified by AR2), an integer value that might result from an analog-to-digital converter (ADC), is converted to its floating-point equivalent, and stored in the extended-precision register R3.

   (b) *Output.* A data value can be sent to a digital-to-analog converter (DAC), using the following program segment:

```
LDI     @OUT_ADDR,AR3
FIX     R0,R1
STI     R1,*AR3
```

The output address (OUT_ADDR) is directly loaded into AR3. Next, the floating-point output value R0 is converted to its integer equivalent R1, which is then stored at the output memory address (specified by AR3). The FIX instruction rounds down the result; for example, 1.5 would become 1 and -1.5 would become -2.

5. **Branch Instructions.** Branches can be unconditional as well as conditional, based on a wide range of conditions. A standard branch, conditional or not, executes in four cycles. Delayed branches, conditionally and unconditionally, are also available. The delayed branch instruction allows the subsequent three instructions to be fetched

before the program counter PC is modified, effectively resulting in a single-cycle branch. For example, consider the following program segment:

```
BD     LOOP
ADDF   R0,R1
FIX    R1
STI    R1,*AR3
```

The unconditional branch to LOOP (a label that specifies the address where to branch)takes place *after* the STI instruction. The floating-point value R0 is added to R1, and the result is stored in R1. R1 is next converted from a floating-point to a fixed-point value, then stored in a memory location specified by AR3 (the auxiliary register AR3 might have first been loaded with an output address).

An instruction to decrement a register and branch (conditionally or unconditionally) with delay is also available:

```
DBNZD  AR2,LOOP
ADDF   R0,R1
FIX    R1
STI    R1,*AR3
```

AR2 is decremented by 1, and the delayed branch to LOOP is performed after the STI instruction, until AR2 < 0. Note that branching would continue as long as AR2, which could be used as a loop counter, is greater than or equal to zero.

6. **Repeat Instructions.** The RPTS instruction allows the subsequent single instruction to be repeated a number of times without the penalty for looping. A block of instructions can be repeated a number of times using RPTB. With the RPTB instruction, the starting address is loaded into the special repeat start address register (RS) and the ending address is loaded into the special repeat end address register (RE). A special repeat counter register (RC) must first be loaded with the number of times the block of code is to be repeated. For example, consider the program segment:

```
            LDI    9,RC
            RPTB   END_ADDR
            CALL   FILTER
            FIX    R0
END_ADDR    STI    R0,*AR3
```

The repeat counter register RC is loaded with 9, causing the block of code starting with the CALL FILTER instruction to be repeated (after the first time) nine times. Hence, this block of code is executed a total of 10 times. The label END_ADDR specifies the end of the block of code to be repeated. Within the block of code to be repeated, a subroutine FILTER is called. Upon returning from the FILTER routine, a floating-point value in R0 is converted from floating-point to integer, then stored in memory, with the memory address in AR3.

The RPTS instruction is not interruptable. If an interrupt is allowed to occur within a loop controlled by a repeat command, the RPTS instruction must be replaced by the RPTB instruction. The RPTS instruction will be illustrated next in conjunction with a set of parallel instructions.

7. *Parallel Instructions.* A number of parallel instructions are available, such as load and store, multiply and add. Such parallel instructions make the TMS320C30 capable of performing up to 33.3 MIPS. Consider the following program segment, which uses the single repeat instruction in conjunction with the set of parallel instructions MPYF and ADDF:

```
           LDF    0,R0
           LDI    29,AR2
           RPTS   AR2
           MPYF   *AR0++,*AR1++,R0
||         ADDF   R0,R2,R2
           ADDF   R0,R2
```

(a) Note the parallel symbol || used with the first ADDF instruction. AR2 is loaded with 29, causing the next single MPYF instruction to be executed a total of 30 times. However, this single MPYF instruction is in parallel with the first ADDF instruction. Hence, the MPYF and ADDF pair of instructions is executed 30 times.

(b) The floating-point value contained in memory specified (pointed) by AR0 is multiplied by the floating-point value contained in memory specified by AR1, and the result is stored in R0. At the same time, R0 is added with R2, with the result stored in R2. The first R0 value is *not* the resulting product of the most recent multiplication, because the ADDF and MPYF instructions are performed in parallel. The first R0 value is 0, initialized with the instruction LDF 0,R0.

(c) Both AR0 and AR1 are then incremented to point at the next-higher memory addresses. A second multiplication operation is then performed, and the result stored in R0. At the same time, R0, which resulted from the *first* multiplication, is added to R2, with the accumulation of the result again into R2.

(d) After the 30th and last multiplication operation, R0, containing the result from the 29th multiplication, is accumulated into R2. The second ADDF instruction, ADDF R0,R2, is executed only once, in order to accumulate the result of the 30th multiplication operation. Note that the second R2 can be omitted from the ADDF R0,R2,R2 instruction, because it is implied. The destination or result of the MPYF multiplication operation must be in either R0 or R1, and the destination or result of the ADDF addition instruction must be in either R2 or R3.

8. *Instructions Using Circular Buffering.* Circular buffering will be used quite often, within routines that implement a convolution equation representing a digital filter. Consider the following program segment:

```
LENGTH    .SET    30
          LDI     LENGTH,BK
          LDF     0,R0
          RPTS    LENGTH-1
          MPYF    *AR0++%,*AR1++%,R0
||        ADDF    R0,R2,R2
          ADDF    R0,R2,R2
```

The size of the circular buffer BK is loaded with 30, the LENGTH previously set to 30. The MPYF and the first ADDF operations are executed 30 times. Two circular buffers are used in conjunction with AR0 and AR1. If AR0 and AR1 are pointing initially to memory addresses X (first circular buffer) and Y (second circular buffer), respectively, when performing the 30th multiplication operation, AR0 and AR1 would be pointing at memory addresses X+29 and Y+29, respectively. After being postincremented, AR0 and AR1 will point "back" to the initial memory (top or lower memory) addresses X and Y, respectively.

Several programming examples will be discussed later in this chapter in order to illustrate many of the TMS320C30 instructions, providing at the same time background materials for more advanced topics.

2.5 DATA AND FLOATING-POINT FORMATS

Three types of data—integer (or, equivalently, signed integer), unsigned integer, and floating-point—are available on the TMS320C30. Short and single-precision formats are supported for signed- and unsigned-integer data and single- and extended-precision formats for floating-point data.

The floating-point format is specific to the TMS320C30, which is not the same as the IEEE floating-point format. The TMS320C30 floating-point format consists of an exponent field (E), a sign bit (S), and a fraction field (F). This floating-point representation, while not a standard, provides effectively

2.5 Data and Floating-Point Formats   29

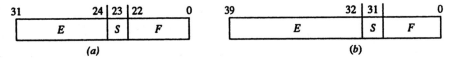

FIGURE 2.3. Floating-point formats: (*a*) single-precision; (*b*) extended-precision

an additional bit of precision from an implied bit, enhancing the efficiency of the TMS320C30.

During processing, an integer input value obtained from an ADC can be converted to a floating-point format, using the FLOAT instruction. Processing is done in floating-point format in order to achieve greater accuracy. A floating-point result can then be converted to an integer format, using the FIX instruction, before sending it to a DAC. Format conversions, for example, from short to extended-precision, are done automatically in hardware with no penalty.

The extended-precision floating-point format takes advantage of the eight 40-bit extended-precision registers R0-R7. Floating-point multiplications are performed with the operands in the single-precision format to yield a floating-point result in extended-precision. The floating-point multiplication operation is a single-cycle instruction. A floating-point number in the extended-precision format, is represented by an 8-bit (bits 32–39) exponent field (E), a single-bit (bit 31) sign field (S), and a 31-bit fractional field (F). A floating-point number in the single-precision format is represented by an 8-bit (bits 24–31) exponent field (E), a single-bit (bit 23) sign field (S), and a 23-bit fractional field (F). Figure 2.3 shows a single-precision floating-point format and an extended-precision floating-point format. The range of the floating-point format is from $\simeq -2^{128}$ to $\simeq 2^{128}$.

Because the simulator memory window displays the memory contents in floating-point format, several short examples will show how to convert those values into decimal. Fortunately, these floating-point values would also be stored in the extended-precision registers that display the equivalent decimal values.

A number N can be represented as

$$N = \begin{cases} 01.\text{F} \times 2^{\text{E}}, & \text{if the sign bit S} = 0 \\ 10.\text{F} \times 2^{\text{E}}, & \text{if the sign bit S} = 1 \end{cases}$$

Consider the following 32-bit floating-point hex values:

1. Zero value = 00000000
   E = 00 (bits 24–31), S = 0 (bit 23), F = 0 (bits 0–22)
   Because S = 0, $N = 01.\text{F} \times 2^{\text{E}} = 01.0 \times 2^0 = 1$ in decimal. Note that trailing zeros are deleted.

2. Positive value: 02400000
   E = 0000 0010 = 2, S = 0, F = 100 0000 0000 0000 0000 0000
   $N = 01.10 \times 2^2 = (0110)_b$, in binary, deleting trailing zeros and shifting right two places, or $N = 6$.
3. Negative value: 03A00000
   E = 0000 0011 = 3, S = 1, F = 010, deleting trailing zeros.
   $N = 10.010 \times 2^3 = (10010)_b$
   A value of S = 1 indicates a negative two's complement number. Because this is a two's complement number, one can take the one's complement of $N = (10010)_b$ and add 1 to the resulting number, or $N = [(01101)_b + 1) = -14$.
4. Fractional value: FC200000
   E = 1111 1100, S = 0, F = 010, deleting trailing zeros
   Because this is a fractional value, determined from the most significant bit being a 1, the one's complement of E is taken to yield $(0000\,0011)_b + 1 = -4$. Then,

$$N = 01.010 \times 2^{-4} = (0.0001010)_b = 2^{-4} + 2^{-6} = \tfrac{5}{64}$$

2.6 PROGRAMMING EXAMPLES USING BOTH C AND TMS320C30 CODE WITH THE SIMULATOR

Several programming examples are discussed in this section, to provide more knowledge of the instructions of the TMS320C30. Examples using C and mixed mode (C and assembly) are presented. These examples will provide more familiarity with the development tools. The best way to learn that is through examples.

Example 2.1 *Addition of Five Values Using TMS320C30 Code.* Figure 2.4 shows the program listing ADDF.ASM for adding five floating-point values 0, 1, 2, 3, 4. The associated command file is on the accompanying disk. The command file listed in Example 2.4 (Figure 2.10, used with PREFIR.CMD) can be modified to obtain ADDF.CMD, that is, the IO memory definition is not needed, and the IN_PORT, OUT_PORT, and XN_BUFFER definitions (in the SECTIONS) also can be deleted. This command file is quite general and can be used in future examples with minor modifications, for both simulation and real-time processing. Note how the various sections are allocated in memory. The memory organization of the TMS320C30 in Figures 2.2 and B.5 (Appendix B) shows that the RESET vector is at memory location 0 and that memory locations 1–3Fh are either reserved or for interrupts and traps. The command file shows the VECS section starts at memory location 0, with a length of 40h. The text section is specified in SRAM, which has a length of 16K. This memory section is on the EVM, external to the TMS320C30. The

2.6 Programming Examples Using both C and TMS320C30 Code

```
;ADDF.ASM  - ADD 5 FLOATING-POINT VALUES
           .TITLE   "ADDF.ASM"       ;PROGRAM TITLE
           .GLOBAL  BEGIN            ;REF/DEF SYMBOLS
VAL_ADDR   .WORD    VALUES           ;SPACES FOR VALUES
VALUES     .FLOAT   0,1,2,3,4        ;5 VALUES TO BE ADDED
BEGIN      LDI      @VAL_ADDR,AR0    ;LOAD ADDR OF VALUES->AR0
           LDF      0.0,R0           ;INIT R0=0
           CALL     ADDF             ;CALL SUBROUTINE ADDF
WAIT       BR       WAIT             ;WAIT
;SUBROUTINE ADDF
ADDF       RPTS     4                ;NEXT INSTRUCTION 5 TIMES
           ADDF     *AR0++,R0        ;ACCUMULATE INTO R0
           RETS                      ;RETURN FROM SUBROUTINE
           .END                      ;END
```

FIGURE 2.4. Program listing for addition of five values (ADDF.ASM)

data section is specified in internal memory RAM of the TMS320C30, starting at 809800h (see Figure 2.2), with a length of 2K. This command file can be used throughout the text, changing the specification of the object and output files (and possibly changing the text and bss sections).

Consider the addition program in Figure 2.4. The .GLOBAL is not an instruction and does not show in the code window of the simulator. It is an assembler directive that can be used as either of the following:

1. .DEF to identify a symbol that is defined in the current module and can be accessed by another file
2. .REF to identify a symbol defined in another module but used in the current module

The .WORD in the program listing (with the label VAL_ADDR) is also an assembler directive, used to reserve the memory locations for the five values.

The .FLOAT is also an assembler directive to place one or more of the floating-point constants into successive memory locations within the specified section. The values 0 through 4 are converted into their equivalent 32-bit floating-point representations with the .FLOAT assembler directive. Proceed with the following:

1. Type ASM30 ADDF to assemble the program ADDF.ASM and create ADDF.OBJ.
2. Type LNK30 to link, using ADDF.CMD as the command file. Press <ENTER> twice, because the object and output files are already specified within the command file. You may wish to obtain a MAP file, by typing -M ADDF.MAP as an option during the linking process, or within

the command file. The map file provides debugging information such as where code or data reside in memory, showing the addresses of instructions with labels.

3. Type the commands SIM30 L 0, then ADDF.OUT to access the simulator, and load the linked COFF output file
4. Type D M 40h to display the memory starting at address 40h
5. The five values (0, 1, 2, 3, 4) to be added are displayed in floating-point format as 80000000, 00000000, 01000000, 01400000, and 02000000, respectively, starting at memory addresses 41h–45h. Do not worry too much about these floating-point values, because these floating-point values are accumulated in R0, which displays the equivalent decimal number.
6. Press F1 to access the code window and F3 or the down-arrow key to line 46h (the first column), which corresponds to the memory address. This program was loaded in memory starting at 40h, as specified in the command file. The code for the instructions are in the second column of the code window, displayed at memory addresses 46h through 4Ch.
7. Press F1 twice to access the CPU status window. Press E to edit, and type 00000046 to set the program counter PC to 46h. Note that it is not necessary to follow this procedure (and the previous step) if the command file includes the option -E BEGIN. Press X I 1, to execute with a repeat count of 1, effectively single-stepping.

Single-step, and observe the following:

1. The starting address of the values to be added, 41h, is loaded into auxiliary register AR0, using the instruction LDI @VAL_ADDR,AR0.
2. The extended-precision register R0 is initialized to zero.
3. The subroutine ADDF is called.
4. The instruction ADDF *AR0++,R0 is repeated four times (executed five times). As you single-step through the addition instruction, observe R0, where both the result in floating-point and its equivalent decimal value are displayed. Also observe AR0 being incremented to point at the next-higher memory address.
5. Execution is returned from the subroutine ADDF to the instruction following the subroutine call.
6. The display of the clock CLK from the CPU status window section. The execution time of this program can be obtained by multiplying the CLK value by the instruction cycle time of 60 ns.

Figure 2.5 shows the simulator screen display for the addition example.

2.6 Programming Examples Using both C and TMS320C30 Code

```
Simulator: Display eXecute Reg Mem Bpoint Eval Config Trace Load Save ->
                                              HALTED
═══════════════════code═══════════════════  ═══extended precision registers═══
    000044  01400000  ADDC  *+AR0(0),R0      R0 03 20000000 = 1.00000000e+01
    000045  02000000  ADDI  R0,R0            R1 00 00000000 = 1.00000000e+00
    000046  XXXXXXXX  BEGIN:                 R2 00 00000000 = 1.00000000e+00
    000046  08280040  LDI   .text,AR0        R3 00 00000000 = 1.00000000e+00
    000047  07608000  LDF   0.0,R0           R4 00 00000000 = 1.00000000e+00
    000048  6200004A  CALL  04aH             R5 00 00000000 = 1.00000000e+00
-r- 000049  60000049  BR    049H             R6 00 00000000 = 1.00000000e+00
    00004A  13FB0004  RPTS  4                R7 00 00000000 = 1.00000000e+00
    00004B  01C02001  ADDF  *AR0++(1),R0     ═══════════cpu status═══════════
    00004C  78800000  RETSU                  PC     0000004A   CLK   0000001A
═══════════════════display═══════════════  ST     00000000   DP    00000000
    000040  00000041 80000000 00000000 01000000   SP     00000000   BK    00000000
    000044  01400000 02000000 08280040 07608000   AR0    00000046   AR1   00000000
    000048  6200004A 60000049 13FB0004 01C02001   AR2    00000000   AR3   00000000
    00004C  78800000 00000000 00000000 00000000   AR4    00000000   AR5   00000000
    000050  00000000 00000000 00000000 00000000   AR6    00000000   AR7   00000000
    000054  00000000 00000000 00000000 00000000   IR0    00000000   IR1   00000000
    000058  00000000 00000000 00000000 00000000   IE     00000000   IF    00000000
    00005C  00000000 00000000 00000000 00000000   IOF    00000000   RS    0000004B
    000060  00000000 00000000 00000000 00000000   RE     0000004B   RC    00000000
F1Inspect                       F8Command F9Update F10Help
```

FIGURE 2.5. Simulator screen display for addition example

Example 2.2 Multiplication of Two Arrays Using TMS320C30 Code. Figure 2.6 lists the main program `MULT.ASM` to multiply two arrays, each containing four values. This program calls the subroutine `MULTSUB.ASM` (listed in Figure 2.7), which performs the multiplication task. The appropriate command file (`MULT.CMD`) is similar to the previous addition example command

```
;MULT.ASM - MULTIPLY TWO ARRAYS
            .TITLE  "MULT.ASM"     ;PROGRAM TITLE
            .GLOBAL BEGIN,MULT     ;REF/DEF SYMBOLS
            .DATA                  ;ASSEMBLE INTO DATA SECTION
H_ADDR      .WORD   HN             ;STARTING ADDRESS OF HN ARRAY
X_ADDR      .WORD   XN             ;STARTING ADDRESS OF XN ARRAY
HN          .FLOAT  1,2,3,4        ;ARRAY OF HN VALUES
XN          .FLOAT  2,3,4,5        ;ARRAY OF XN VALUES
            .TEXT                  ;ASSEMBLE INTO TEXT SECTION
BEGIN       LDP     H_ADDR         ;INIT DATA PAGE
            LDI     @H_ADDR,AR0    ;ADDRESS OF FIRST HN VALUE->AR0
            LDI     @X_ADDR,AR1    ;ADDRESS OF FIRST XN VALUE->AR1
            LDI     3,RC           ;REPEAT COUNT REG RC=3
            CALL    MULT           ;GO TO SUBROUTINE MULT
WAIT        BR      WAIT           ;WAIT
```

FIGURE 2.6. Program listing for multiplication of two arrays (`MULT.ASM`)

```
;MULTSUB.ASM - SUBROUTINE FOR MULTIPLICATION PROGRAM
         .TITLE   "MULTSUB.ASM"   ;MULTIPLY SUBROUTINE
         .GLOBAL  MULT            ;REF/DEF SYMBOL
         .TEXT                    ;ASSEMBLE INTO TEXT SECTION
MULT     LDF      0,R0            ;INIT R0=0
         LDF      0,R2            ;INIT R2=0
         RPTS     RC              ;EXEC NEXT 2 INSTR RC+1 TIMES
         MPYF     *AR0++,*AR1++,R0   ;(AR0)*(AR1)->R0
||       ADDF     R0,R2           ;IN // WITH ACCUM INTO R2
         ADDF     R0,R2           ;LAST MULT RESULT ADDED TO R2
         RETS                     ;RETURN FROM SUBROUTINE
         .END                     ;END
```

FIGURE 2.7. Subroutine program listing for multiplication of two arrays (MULTSUB.ASM)

file. The files to be linked and the output file can be replaced by using the following:

```
-E BEGIN          /*specifies the entry point*/
MULT.OBJ          /*the main program*/
MULTSUB.OBJ       /*the subroutine*/
-O MULT.OUT       /*the linked COFF output file*/
```

First assemble the two source files MULT.ASM and MULTSUB.ASM in order to create the two files MULT.OBJ and MULTSUB.OBJ. Link using LNK30 MULT.CMD to create the output file MULT.OUT. Download this output file into the simulator, as in the previous example, using SIM30 MULT.OUT. Note that by adding the -E linker option, the program counter is automatically set to 40h, specified by BEGIN, and the code window displays the code starting at 40h. Display the memory addresses (using D then M), starting at 809800h. The first set of floating-point values (HN) start at memory location 809802h and the second set of floating-point values (XN) start at 809806h. Press (enter separately) X I 1 to single-step, and note the following:

1. The data page is loaded and initialized. The data page register DP is loaded with 128. The memory map of the TMS320C30 contains 256 pages, each with 64K words long, yielding a total of 16M of memory. DP is initialized to the start of the middle section of the total memory space, because H_ADDR is assembled into the data section, in RAM, as specified in the command file (809800h corresponds to page 128).

2. The starting addresses of the HN and XN arrays are 809802h and 809806h, respectively. These addresses are within the data section, specified by using the .DATA assembler directive. This data section, specified within the command file, starts at 809800h (the starting address of internal RAM).

2.6 Programming Examples Using both C and TMS320C30 Code

3. The .TEXT directs the assembler to assemble the code into the text section, starting at 40h, as specified in the command file.
4. AR0 is loaded with 809802h, the starting address of the HN array, and AR1 is loaded with 809806h, the starting address of the XN array. The repeat counter register RC is loaded with 3, and the program flow proceeds to the subroutine MULT. Note that in this example, the subroutine MULT, within the separate file MULTSUB.ASM was linked with the main program.
5. In the subroutine MULT, R0 and R2 are initialized to zero. The two parallel instructions MPYF (multiply) and ADDF (add) are executed a total of four times. After each multiply operation, both AR0 and AR1 are incremented. Monitor R0 in the simulator window for the result of each multiplication, and R2 for the accumulation of the intermediate and final results. Note that when the ADDF instruction is executed the first time, R0 is zero (initialized previously) and not the product that resulted from the first multiplication operation. The second ADDF instruction accumulates the last multiplication product. Monitor R2 and verify that it becomes subsequently equal to 0, 2, 8, 20, 40.
6. Execution is next returned to the main program, using the return from subroutine instruction RETS. The BR WAIT instruction causes a branch to the address specified by the WAIT label, effectively causing the processor to execute the same instruction indefinitely (waits).

Example 2.3 Circular Buffer and Data Transfer Using TMS320C30 Code. This example demonstrates the circular mode of addressing and the transfer of data from external memory SRAM to internal memory RAM, block 0.

The program listing for this example CIRC.ASM is shown in Figure 2.8. A similar command file as with the previous multiplication example can be used. In the command file, both the data and the text sections are specified in the SRAM (not RAM), starting at memory address 40h. The bss section is specified in RAM, starting at 809800h (block 0).

Note the following initialization:

1. H_ADDR   .WORD HN   the address of the first HN values in external SRAM
2. XSRAM_ADDR   .WORD XN   the address of the first XN values in SRAM
3. XRAM_ADDR   .WORD X3   the starting address reserved (using .BSS) for the XN values in internal RAM; the BSS section is specified in RAM within the command file.

```
;CIRC.ASM - DEMONSTRATES CIRCULAR BUFFER & DATA TRANSFER
            .TITLE  "CIRC.ASM"      ;PROGRAM TITLE
            .GLOBAL BEGIN            ;REF/DEF SYMBOLS
            .DATA                    ;ASSEMBLE INTO DATA SECTION
HN          .FLOAT  1                ;1ST VALUE IN HN ARRAY (H3)
            .FLOAT  2                ;2ND HN VALUE (H2)
            .FLOAT  0                ;3RD HN VALUE (H1)
H0          .FLOAT  3                ;4TH HN VALUE (H0)
            .BSS    X3,LENGTH        ;4 SPACES FOR XN IN BSS(INTERNAL RAM)
XN          .FLOAT  2                ;1ST XN VALUE (X3)
            .FLOAT  3                ;2ND XN VALUE (X2)
            .FLOAT  4                ;3RD XN VALUE (X1)
X0          .FLOAT  0                ;4TH XN VALUE (X0)
H_ADDR      .WORD   HN               ;STARTING ADDR OF HN ARRAY
XSRAM_ADDR  .WORD   XN               ;STARTING ADDR OF XN IN SRAM
XRAM_ADDR   .WORD   X3               ;START ADDR RESERVED FOR XN IN RAM
X0_ADDR     .WORD   X3+LENGTH-1      ;BOTTOM SAMPLE ADDR OF X0 IN RAM
LENGTH      .SET    H0-HN+1          ;# OF HN VALUES
;           +--------+                +------------+
; LOW ADDR  | H3 = 1 |                | X3 = 2     |
;           +--------+                +------------+
;           | H2 = 2 |                | X2 = 3     |
;           |--------+                +------------+
;           | H1 = 0 |                | X1 = 4     |
;           +--------+                +------------+
; HIGH ADDR | H0 = 3 |                | X0 = 0     |
;           +--------+                +------------+
            .TEXT                    ;ASSEMBLE INTO TEXT SECTION
BEGIN       LDP     H_ADDR           ;INIT DATA PAGE (0)
            LDI     @XSRAM_ADDR,AR0  ;XN ADDR IN SRAM -> AR0
            LDI     @XRAM_ADDR,AR1   ;BEGINNING ADDR IN RAM 0->AR1
            LDI     LENGTH,BK        ;BK = 4 (LENGTH OF CIRC BUFFER)
            LDI     LENGTH-1,RC      ;REPEAT COUNTER REG RC = 3
            RPTB    XFER             ;BLOCK REPEAT TIL XFER
            LDF     *AR0++,R0        ;TO TRANSFER SAMPLES XN
XFER        STF     R0,*AR1++%       ;FROM SRAM TO RAM 0(START @ 809800H)
            LDI     @H_ADDR,AR0      ;START ADDR OF HN (H3) -> AR0
            LDI     @X0_ADDR,AR1     ;BOTTOM ADDR(X0)->AR1 (LAST XN ADDR)
            LDF     6,R5             ;SET R5=6
            STF     R5,*AR1++%       ;STORE R5 @ X0 ADDR THEN AR1->X3
            LDF     0,R0             ;INIT R0=0
            LDF     0,R2             ;INIT R2=0
            RPTS    LENGTH-1         ;EXECUTE NEXT 2 INSTR 4 TIMES
            MPYF    *AR0++,*AR1++%,R0  ;(AR0)*(AR1) -> R0
||          ADDF    R0,R2            ;IN PARALLEL WITH ACC INTO R2
            ADDF    R0,R2            ;LAST ACC -> R2=H3*X3+...+H0*X0
WAIT        BR      WAIT             ;WAIT
            .END                     ;END
```

FIGURE 2.8. Program listing to demonstrate circular buffering and data transfer (CIRC.ASM)

2.6 Programming Examples Using both C and TMS320C30 Code

4. `X0_ADDR   .WORD X3 + LENGTH- 1` specifies the "bottom" (last) address of the XN values (X0)
5. The length of the HN array is found by using the difference between the "bottom" higher-memory address (H0) of the HN array and the "top" lower-memory address (HN), plus 1
6. Whereas XN is the starting address of the XN values in external SRAM, X3 is the starting address in internal RAM reserved for the XN values.

Assemble, link, and download this program into the simulator. Display the memory addresses starting at 40h, and verify that the four HN data values (1, 2, 0, 3) and the four XN data values (2, 3, 4, 0), in floating-point representation, are in memory addresses 40h through 47h. Single-step through the program, and observe the following:

1. The data page register DP is initialized to zero (because H_ADDR is in page 0), and the text section is specified in SRAM. AR0 is loaded with the starting address of the XN values in SRAM, and AR1 is loaded with the starting address in internal RAM (specified by using the .BSS assembler directive) reserved for the XN values. The size of the circular buffer, LENGTH = 4, is loaded into register BK. The repeat counter register RC is loaded with 3, in order to use the repeat block of code instruction RPTB.
2. The two instructions following the repeat block instruction are executed four times. Display the memory window starting at 809800h to verify that the XN values are transferred from memory locations 44h–47h in external SRAM to memory locations 809800h–809803h in internal RAM.
3. After the data transfer, AR1 points back to the top lower-memory address 809800h, due to the circular mode of addressing, specified by the modulo operator % in *AR1 + + %.
4. The starting address (40h) of HN is loaded into AR0. The "bottom" address (809803h) of the XN values is loaded into AR1. Register R5 is initialized to the value 6 and its floating-point equivalent is stored in 809803h. Then AR1 is postincremented in a circular fashion, to point "back" at 809800h.
5. The remainder of the code is similar to the previous example, which multiplies two arrays HN and XN. Step through this section of code and verify that R2 = 2 + 6 + 0 + 18 = 26 (X0 = 6). Note that AR1 then points to the top of the block of memory reserved for XN.

It is coincidental that the circular buffer section is properly specified or aligned and performs as expected. However, it is desirable to ensure that the data is properly aligned within the circular buffer. It would not be so if the first data value were at 809801h instead of 809800h. Instead of using the assembler directive .BSS X3,LENGTH to reserve four spaces for XN in the bss section (specified within the command file as internal RAM), using the alternative directives

```
.ALIGN
.FLOAT    0,0,0,0
```

would align for circular buffering (with .ALIGN) and reserve four spaces for XN. This method would cause alignment on a 32-word boundary and guarantee circular buffering. NOP (no operation) instruction would be placed in the remainder of the memory locations to guarantee that the data is properly aligned on a 32-word boundary. Note that if the size of the arrays were 30, for example, 30 zeros would be included with the .FLOAT directive. A more convenient and preferable method is to specify an ALIGN command within the command file, as discussed in the next example.

Example 2.3, along with Example 2.4, provides the background for coding a digital filter, as discussed in Chapter 4.

Example 2.4 *Background Programming for Digital Filtering Using TMS320C30 Code.* This very useful example builds upon the previous two examples, to provide the background necessary for implementing digital filters using TMS320C30 code. An input file (NOISE4) representing four samples of "noise" is used. Figure 2.9 shows the program listing PREFIR.ASM and Figure 2.10 shows the command file listing PREFIR.CMD. Be careful when modifying the command file, because the commands listed are *case sensitive*. In the command file, the input and output port addresses are specified (804000h, 804002h). For proper alignment of the circular buffer, the buffer XN_BUFFER is specified in the main program with the .USECT assembler directive. The .USECT is to reserve space for variables in a user-defined section (the assembler directives .DATA and .TEXT correspond to unnamed sections). This named section appears under the double quote in the main program followed by the size of the buffer. Although the buffer may be properly aligned (by luck) for circular addressing, the ALIGN statement within the command file guarantees it. This buffer section must be aligned on an n-word boundary, where n is a power of 2. In this example, we chose a 64-word boundary, even though we needed only four. For five samples, we would need a minimum of an eight-word boundary. We can change this boundary to 512, if we wanted to implement a filter with 500 samples.

Create the noise input file NOISE4, with the four values that represent ±4,096:

```
0X00001000
0XFFFFF000
0X00001000
0XFFFFF000
```

```
;PREFIR.ASM - BACKGROUND FOR FILTER,NOISE(4 SAMPLES)INPUT
            .TITLE    "PREFIR.ASM"         ;BACKGROUND FOR CONVO
            .GLOBAL   BEGIN,MAIN           ;REF/DEF SYMBOLS
            .DATA                          ;ASSEMBLE INTO DATA SECTION
HN          .FLOAT    1                    ;1ST VALUE IN HN ARRAY (H3)
            .FLOAT    2                    ;2ND HN VALUE (H2)
            .FLOAT    0                    ;3RD HN VALUE (H1)
H0          .FLOAT    3                    ;4TH HN VALUE (H0)
H_ADDR      .WORD     HN                   ;STARTING ADDR OF HN ARRAY
X0_ADDR     .WORD     XN+LENGTH-1          ;ADDR OF BOTTOM SAMPLE X(n)
IN_ADDR     .WORD     IN                   ;INPUT PORT ADDR
OUT_ADDR    .WORD     OUT                  ;OUTPUT PORT ADDR
LENGTH      .SET      H0-HN+1              ;# OF HN VALUES
XN          .USECT    "XN_BUFFER",LENGTH   ;BUFFER SIZE OF SAMPLES XN
            .SECT     "VECTORS"            ;ASSEMBLE INTO VECT SECTION
MAIN        .WORD     BEGIN                ;START OF CODE
IN          .USECT    "IN_PORT",1          ;1 SPACE FOR INPUT PORT ADDR
OUT         .USECT    "OUT_PORT",1         ;1 SPACE FOR OUTPUT PORT ADDR
;                   +---------+          +------------+
; LOW ADDR  | H3 = 1 |          |  X(n-3)  |
;                   +---------+          +------------+
;                   | H2 = 2 |          |  X(n-2)  |
;                   |---------+          +------------+
;                   | H1 = 0 :          |  X(n-1)  |
;                   +---------+          +------------+
; HIGH ADDR | H0 = 3 |          |  X(n)    |
;                   +---------+          +------------+
            .TEXT                          ;ASSEMBLE INTO TEXT SECTION
BEGIN       LDP       IN_ADDR              ;INIT DATA PAGE (0)
            LDI       @X0_ADDR,AR1         ;BOTTOM SAMPLE X(n) ADDR->AR1
            LDI       LENGTH,BK            ;BK = 4 (LENGTH OF CIRC BUFFER)
            LDF       0,R0                 ;INIT R0=0
            RPTS      LENGTH-1             ;NEXT INST. LENGTH TIMES
            STF       R0,*AR1--%           ;INIT SAMPLES=0,AR1->X(n) AGAIN
            LDI       @IN_ADDR,AR5         ;INPUT PORT ADDR -> AR5=804000h
            LDI       @OUT_ADDR,AR6        ;OUTPUT PORT ADDR-> AR6=804002h
            LDI       LENGTH,R4            ;R4=4,LOOP COUNTER
LOOP        FLOAT     *AR5,R3              ;INPUT FROM PORT ADDR->R3=4096
            STF       R3,*AR1++%           ;NEWEST SAMPLE @ 809803
            LDI       @H_ADDR,AR0          ;START ADDR OF HN->AR0
            CALL      FILTER               ;GO TO SUBROUTINE FILTER
            FIX       R2,R1                ;R1=INTEGER(R2)
            STI       R1,*AR6              ;OUTPUT FROM @ 804002H
            SUBI      1,R4                 ;DECREMENT R4
            BNZ       LOOP                 ;BRANCH BACK UNTIL R4=0
WAIT        BR        WAIT                 ;WAIT INDEFINITELY
;SUBROUTINE FILTER
FILTER      LDF       0,R0                 ;INIT R0=0
            LDF       0,R2                 ;INIT R2=0
            RPTS      LENGTH-1             ;EXECUTE NEXT 2 INSTR 4 TIMES
            MPYF      *AR0++,*AR1++%,R0    ;(AR0)*(AR1) -> R0
||          ADDF      R0,R2                ;IN PARALLEL WITH ACC IN R2
            ADDF      R0,R2                ;LAST ACC->R2=H3*X(n-3)+..+H0*X(n)
            RETS                           ;RETURN FROM SUBROUTINE
            .END                           ;END
```

FIGURE 2.9. Program listing as background for digital filtering (PREFIR.ASM)

```
/*PREFIR.CMD          COMMAND FILE              */
-E BEGIN              /*SPECIFIES ENTRY POINT   */
PREFIR.OBJ            /*OBJECT CODE             */
-O PREFIR.OUT         /*LINKED COFF OUTPUT FILE*/
MEMORY
{
 VECS: org = 0          len = 0x40    /*VECTOR LOCATIONS */
 SRAM: org = 0x40       len = 0x3FC0  /*PRIMARY BUS (16K)*/
 RAM : org = 0x809800   len = 0x800   /*2K INTERNAL RAM  */
 IO  : org = 0x804000   len = 0x2000  /*8K -IOSTRB       */
}
SECTIONS
{
 .data:    {} > SRAM              /*DATA SECTION IN SRAM       */
 .text:    {} > SRAM              /*CODE                       */
 .cinit:   {} > SRAM              /*INITIALIZATION TABLES      */
 .stack:   {} > RAM               /*SYSTEM STACK               */
 .bss:     {} > RAM               /*BSS SECTION IN RAM         */
 vecs:     {} > VECS              /*RESET & INTERRUPT VECTORS  */
 IN_PORT   804000h : {} > IO      /*INPUT PORT ADDRESS         */
 OUT_PORT  804002h : {} > IO      /*OUTPUT PORT ADDRESS        */
 XN_BUFFER ALIGN(64): {} > RAM    /*ALIGN CIRCULAR BUFFER IN RAM*/
}
```

FIGURE 2.10. Command file for digital filtering background example (PREFIR.CMD)

Assemble and link the PREFIR program. As in Chapter 1, configure the memory by loading the memory configuration file M.CFM created in Chapter 1. Configure the port map (toggle with F1) to specify the input file NOISE4 at address 804000h and the output file PREFIR.DAT at address 804002h.

Single-step through the program, and observe the following:

1. The data page is initialized to zero, AR1 is loaded with the internal RAM address 809803h. BK is initialized with the size of the circular buffer, and R0 is initialized to zero. Display the memory window starting at address 809800h, and note that the memory locations 809803h through 809800h, reserved for the sample values XN, are initialized to zero (in floating-point format). Upon reaching the "top" (lower-memory) XN address, AR1 is postdecremented to point again at the "bottom" address (higher-memory) 809803h.

2. AR5 and AR6 are loaded with the input and output port addresses 804000h and 804002h, respectively. Register R4, used as a loop counter, is loaded with the value 4.

3. The FLOAT instruction (note the assembler directive is .FLOAT) inputs the first noise sample value 1000h (or 4,096) from memory port

2.6 Programming Examples Using both C and TMS320C30 Code

address 804000h as a floating-point number and loads it into the extended-precision register R3. This value is then stored in memory location 809803h, pointed by AR1. AR1 is next postincremented to point at 809800h.

4. The starting address (40h) of HN is loaded into AR0, and the program flow continues to the subroutine FILTER.
5. Both R0 and R2 are initialized to zero. The first time through the FILTER subroutine, monitor R0 and, especially, R2, to verify that

$$R2 = H3*X(n-3) + H2*X(n-2) + H1*X(n-1) + H0*X(n)$$
$$= 1(0) + 2(0) + 0(0) + 3(4,096)$$
$$= 12,288$$

6. Execution returns next to the main program, where the floating-point value in R2 is converted to integer. This value is then stored in the output memory port address 804002h and also in the file PREFIR.DAT. In a real-time environment, the input would come from an analog-to-digital converter ADC, and the output sent to a digital-to-analog converter DAC.
7. The loop counter R4 is decremented and the program flow is back to the address specified by LOOP. A new input noise value FFFF000h (or −4,096) is loaded into R3 and stored in memory location 809800h. AR1 then points at 809801h, which contains X(n-2). Steps 4–6 are next repeated. The second time through the FILTER subroutine,

$$R2 = H3*X(n-2) + H2*X(n-1) + H1*X(n) + H0*X(n-3)$$
$$= 1(0) + 2(0) + 0(4,096) + 3(-4,096)$$
$$= -12,288$$

Note that the last multiplication operation involves the last coefficient H0 and the newest input sample.

8. Continue to single-step through and verify that the third input is 4,096 and is stored at 809801h as X(n-2). After processing the FILTER subroutine a third time,

$$R2 = H3*X(n-1) + H2*X(n) + H1*X(n-3) + H0*X(n-2)$$
$$= 1(0) + 2(4,096) + 0(-4,096) + 3(4,096)$$
$$= 20,480$$

After processing the filter subroutine a fourth and last time,

$$R2 = H3*X(n) + H2*X(n-3) + H1*X(n-2) + H0*X(n-1)$$
$$= 1(4,096) + 2(-4,096) + 0(4,096) + 3(-4,096)$$
$$= -16,384$$

```
;MATRIX.ASM-MATRIX MULTIPLICATION IN ASSEMBLY CODE
        .GLOBAL  BEGIN
        .DATA                        ;ASSEMBLE INTO DATA SECTION
A       .FLOAT   1,2,3,4,5,6,7,8,9   ;VALUES FOR MATRIX A
B       .FLOAT   1,2,3               ;VALUES FOR MATRIX B
A_ADDR  .WORD    A                   ;ADDR OF MATRIX A
B_ADDR  .WORD    B                   ;ADDR OF MATRIX B
IO_OUT  .WORD    804001H             ;ADDR OF OUTPUT PORT
        .TEXT                        ;ASSEMBLE INTO TEXT SECTION
BEGIN   LDI      @A_ADDR,AR0         ;AR0=START ADDR OF A
        LDI      @B_ADDR,AR1         ;AR1=START ADDR OF B
        LDI      @IO_OUT,AR2         ;AR2=OUTPUT PORT ADDR
        LDI      3,R4                ;R4 IS LOOPI COUNTER
LOOPI   LDF      0,R0                ;INITIALIZE R0
        LDI      2,AR4               ;AR4 IS LOOPJ COUNTER
LOOPJ   MPYF     *AR0++,*AR1++,R1    ;A[I,J]*B[J]=R1
        ADDF     R1,R0               ;R0 ACCUMULATES RESULT
        DB       AR4,LOOPJ           ;DEC AR4 & BRANCH TIL AR4<0
        FIX      R0                  ;FLOAT TO INTEGER
        STI      R0,*AR2             ;OUTPUT RESULT TO IO_OUT
        LDI      @B_ADDR,AR1         ;RELOAD START ADDR OF B
        SUBI     1,R4                ;DECREMENT R4
        BNZ      LOOPI               ;BRANCH WHILE  R4 <> 0
WAIT    BR       WAIT                ;WAIT INDEFINITELY
```

FIGURE 2.11. Matrix multiplication program (MATRIX.ASM)

The circular buffering is a procedure used to model the delays in a difference or convolution equation. We will see later how digital filters can be implemented, modeling the delay samples using this technique. Each time the FILTER subroutine is processed, it corresponds to incrementing *n*, which represents time, by 1. View and verify the file PREFIR.DAT with the four output values in hex.

Example 2.5 *Matrix Multiplication Using TMS320C30 Code.* A matrix multiplication program using TMS320C30 assembly code is listed in Figure 2.11. This program is equivalent to the C version in Example 1.1, and was introduced and executed in Example 1.2. Although this assembly source program is longer and looks more difficult than its C version, it executes faster. This can be verified from the clock cycles displayed by CLK in the simulator after the program execution. Assemble the matrix program MATRIX.ASM and link it using the following command file:

```
-E    BEGIN
MATRIX.OBJ
-O    MATRIX.OUT
```

2.6 Programming Examples Using both C and TMS320C30 Code

Configure the memory (by loading M.CFM) and port maps to create an output file MATRIX.DAT at 804001h. Load the linked output COFF file MATRIX.OUT into the simulator. Observe the following:

1. AR0, AR1, and AR2 are loaded with the addresses of the A and B arrays and the output port address, respectively. R4 is the outer loop counter (starting with LOOPI).
2. Within the inner loop starting at LOOPJ (nestled within LOOPI), the multiplications of the first row numbers (from A) times the column numbers (from B) are performed, with each resulting value accumulated in R0.
3. AR1 is re-initialized to the beginning address of the B array before execution returns to the outer loop (starting with LOOPI) to repeat the procedure using the next two row numbers in the A array.

This example is specific to a $(3 \times 3)(3 \times 1) = (3 \times 1)$ matrix multiplication example; however, it can be extended to handle different sizes of matrices. The result from file MATRIX.DAT is identical with the result obtained with the C version (MATRIXC.DAT).

Example 2.6 Input and Output Using C Code. This example demonstrates input from a port and output to a port using C code. Compile, assemble the source program listed in Figure 2.12. Link with a command file INOUTC.CMD similar to the command file MATRIXC.CMD used in Example 1.1, or

```
CL30 -o2 INOUTC -z INOUTC.CMD
```

Create a file INOUTC.IN with the following values:

```
0x00000001
0x00000002
0x00000003
0x00000004
0x00000005
```

as the input file to port 804000h. Download the linked output file INOUTC.OUT into the simulator. Configure the memory and port maps as shown in the simulator screen, Figure 2.13. The output file INOUTC.DAT is specified at port address 804001h. Execute the program and verify (view

```
/*INOUTC.C-DEMONSTRATES INPUT AND OUTPUT         */
main( )
{
  volatile int *IO_INPUT=(volatile int *) 0x804000;
  volatile int *IO_OUTPUT=(volatile int *)0x804001;
  int count,value;
  for (count=0; count <=4; ++count)
    {
      value = *IO_INPUT;              /*input -> value*/
      *IO_OUTPUT = 2 * value;         /*output=2*input*/
    }
}
```

FIGURE 2.12. Input/output program using C code (INOUTC.C)

INOUTC.DAT) that the output is twice the input, or

0x00000002
0x00000004
0x00000006
0x00000008
0x0000000a

The use of the volatile declaration tells the compiler not to optimize any references to IO_INPUT and IO_OUTPUT, because they are pointing at the input and output addresses 804000h and 804001h, respectively. Note the use of *IO_INPUT and *IO_OUTPUT as pointers to these memory addresses.

```
Port Map:  Load  Save  Reset  Add  Delete
=================================device mode=================================
                         MC/MP = 1 (microComputer)
===================memory map===================|===============port map===============
ID#      from        to       #wait  typ  | ID#  type    address    filename
 1   0000000h - 0000FFFh        0    ROM  |  1   IO_IN   0804000h   inoutc.in
 2   0804000h - 08043FFh        0    EXT  |  2   IO_OUT  0804001h   inoutc.dat
 3   0808000h - 08083FFh        0    MMR  |
 4   0809800h - 0809FFFh        0    RAM  |




F1configMem           F8Command  F10Help
```

FIGURE 2.13. Memory and port maps for input/output program example

2.6 Programming Examples Using both C and TMS320C30 Code

```
/*ADDM.C-PROGRAM IN C CALLING A FUNCTION IN ASSEMBLY*/
extern int addmfunc();   /*external assembly function*/
int temp = 10;           /*global C variable          */
main()
{
 volatile int *IO_OUTPUT=(volatile int *) 0x804001; /*output port addr*/
 int count;
 for (count = 0; count < 5; ++count)
   {
    *IO_OUTPUT=addmfunc(count); /*calls assembly function*/
   }
}
```

FIGURE 2.14. Add program in C calling an assembly function (ADDM.C)

Example 2.7 Matrix Multiplication Using C Code. Consider again the matrix example in Chapter 1 (Example 1.1). Before running this matrix program again, delete the simulator output file MATRIXC.DAT since the second output result will be appended to the first one. Download MATRIXC.OUT into the simulator. Configure the memory and port maps, specifying the simulator output file MATRIXC.DAT at 804001h. Single-step through the program. Note the value of CLK at the end of the program execution, and verify that the C version of the matrix program executes slower than its assembly counterpart even though both the C and the assembly version programs can be made more efficient. The execution time is the CLK value multiplied by 60 ns. The execution time of this matrix program can be reduced further by specifying the memory sections in the command file.

Example 2.8 Addition with Both C and C-Called Function Using TMS320C30 Code. This example illustrates the use of mixed mode, using a C program, ADDM.C, listed in Figure 2.14, that calls an assembly function ADDMFUNC.ASM, listed in Figure 2.15. First assemble this function, using

ASM30   ADDMFUNC

Then compile and assemble ADDM.C and link (using a command file similar to that in the previous example), or

CL30 -o2 ADDM -z ADDM.CMD

Consider the assembly function ADDMFUNC.ASM The frame pointer FP is set in auxiliary register AR3. All C identifiers are referenced in assembly code with underscores, such as _addfunc and _addtemp. The frame pointer, offset appropriately, is used for passing the address of an argument from the C program to the assembly function. The address of the only argument, count, is at *-FP(2), because the old frame pointer FP is at the first location in the stack. The next example (matrix multiplication with mixed mode) shows how the address of an array can be passed into an assembly function.

```
*ADDMFUNC.ASM-ASSEMBLY FUNCTION CALLED FROM C PROGRAM
FP       .set    AR3             ;FRAME POINTER IN AR3
         .global _addmfunc       ;GLOBAL REF/DEF
         .global _temp           ;GLOBAL REF/DEF
_addmfunc                        ;FUNCTION IN ASSEMBLY
         PUSH    FP              ;SAVE FP ONTO STACK
         LDI     SP,FP           ;POINT TO START OF STACK
         LDI     *-FP(2),R0      ;1ST ARGUMENT INTO R0
         ADDI    @_temp,R0       ;ADD GLOBAL VARIABLE WITH R0
         POP     FP              ;RESTORE FP
         RETS                    ;RETURN FROM SUBROUTINE
```

FIGURE 2.15. C-called assembly function for add program (ADDMFUNC.ASM)

Certain registers are dedicated and any of the registers R4–R7, AR4–AR7, SP, DP, and FP (used as AR3), which are modified by the assembly function, must be preserved on the stack, using PUSH or PUSHF and POP or POPF to save and restore them, respectively. These registers must be saved as integers, except R6 and R7. The returned value from an assembly function must be in R0.

Download the linked COFF output file ADDM.OUT into the simulator. Configure the memory and the port maps to create the simulator output file ADDM.DAT at port address 804001h. Run this program and verify that the output file is

```
0x0000000a
0x0000000b
0x0000000c
0x0000000d
0x0000000e
```

Example 2.9 Matrix Multiplication with Both C and C-Called Function Using TMS320C30 Code. This example illustrates how the address of an array can be passed from a C program into an assembly function. Figure 2.16 shows the matrix multiplication program listing MATRIXM.C, in C, which calls the assembly function, MATRIXMF.ASM listed in Figure 2.17. Note how the addresses of the A and the B arrays, and the output address are passed into the assembly function using the frame pointer FP, with offsets of 2, 3, and 4, respectively. Assemble the TMS320C30 assembly function MATRIXMF.ASM. Compile and assemble the C program MATRIXM.C, and link with the object code of the assembly function using the command file

```
-c
MATRIXM.OBJ
MATRIXMF.OBJ
-L RTS30.LIB
-O MATRIXM.OUT
```

2.6 Programming Examples Using both C and TMS320C30 Code

```c
/*MATRIXM.C MATRIX MULTIPLICATION.CALLS ASSEMBLY FUNCTION*/
volatile int *IO_OUTPUT=(volatile int *) 0x804001;
extern void matrixmf (float *, float *, int *);
main()
{
 float A[3][3] =  {{1,2,3},
                   {4,5,6},
                   {7,8,9}};
 float B[3] =     {1,2,3};
  matrixmf ((float *) A, (float *)B, (int *) IO_OUTPUT);
}
```

FIGURE 2.16. Matrix multiplication program in C calling an assembly function (MATRIXM.C)

```
*MATRIXMF.ASM-ASSEMBLY FUNCTION CALLED FROM C PROGRAM
FP        .SET     AR3            ;FRAME POINTER IN AR3
          .GLOBAL  _matrixmf      ;GLOBAL REF/DEF
_matrixmf                         ;FUNCTION IN ASSEMBLY
          PUSH     FP             ;SAVE OLD FRAME POINTER
          LDI      SP,FP          ;POINT TO START OF STACK
          PUSH     R4             ;R4 IS A DEDICATED C REGISTER
          PUSH     AR4            ;AR4 IS A DEDICATED C REGISTER
          LDI      *-FP(2),AR0    ;A ARRAY
          LDI      *-FP(3),AR1    ;B ARRAY
          LDI      *-FP(4),AR2    ;POINTER TO IO_OUTPUT
          LDI      3,R4           ;R4 IS LOOPI COUNTER
LOOPI     LDF      0,R0           ;INITIALIZE R0
          LDI      2,AR4          ;AR4 IS LOOPJ COUNTER
LOOPJ     MPYF     *AR0++,*AR1++,R1  ;A[I,J] * B[J] = R1
          ADDF     R1,R0          ;R0 ACCUMULATES RESULT
          DB       AR4,LOOPJ      ;DEC AR4 AND BRANCH TIL AR4<0
          FIX      R0             ;FLOAT TO INTEGER
          STI      R0,*AR2        ;OUTPUT RESULT TO IO_OUT
          LDI      *-FP(3),AR1    ;RELOAD START ADDR OF B ARRAY
          SUBI     1,R4           ;DECREMENT R4
          BNZ      LOOPI          ;BRANCH WHILE  R4 <> 0
          POP      AR4            ;RESTORE AR4
          POP      R4             ;RESTORE R4
          POP      FP             ;RESTORE FP
          RETS                    ;RETURN FROM SUBROUTINE
```

FIGURE 2.17. C-called assembly function for matrix program (MATRIXMF.ASM)

Run the mixed-mode executable file MATRIXM.OUT through the simulator, and verify the same result as using the C or assembly version of the matrix multiplication examples.

Exercise 2.1. Implement the PREFIR example using five values for the coefficients HN and five values for the samples XN. Let H0 = 1, H1 = 2, H2 = 3, H3 = 4, and H4 = 5. Use the XN noise sample values as before and let X(n-4) = 1000h. Use two circular buffers, one for XN and one for HN. Note that a second ALIGN statement associated with HN can be specified within the command file. Find R2 for each time ($n = 0, 1, \ldots, 4$).

Exercise 2.2. Generalize the matrix program example in order to implement an (NxM)(MxK) = (NxK) matrix. Verify your result using the values in the previous (3 × 3) matrix example.

REFERENCES

[1] D. W. Horning, An undergraduate digital signal processing laboratory, in *Proceedings of the 1987 ASEE Annual Conference*, Vol. 3, June 1987.

[2] R. Chassaing, Applications in digital signal processing with the TMS320 digital signal processor in an undergraduate laboratory, in *Proceedings of the 1987 ASEE Annual Conference*, June 1987.

[3] R. Chassaing, A senior project course in digital signal processing with the TMS320, *IEEE Transaction on Education*, 32, May 1989, pp. 139–145.

[4] K. S. Lin (editor), *Digital Signal Processing Applications with the TMS320 Family: Theory, Algorithms and Implementations*, Vol. 1, Prentice-Hall, Englewood Cliffs, N.J., 1988.

[5] *TMS320C2X User's Guide*, Texas Instruments, Inc., Dallas, Tex., 1989.

[6] R. Chassaing and D. W. Horning, *Digital Signal Processing with the TMS320C25*, Wiley, New York, 1990.

[7] R. Chassaing and D. W. Horning, Digital filtering with fixed- and floating-point processors, in *Proceedings of the 1990 ASEE Annual Conference*, June 1990.

[8] R. Chassaing and P. Martin, Digital filtering with the floating point TMS320C30 digital signal processor, in *Proceedings of the 21st Annual Pittsburgh Conference on Modeling and Simulation*, May 1990.

[9] R. Chassaing, P. Martin, and R. Thayer, Multirate filtering using the TMS320C30 floating-point digital signal processor, in *Proceedings of the 1991 ASEE Annual Conference*, June 1991.

[10] P. Papamichalis (editor), Digital Signal Processing Applications with the TMS320 Family: Theory, Algorithms, and Implementations, Vol. 3, Texas Instruments, Inc., Dallas, Tex., 1990.

[11] R. Chassaing, B. Bitler, and D. W. Horning, Real-time digital filters in C, in *Proceedings of the 1991 ASEE Annual Conference*, June 1991.

[12] *TMS320C3X User's Guide*, Texas Instruments, Inc., Dallas, Tex., 1991.

[13] *TMS320 Floating-Point DSP Optimizing C Compiler User's Guide*, Texas Instruments, Inc., Dallas, Tex., 1991.

[14] *TMS320 Floating-Point DSP Assembly Language Tools User's Guide*, Texas Instruments, Inc., Dallas, Tex., 1991.

[15] *TMS320C30 Simulator User's Guide*, Texas Instruments, Inc., Dallas, Tex., 1989.

3
Alternative Input / Output and Extended Development System

CONCEPTS AND PROCEDURES
- I / O *programming with the* TI *analog interface chip* (AIC)
- *Use of interrupts*
- I / O *programming with the Burr–Brown two-channel* DSP102 / 202 *analog fixture*
- *Alternative DSP development system using the* TI XDS1000 *emulator*
- *Development of an application board for the* XDS1000

This chapter discusses input and output alternatives using the following:
1. *the two-input* TLC32044 *analog interface chip* (AIC) *module; the* AIC *provides an inexpensive ($30) alternative for analog interfacing, with two-input capability;*
2. *the analog evaluation fixture* DSP102/202 *through serial port 1 of the* TMS320C30, *available through a connector on the* EVM; *the reasonably priced ($200)* DSP102/202, *is a two-channel analog fixture with two inputs and two outputs, input and output filters, and a maximum sampling rate of 200 kHz.*

A powerful, but rather expensive, DSP development system XDS1000 is also discussed along with an application board designed and developed as a target system. This application board includes the TMS320C30 and 16K of user SRAM. [Interfacing the XDS1000 to the Texas Instruments analog interface board (AIB) and the analog interface chip are discussed in Appendix E.] The AIB contains 16-bit analog-to-digital and digital-to-analog converters, with a maximum sampling rate of 58 kHz. There is also a TLC32040 AIC on the AIB. Loop programs, with appropriate communication routines for interfacing these various I/O alternatives, are also presented. These loop programs serve as sample programs for developing other program examples.

3.1 INTRODUCTION

The TMS320C30 has two serial ports, two timers, and an on-chip direct memory access (DMA) controller for I/O. These features are controlled through memory-mapped registers, to communicate or to interface with the outside world. The serial ports are interrupt-driven, with interrupt vectors at specific memory locations.

Input and Output Signals

A typical DSP application requires at least an analog input and an analog output as illustrated in Figure 3.1. Along the input path is a filter for suppressing frequencies greater than one-half the sampling frequency. This filter acts as an antialiasing filter, to eliminate frequencies above the Nyquist frequency (defined as one-half the sampling frequency). Otherwise, aliasing can occur, in which case a higher-frequency signal can be disguised as a lower-frequency signal. The analog input signal is then converted to a digital representation by an analog-to-digital converter (ADC). The maximum level of the input signal for conversion is based on the specific ADC. For example, the input signal level to the analog interface chip (AIC) used on board the EVM is limited to ± 1.5 V, whereas the range of the input signal to the TI analog interface board (AIB) is ± 10 V. This voltage range is decomposed into discrete levels. The number of these discrete levels is based on the number of bits of the ADC. The ADC on the Burr–Brown analog fixture has 18-bit conversion capability, whereas the AIC uses a 14-bit ADC.

After processing, communication is again needed with the outside world. Along the output signal path is a digital-to-analog converter (DAC). The DAC performs the reverse operation of the ADC. Different output levels from the DAC are produced based on the digital word on its input. This

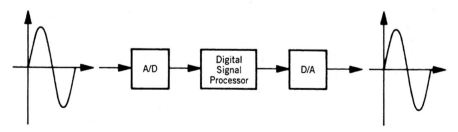

FIGURE 3.1. A DSP system with input and output

step-level signal is passed through a lowpass output filter. This filter smoothes out the steps and reconstructs an equivalent analog signal based on the step levels.

3.2 THE ANALOG INTERFACE CHIP

This section describes the analog interface chip and a way to connect an additional AIC to the TMS320C30 [1–5]. This additional AIC provides the user with an alternative and inexpensive analog interface with two inputs. The AIC functional block diagram is shown in Figure 3.2. The AIC is the TLC32044, a member of the TLC32040 family of voice-band analog interface circuits, which is on board the evaluation module (EVM). The AIC supports two input channels and one output channel. However, the specific AIC on board the EVM has one of its input connected to ground. Several programmable sampling rates can be obtained with the AIC, with a maximum sampling rate of 19.2 kHz. The AIC includes an ADC, DAC, and switched-capacitor antialiasing input and reconstruction output filters, constructed on a single C-MOS chip. Both the ADC and the DAC have 14 bits of resolution. The lowpass and highpass input filters are programmable, creating an effective input bandpass filter. The output filter is fixed.

The TLC32046 AIC is a recent addition to the TI TLC32040 family. This version allows for an input bandpass filter of greater bandwidth and a maximum sampling rate of 25 kHz (as compared to 19.2 kHz with the TLC32044).

AIC Control

Data transmission occurs through two of the AIC's serial port registers, the data receive register (DRR) and the data transmit register (DXR). The AIC is controlled through the data transmit register. The two least significant bits (LSBs) are used for communication functions. Normal transmission occurs with the two LSBs being 0s, and secondary communication is initiated with the two LSBs being 1s. The secondary communication mode initializes and controls the AIC, allowing one secondary transmission before switching back to the communication mode. Control functions are initialized by writing to several registers in the AIC. The control register specifies the input port, the input filter, the input voltage level, and the activation of a loopback test. Figure 3.3 shows the internal timing configuration of the AIC. The letters **A** and **B** designate the location of control, with **A** representing a filter control register and **B** the analog-to-digital (A/D) and digital-to-analog (D/A) control. The TA and RA registers divide the system clock to the switched capacitor filter frequency, and TB and RB further divide the switched capacitor filter frequency down to the sampling rate frequency. The bit locations for

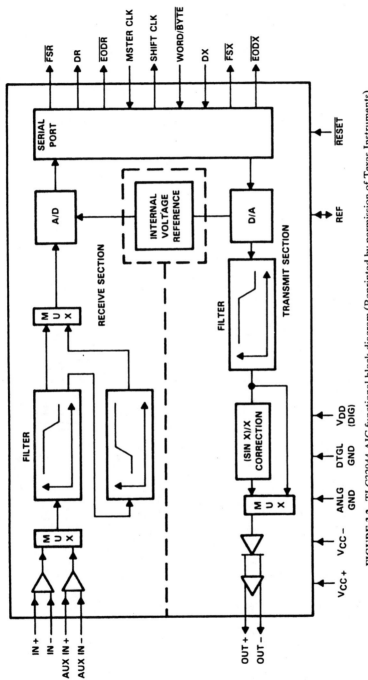

FIGURE 3.2. TLC32044 AIC functional block diagram (Reprinted by permission of Texas Instruments)

54 Alternative Input/Output and Extended Development System

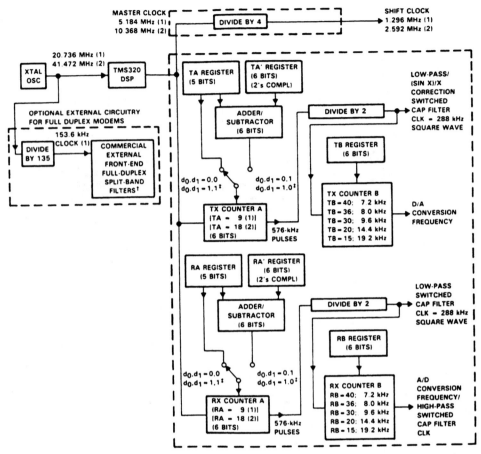

FIGURE 3.3. AIC internal timing configuration (Reprinted by permission of Texas Instruments)

the transmit and receive registers TA and RA follow:

0,0	bis 0 and 1
RA	bits 2–6
don't care	bits 7–8
TA	bits 9–13
don't care	bits 14–15

The bit locations for the transmit and receive registers TB and RB follow:

0,1	bits 0 and 1
RB	bits 2–7
don't care	bit 8
TB	bits 9–14
don't care	bit 15

3.2 The Analog Interface Chip

secondary DX serial communication protocol

x x	← to TA register →	x	x	← to RA register →		0	0	d13 and d6 are MSBs (unsigned binary)
x	← to TA' register →	x	← to RA' register	→		0	1	d14 and d7 are 2's complement sign bits
x	← to TB register →	x	← to RB register	→		1	0	d14 and d7 are MSBs (unsigned binary)
x x x x x x d9 x	d7 d6 d5 d4 d3 d2 ←CONTROL REGISTER→					1	1	d2 = 0/1 deletes inserts the A D high-pass filter d3 = 0/1 disables/enables the loopback function d4 = 0/1 disables/enables the AUX IN + and AUX IN − pins d5 = 0/1 asynchronous/synchronous transmit and receive sections d6 = 0/1 gain control bits (see Gain Control Section) d7 = 0/1 gain control bits (see Gain Control Section) d9 = 0/1 delete insert on-board second-order (sin x)/x correction filter

FIGURE 3.4. AIC secondary communication protocol (Reprinted by permission of Texas Instruments)

Figure 3.4 shows the AIC secondary communication protocol. Table 3.1 shows the AIC data configuration, which is very useful when using the AIC for I/O. This will be demonstrated later in this chapter with a programming example.

The AIC can be configured for a specific sampling rate and filter bandwidth. This is accomplished by requesting an AIC secondary communication, loading a 1 in the first two LSBs. For example, to set a sampling rate of 10 kHz with a filter bandwidth of 4,700 Hz, the following sequence of data need to be loaded to the serial port data transmit register at address 0x808048:

0x00000003	to request secondary AIC communication
0x00001428	for TA and RA registers
0x00000003	to request secondary AIC communication (again)
0x00004A96	for TB and RB registers
0x00000003	to request secondary AIC communication
0x00000067	to configure the control register

TABLE 3.1 AIC Data Configuration

Register	Address	Command
Timer 0 period	0x808028	Load 0x00000001
Timer 0 global control	0x808020	Load 0x000002C1
I/O flag	IOF	Load 0x00000002
SP0 transmit port control	0x808042	Load 0x00000111
SP0 receive port control	0x808043	Load 0x00000111
SP0 global control	0x808040	Load 0x0E970300
SP0 data transmit	0x808048	Load 0x00000000
I/O flag	IOF	Load 0x00000006
Interrupt flag	IF	Load 0x00000000
Interrupt enable	IE	OR with 0x00000010
Status register	ST	OR with 0x00002000

For example, for a switched capacitor filter (SCF) frequency of 288 kHz and a sampling rate of 10 kHz,

$$A = \frac{(\text{master clock frequency})/4}{2 \times \text{SCF}} = \frac{(30 \text{ MHz})/4}{2 \times 288 \text{ kHz}}$$
$$\simeq 13$$
$$= (01101)_b$$

where the master clock frequency is 30 MHz on the EVM, and

$$B = \frac{\text{SCF}}{F_s} = \frac{288 \text{ kHz}}{10 \text{ kHz}} \simeq 29 = (011101)_b$$

This produces a (lowpass) filter bandwidth of $\simeq 3.6$ kHz (see Example 3.6). Different filter bandwidth can be obtained as shown later. There is an additional set of registers TA' and RA' that can be used for further adjustment (fine-tuning) of the sampling rate and filter bandwidth.

Example 3.6 and Exercises 4.2–4.4 illustrate further how to obtain a desired sampling frequency and input filter bandwidth using the AIC. An interactive oscilloscope application program [3, 4], available with the EVM package, illustrates the effects of a desired sampling frequency on the A and B registers, and on the input and output lowpass and input highpass filter bandwidths.

3.3 INTERRUPTS

The TMS320C30 supports a nonmaskable external RESET signal, as well as internal and external interrupts, which can be used to interrupt the CPU or the DMA. Figure B.6 (in Appendix B) shows the peripheral addresses of the serial ports and timers. The global interrupt register enable bit (GIE) controls all CPU interrupts, and is located in the status register ST. The DMA global interrupt enable bit, independent of ST, controls DMA interrupts. To enable an interrupt, set the GIE bit to 1. To disable an interrupt, disable the interrupt enable (IE) register, setting it to zero, then set the global interrupt enable bit GIE also to zero. Figure B.5 (Appendix B) shows that the memory locations 1 through Bh are used on the TMS320C30 for interrupts.

Interrupt handlers in C are such that no arguments should be passed to or from a function. An interrupt function should be declared with the following format so that the C compiler can use a specified register preservation requirement:

```
c_intnm
void c_int05 ( )
```

where nm is a two-digit number between 00 and 99, and c_int05 ( ) is an example of an interrupt function. The C compiler saves only the registers used by the interrupt function, unless the function calls another C function.

In that case, the C compiler saves all the registers, which results in a significant increase in execution time.

The interrupt handler or service routine is installed in the interrupt vectors located at memory addresses starting at 0h. An interrupt function should not be called directly. Instead, the address of the function should be installed in the appropriate interrupt vector location. On interrupt, this function can be serviced by the TMS320C30. The following instructions can be used to install an interrupt function:

1. #define VEC_ADDR (volatile int *)0x00;
2. volatile int *INTVEC = VEC_ADDR;
3. INTVEC[5] = (volatile int) c_int05;

where the beginning address of the vectors is defined in the instruction. The second instruction uses a pointer to install the function, and the third instruction is to install the address of the function at 0x05.

Timers

Two timers are supported by the TMS320C30. Those timers can be used to count external events. They provide the timing necessary to signal an ADC to start a conversion. Figure B.7 (Appendix B) shows the memory-mapped timer locations. The timer global control register monitors the timer status. The timer period register specifies the timer's frequency. The timer counter register contains the value of the incrementing counter. When the value of the timer period register equals that of the timer counter register, the counter register resets to zero. At reset, both the timer counter and period registers are set to zero.

Serial Ports

The TMS320C30 has two serial ports, which are totally independent, each with a set of control registers. The memory-mapped registers for the serial ports are shown in Figure B.9 (Appendix B).

Several examples using both C and assembly code illustrate the interrupt structure of the TMS320C30. Additional examples illustrate the use of the AIC using serial ports 0 and 1.

3.4 PROGRAMMING EXAMPLES

Example 3.1 Output Rate Controlled by an Interrupt, Using TMS320C30 Code. This example illustrates the use of an interrupt to control the output rate. Figure 3.5 shows the program listing INTERINC.ASM, using TMS320C30 code. On interrupt, execution of the program proceeds to the address specified by the appropriate interrupt vector. The RESET vector is located at

58 Alternative Input/Output and Extended Development System

address 0, and the next eight vectors are skipped (in order to select timer interrupt 0). Timer interrupt 0 (TINT0) is specified at address 9, as shown in Figure B.5 (Appendix B). On interrupt, execution proceeds to the address specified with the label TIM_INT. The interrupt rate is found using

$$\text{rate} = \frac{(\text{master clock frequency})/4}{2 \times \text{period}}$$

A PERIOD value of 30h is set, which corresponds to an interrupt rate of 78.125 kHz. This frequency can be verified with an oscilloscope on pin 4 (TCLK0) of the TMS320C30.

Figure B.2 shows the interrupt enable register (IE). To select timer interrupt 0, bit 8 is set to 1, enabling ETINT0. This is accomplished by loading 100h in the IE register.

Assemble and link the program in Figure 3.5. Configure the memory and port maps in order to obtain the output file INTERINC.DAT at port address 804000h. Load the executable file INTERINC.OUT into the simulator, and display the memory window starting at address 808020h. A description of the program in Figure 3.5 follows (single-step through):

1. The data page is initialized to page 128. AR0 is loaded with the period address 808028h, and R0 is loaded with 30h. This small value for PERIOD is chosen so that one can single-step quickly through to demonstrate the occurrence of an interrupt. This occurs when the value at 808024h is incremented from 0 and reaches 30h. R0, with a value of 30h is stored at the address of the period register 808028h. Then AR0 is decremented (offset by 8) to point at 808020h.
2. R0 is loaded with the control register value 2C1h and stored at the address of the timer global control register 808020h.
3. An initial output value of 1000h is loaded into R0. The IE register (displayed from the CPU status window) is set to 100h (bit 8 set to 1). The status register is set to 2000h so that the global interrupt enable register GIE (bit 13) is set. If GIE = 0, the CPU will not respond to an enabled interrupt.
4. Execution is between the IDLE instruction (until interrupt) and the branching instruction (to return back to the IDLE instruction), as the content of the timer counter register at address 808024h is incremented.
5. When the value in the timer counter register (at 808024h) equals the timer period register value of 30h (set at 808028h), execution proceeds to the interrupt vector address specified by TIM_INT. AR0 is then loaded with the output address of 804000h, where the initial output value of 1000h is stored. R0 is then incremented by one (to 1001h), and

3.4 Programming Examples 59

```
;INTERINC.ASM-DEMONSTRATES THE USE OF INTERRUPT
          .TITLE    "INTERINC.ASM"  ;INTERRUPT PROGRAM
          .SECT     "VECTORS"       ;ASSEMBLE INTO VECTOR SECTION
RESET     .WORD     BEGIN           ;RESET VECTOR
          .SPACE    8               ;SKIP 8 WORDS
TIMER0    .WORD     TIM_INT         ;TINT0 VECTOR LOCATION @ 9h
          .SPACE    54              ;REMAINDER OF VECTOR SECTION
          .DATA                     ;ASSEMBLE INTO DATA SECTION
STACKS    .WORD     809F00H         ;INIT STACK POINTER DATA
PERIOD    .WORD     30H             ;INTERRUPT RATE=7.5MHz/(2*PERIOD)
IE_REG    .WORD     100H            ;ENABLE TIMER 0 (TINT0)INTERRUPT
PER_ADDR  .WORD     808028H         ;(TINT0) PERIOD REG LOCATION
TCNTL     .WORD     2C1H            ;CONTROL REGISTER VALUE
ST_REG    .WORD     2000H           ;SET STATUS REG
IO_ADDR   .WORD     804000H         ;I/O ADDRESS
OUTPUT    .WORD     01000H          ;INITIAL OUTPUT VALUE
          .TEXT                     ;ASSEMBLE INTO TEXT SECTION
BEGIN     LDP       STACKS          ;INIT DATA PAGE
          LDI       @STACKS,SP      ;SP-> 809F00H
          LDI       @PER_ADDR,AR0   ;TINT0 PERIOD REG =>AR0
          LDI       @PERIOD,R0      ;PERIOD VALUE => R0
          STI       R0,*AR0--(8)    ;SET TINT0 PERIOD @ 808028H
          LDI       @TCNTL,R0       ;CONTROL REGISTER VALUE =>R0
          STI       R0,*AR0         ;SET TINT0 GLOBAL CNTRL @ 808020H
          LDI       @OUTPUT,R0      ;R0 = OUTPUT VALUE
          LDI       @IE_REG,IE      ;ENABLE TINT0 INTERRUPT(BIT 8)
          OR        @ST_REG,ST      ;SET STATUS REG (ENABLE GIE BIT)
WAIT      IDLE                      ;WAIT FOR INTERRUPT
          BR        WAIT            ;BRANCH TO WAIT TIL INTERRUPT
; INTERRUPT VECTOR
TIM_INT   LDI       @IO_ADDR,AR0    ;1ST ADDRESS OF EXP BUS=>AR0
          STI       R0,*AR0         ;OUTPUT R0 @ 804000H
          ADDI      1,R0            ;INCREMENT OUTPUT VALUE BY 1
          RETI                      ;RETURN FROM INTERRUPT
          .END                      ;END
```

FIGURE 3.5. Program listing to demonstrate interrupt using TMS320C30 code (INTERINC.ASM)

execution returns to the IDLE or BRANCH instruction, until the next interrupt. Run this program and obtain an output file that contains 1000h, 1001h, 1002h,... (until execution is stopped).

Example 3.2 Pseudorandom Noise Generator with Interrupt, Using TMS320C30 Code. This program example generates a pseudorandom noise (PRN) using TMS320C30 code. The output rate of each noise sample is controlled by interrupt. A 32-bit random noise sequence is implemented using the program listed in Figure 3.6. The noise generation uses the following scheme:

1. A 32-bit initial seed value is chosen (for example, 7E521603h).
2. A modulo 2 summation is performed between bits 17, 28, 30, and 31.
3. The least significant bit (LSB) of the resulting summation is selected. This bit is either a 0 or a 1 and is scaled accordingly to a positive or a negative value.
4. The seed value is shifted left by one, and the resulting bit from the previous step is placed in the LSB position and the process repeated with the new (shifted by one) seed value (as shown in the following noise generator diagram).

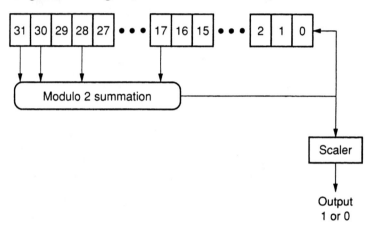

Assemble and link this program. Use the simulator to verify the following:

1. Interrupt is enabled in a fashion similar to that in the previous example, with proper initialization of the timer period and control registers as well as the interrupt enable and status registers.
2. On interrupt, execution proceeds to the interrupt vector address, located at 9h. This address is specified by the label TIM_INT, where the noise generation routine starts.
3. Register R2 is loaded with the initial seed value of 7E521603h and this value is shifted *Right* by 17, using LSH -17,R2. This places bit 17 into the LSB position, where the addition is meaningful.
4. R2 is then shifted right by 11 to place bit 28 in the LSB position. Note that it was previously shifted right by 17. Then the summation takes place, effectively adding bit 17 with bit 28.
5. The procedure continues, adding bits 17, 28, 30, and 31. The LSB is selected by ANDing the 32-bit value in R4 by 1. This bit, a 1 or a 0, represents the output noise sample.

```
;NOISEINT.ASM-PSEUDORANDOM NOISE GENERATOR USING INTERRUPT
          .TITLE    "NOISEINT.ASM"  ;INTERRUPT WITH NOISE PROGRAM
          .SECT     "VECTORS"       ;ASSEMBLE INTO VECTOR SECTION
RESET     .WORD     BEGIN           ;RESET VECTOR
          .SPACE    8               ;SKIP 8 WORDS
TIMER0    .WORD     TIM_INT         ;TINT0 VECTOR LOCATION @ 9h
          .SPACE    54              ;REMAINDER OF VECTOR SECTION
          .DATA                     ;ASSEMBLE INTO DATA SECTION
STACKS    .WORD     809F00H         ;INIT STACK POINTER DATA
PERIOD    .WORD     30H             ;INTERRUPT RATE=7.5MHz/(2*PERIOD)
IE_REG    .WORD     100H            ;ENABLE TIMER 0 (TINT0)INTERRUPT
PER_ADDR  .WORD     808028H         ;(TINT0) PERIOD REG LOCATION
TCNTL     .WORD     2C1H            ;CONTROL REGISTER VALUE
ST_REG    .WORD     2000H           ;SET STATUS REG
IO_ADDR   .WORD     804000H         ;I/O ADDRESS
SEED      .WORD     7E521603H       ;INITIAL SEED VALUE
MINUS     .WORD     0FFFFF000H      ;NEGATIVE LEVEL
PLUS      .WORD     1000H           ;POSITIVE LEVEL
          .TEXT                     ;ASSEMBLE INTO TEXT SECTION
BEGIN     LDP       STACKS          ;INIT DATA PAGE
          LDI       @STACKS,SP      ;SP-> 809F00H
          LDI       @PER_ADDR,AR0   ;TINT0 PERIOD REG =>AR0
          LDI       @PERIOD,R0      ;PERIOD VALUE => R0
          STI       R0,*AR0--(8)    ;SET TINT0 PERIOD
          LDI       @TCNTL,R0       ;CONTROL REGISTER VALUE =>R0
          STI       R0,*AR0         ;SET TINT0 GLOBAL CNTRL @ 808020H
          LDI       @SEED,R0        ;R0 = INITIAL SEED VALUE
          LDI       0,R7            ;INIT R7 (OUTPUT) TO 0
          LDI       @IE_REG,IE      ;ENABLE TINT0 INTERRUPT(BIT 8)
          OR        @ST_REG,ST      ;SET STATUS REG (ENABLE GIE BIT)
WAIT      IDLE                      ;WAIT FOR INTERRUPT
          BR        WAIT            ;BRANCH TO WAIT TIL INTERRUPT
;   INTERRUPT VECTOR
TIM_INT   LDI       0,R4            ;INIT R4=0
          LDI       R0,R2           ;PUT SEED IN R2
          LSH       -17,R2          ;MOVE BIT 17 TO LSB      =>R2
          ADDI      R2,R4           ;ADD BIT (17)            =>R4
          LSH       -11,R2          ;MOVE BIT 28 TO LSB      =>R2
          ADDI      R2,R4           ;ADD BITS (28+17)        =>R4
          LSH       -2,R2           ;MOVE BIT 30 TO LSB      =>R2
          ADDI      R2,R4           ;ADD BITS (30+28+17)     =>R4
          LSH       -1,R2           ;MOVE BIT 31 TO LSB      =>R2
          ADDI      R2,R4           ;ADD BITS (31+30+28+17)  =>R4
          AND       1,R4            ;MASK LSB OF R4
          LDIZ      @MINUS,R7       ;IF R4 = 0, R7 = @MINUS
          LDINZ     @PLUS,R7        ;IF R4 = 1, R7 = @PLUS
          LSH       1,R0            ;SHIFT SEED LEFT BY 1
          OR        R4,R0           ;PUT R4 INTO LSB OF R0
          LDI       @IO_ADDR,AR0    ;1ST ADDRESS OF EXP BUS=>AR0
          STI       R7,*AR0         ;OUTPUT NOISE SAMPLE
          RETI                      ;RETURN FROM INTERRUPT
          .END                      ;END
```

FIGURE 3.6. Pseudorandom noise generator program with interrupt, using TMS320C30 code (NOISEINT.ASM)

0x00001000
0x00001000
0x00001000
0x00001000
0xfffff000
0x00001000
0xfffff000
0x00001000
0x00001000
0x00001000
0x00001000
0xfffff000
0xfffff000
0x00001000
0x00001000
0xfffff000

FIGURE 3.7. Partial pseudorandom noise sequence output (NOISEINT.DAT)

6. This noise sample is scaled to ±4,096 (1000h or FFFFF000h).
7. Before the procedure is repeated, the seed value is shifted by one, and the resulting bit (0 or 1) in R4 is placed in the LSB position of R0, which now contains the new (shifted left by one) seed value.
8. The scaled output noise sample is stored for output. Execution returns to the IDLE or BRANCH instruction until the next interrupt. Upon each interrupt, the noise generator routine is repeated and a new noise sample is obtained. The output pseudorandom noise sequence (NOISEINT.DAT) is partially shown in Figure 3.7. With more and more points, this sequence can be seen to become more random.

Example 3.3 Alternative Pseudorandom Noise Generator Program Using TMS320C30 Code. The program listing in Figure 3.8 shows an alternative method to generate a pseudorandom noise sequence. This program is not interrupt-driven. Assemble this program, link it with an appropriate command file, and verify that the identical pseudorandom noise sequence is generated.

Example 3.4 Sine Generation with Four Points and with Interrupt, Using TMS320C30 Code. This example shows how a sinusoid can be generated using only four points. The program listed in Figure 3.9 is used to generate the sinusoid. The appropriate command file should include

```
SINE_BUFF ALIGN(8): {} > RAM
```

to obtain proper alignment of a circular buffer, in a fashion similar to that in the command file of the PREFIR example included in Chapter 2.

3.4 Programming Examples

```
;NOISEALT.ASM-PSEUDORANDOM NOISE GENERATOR
            .TITLE   "NOISEALT"      ;PSEUDORANDOM NOISE GENERATOR
            .SECT    "VECTORS"       ;ASSEMBLE INTO VECTOR SECTION
RESET       .WORD    BEGIN           ;RESET VECTOR
            .DATA                    ;ASSEMBLE INTO DATA SECTION
STACKS      .WORD    809F00H         ;INIT STACK POINTER DATA
IO_ADDR     .WORD    804000H         ;I/O ADDRESS
SEED        .WORD    7E521603h       ;INITIAL SEED VALUE
MINUS       .WORD    0FFFFF000h      ;NEGATIVE LEVEL
PLUS        .WORD    1000h           ;POSITIVE LEVEL
            .TEXT                    ;ASSEMBLE INTO TEXT SECTION
BEGIN       LDP      STACKS          ;INIT DATA PAGE
            LDI      @STACKS,SP      ;SP-> 809F00H
            LDI      @SEED,R0        ;R0 = INITIAL SEED VALUE
LOOP        LDI      R0,R4           ;PUT SEED IN R4
            LSH      -31,R4          ;MOVE BIT 31 TO LSB       =>R4
            LDI      R0,R2           ;R2=R0=SEED
            LSH      -30,R2          ;MOVE BIT 30 TO LSB       =>R2
            ADDI     R2,R4           ;ADD BITS (31+30)         =>R4
            LDI      R0,R2           ;R2=R0=SEED
            LSH      -28,R2          ;MOVE BIT 28 TO LSB       =>R2
            ADDI     R2,R4           ;ADD BITS (31+30+28)      =>R4
            LDI      R0,R2           ;R2=R0=SEED
            LSH      -17,R2          ;MOVE BIT 17 TO LSB       =>R2
            ADDI     R2,R4           ;ADD BITS (31+30+28+17)=>R4
            AND      1,R4            ;MASK LSB OF R4
            LDIZ     @MINUS,R7       ;IF R4=0, R7 = @MINUS
            LDINZ    @PLUS,R7        ;IF R4=1, R7 = @PLUS
            LSH      1,R0            ;SHIFT SEED LEFT BY 1
            OR       R4,R0           ;PUT R4 INTO LSB OF R0
            LDI      @IO_ADDR,AR0    ;1ST ADDR OF EXP BUS=>AR0
            STI      R7,*AR0         ;OUTPUT R7
            BR       LOOP            ;REPEAT FOR NEXT NOISE SAMPLE
            .END                     ;END
```

FIGURE 3.8. Alternative pseudorandom noise generator program (`NOISEALT.ASM`)

The following is a brief description of this program.

1. The four values used to generate the sinusoid (0, 16,384, 0, −16,384) are set in a memory table, with starting address specified by AR1. These values are initially loaded once in the memory table section.

2. The size of the circular buffer is set in BK. The alignment for the circular buffer is generated through the use of the ALIGN statement in the command file.

```
;SINE4.ASM-GENERATES A SINE WITH 4 POINTS.USE OF INTERRUPTS
            .TITLE    "SINE4.ASM"          ;SINE GENERATION,F=Fs/4
            .SECT     "VECTORS"            ;ASSEMBLE INTO VECTOR SECTION
RESET       .WORD     BEGIN                ;RESET VECTOR
            .SPACE    8                    ;SKIP 8 WORDS
TIMER0      .WORD     TIM_INT              ;TINT0 VECTOR LOCATION @ 9h
            .SPACE    54                   ;REMAINDER OF VECTOR SECTION
SINE_TABLE  .USECT    "SINE_BUFF",LENGTH   ;SINE TABLE SIZE
            .DATA                          ;ASSEMBLE INTO DATA SECTION
STACKS      .WORD     809F00H              ;INIT STACK POINTER DATA
PERIOD      .WORD     177H                 ;INTER RATE=7.5MHz/(2*PERIOD)=10 kHz
IE_REG      .WORD     100H                 ;ENABLE TIMER 0 (TINT0)INTERRUPT
PER_ADDR    .WORD     808028H              ;(TINT0) PERIOD REG LOCATION
TCNTL       .WORD     2C1H                 ;CONTROL REGISTER VALUE
ST_REG      .WORD     2000H                ;SET STATUS REG
IO_ADDR     .WORD     804000H              ;I/O ADDRESS
SINE_ADDR   .WORD     SINE_TABLE           ;ADDRESS OF SINE TABLE
LENGTH      .SET      4                    ;LENGTH OF CIRCULAR BUFFER
            .TEXT                          ;ASSEMBLE INTO TEXT SECTION
BEGIN       LDP       STACKS               ;INIT DATA PAGE
            LDI       @STACKS,SP           ;SP-> 809F00H
            LDI       @PER_ADDR,AR0        ;TINT0 PERIOD REG =>AR0
            LDI       @PERIOD,R0           ;PERIOD VALUE => R0
            STI       R0,*AR0--(8)         ;SET TINT0 PERIOD @ 808028H
            LDI       @TCNTL,R0            ;CONTROL REGISTER VALUE =>R0
            STI       R0,*AR0              ;SET TINT0 GLOBAL CNTRL @ 808020H
            LDI       LENGTH,BK            ;SIZE OF BUFFER -> BK
            LDI       @SINE_ADDR,AR1       ;SINE_ADDR -> AR1
            LDI       0,R0                 ;R0=0, FIRST SINE VALUE
            STI       R0,*AR1++%           ;STORE IN FIRST TABLE ADDRESS
            LDI       16384,R0             ;R0=SECOND VALUE
            STI       R0,*AR1++%           ;STORE SECOND VALUE
            LDI       0,R0                 ;R0=0, THIRD SINE VALUE
            STI       R0,*AR1++%           ;STORE THIRD VALUE
            LDI       -16384,R0            ;R0=FOURTH VALUE
            STI       R0,*AR1++%           ;STORE FOURTH VALUE
            LDI       @IE_REG,IE           ;ENABLE TINT0 INTERRUPT(BIT 8)
            OR        @ST_REG,ST           ;SET STATUS REG (ENABLE GIE BIT)
WAIT        IDLE                           ;WAIT FOR INTERRUPT
            BR        WAIT                 ;BRANCH TO WAIT TIL INTERRUPT
;  INTERRUPT VECTOR
TIM_INT     LDI       @IO_ADDR,AR0         ;I/O BASE ADDRESS -> AR0
            LDI       *AR1++%,R7           ;R7=TABLE VALUE
            STI       R7,*+AR0(2)          ;OUTPUT TO PORT ADDRESS 804002H
            RETI                           ;RETURN FROM INTERRUPT
            .END                           ;END
```

FIGURE 3.9. Sine-generation program with four points, using TMS320C30 code (SINE4.ASM)

3.4 Programming Examples

3. The registers used for interrupt are initialized as in the previous example. A period of 177h corresponds to an output rate of

$$\text{rate} = \frac{7.5 \text{ MHz}}{2 \times \texttt{PERIOD}} = 10 \text{ kHz}$$

4. On interrupt, execution proceeds to the instruction specified by the interrupt vector `TIM_INT`. `AR1` is pointing at the starting address of the table that contains the first value 0. This value is stored at the output port 804002h, the I/O address offset by two.

5. Execution returns to the `IDLE` or `BR WAIT` instruction to wait for the next interrupt. On each interrupt, execution proceeds to output the next value in the table. Note that a circular mode of addressing is used in conjunction with `AR1`.

Run this program through the simulator to create an output file `SINE4.DAT`, with output port address at 804002h. Appendix C describes a utility package from Hyperception, Inc., that can be used for data acquisition and plotting (also for filter design and sine or cosine generation). Using 10 kHz as the sampling frequency, the output is plotted in Figure 3.10. Taking the FFT of the output, using the utility package from Hyperception, Inc., produces the plot in Figure 3.11, showing a "delta" function at the frequency $f = 2.5$ kHz of the sine function, since

$$f = \frac{F_s}{\text{number of points}} = 2.5 \text{ kHz}$$

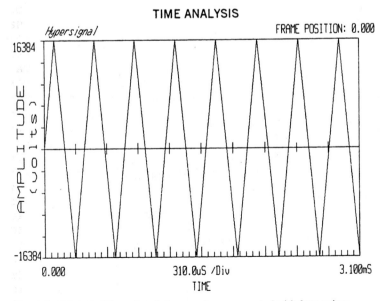

FIGURE 3.10. Time plot of sine waveform generated with four points

Alternative Input/Output and Extended Development System

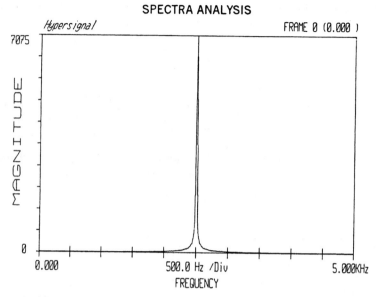

FIGURE 3.11. Frequency plot of sine waveform generated with four points

```
/*SINE4C.C SINE GENERATION WITH 4 POINTS.USE OF INTERRUPT        */
#define period 375                    /*timer 0 period reg value */
#define VEC_ADDR (volatile int *) 0x00    /*addr of interrupt vectors*/
#define TIMER_ADDR (volatile int *) 0x808020 /*addr of timer 0    */
int sin_table[4] = {0, 16384, 0, -16384};    /*values in sine table   */
int loop = 0;                         /*declare global variable  */
volatile int *IO_OUTPUT = (volatile int *) 0x804001; /*output port addr */

void c_int09()                        /*TINT0 Interrupt handler   */
{
 *IO_OUTPUT = sin_table[loop];        /*output value from sine table */
 if (loop < 3) ++loop;                /*increment to next value   */
 else loop = 0;                       /*go to beginning of sine table*/
}

main()
{
  volatile int *INTVEC = VEC_ADDR;    /*pointer to interrupt vector*/
  volatile int *TIMER = TIMER_ADDR;   /*pointer to timer 0 address */
  INTVEC[9] = (volatile int) c_int09; /*Install interrupt 9 handler*/
  TIMER[8] = period;                  /*Set period register       */
  TIMER[0] = 0x2C1;                   /*Set Global Cntrl Register */
  asm ("       LDI    100h,IE");      /*Enable TINT0 interrupt    */
  asm ("       OR     2000h,ST");     /*Enable GIE Bit            */
  for (;;);                           /*Wait for interrupt        */
}
```

FIGURE 3.12. Sine-generation program with four points, using C code (SINE4C.C)

Change the sampling frequency F_s to 5 kHz and verify that a sine function with a frequency of 1,250 Hz is produced.

***Example 3.5** Sine Generation with Four Points and with Interrupt, Using C Code.* This program example is the C-version equivalent of the previous example and is listed in Figure 3.12 as `SINE4C.C`. It uses the interrupt structure supported by C. Example 3.6 illustrates the real-time version of this program. In certain situations, the optimizing compiler reduces a repetitive statement with a variable that changes as a result of an external event, such as an interrupt service routine, to a single `read` statement. In

```
int     index;
:
while (! index);
```

the `while` statement is reduced to a single `read` statement. To prevent this, the following modification should be made:

```
volatile int index;
:
while (! index);
```

This code segment is used in Chapter 6 in conjunction with a real-time FFT routine. The program idles at the `while` statement until the current frame of information is processed and the volatile variable index is set to zero by the interrupt service routine.

Note the following, from the program `SINE4C.C`:

1. A period value of 177h = 375 is set for an interrupt rate of 10 kHz (as in the previous example), which determines the output rate. This value becomes important for real-time implementations. With simulation, the sampling rate can be set later (for example, in the Hypersignal utility package from Hyperception, Inc.).
2. The timer address for serial port 0 is at 0x808020. The reset vector is at memory location 0, and interrupt `TINT0` is at memory location 9. The timer period register at address 0x802028 (timer address offset by eight) is set with the period value 375 and the timer global control register is set with the control value 0x2C1.
3. To enable interrupt, the interrupt enable register `IE` and the global interrupt enable (GIE) bit within the status register must be set. This is accomplished with TMS320C30 code within the C program, using the `asm` statements. Note the format of the assembly code between the quotes. The assembly instructions using `asm` are

```
LDI        100h, IE
OR         2000h, ST
```

68 Alternative Input/Output and Extended Development System

Compile, assemble, and link this program. The loop command file (LOOPALL.CMD) of Chapter 1 can be modified to use with this program. Run this program and verify that the resulting simulator output file is identical to the one generated in Example 3.4 using TMS320C30 code.

Example 3.6 Real-Time Sine Generation with Four Points, Using C Code. This example extends Example 3.5 to include a real-time implementation using the EVM and AIC. Figure 3.13 shows the program listing (SINE4AIC.ASM) for this example. This program calls the AIC communication program AICCOM.C, listed in Figure 3.14. Note that much of the code in AICCOM.C is very similar to the program LOOPALL.C in Example 1.3. Here, the TMS320C30–AIC communication routines are listed in a separate program AICCOM.C. In addition, AICCOM.C includes the function AICSET_I, which is added to be used with an interrupt-driven calling program. The program AICCOM.C is a general AIC communication program (non-interrupt- or interrupt-driven) that is used in conjunction with serial port 0 on the TMS320C30. We will show later a slightly different version, which can be used for either port 0 or 1.

A command file similar to LOOPALL.CMD listed in Figure 1.5 can be used. A program VECSIR.OBJ is to be linked also. The program VECSIR.ASM (on accompanying disk), which contains the reset interrupt vector, follows:

```
.asect     ``vecs´´,0h      ;absolute section defined
.ref       _c_int00         ; C init reset
.word      _c_int00
```

From the main program SINE4AIC.C listed in Figure 3.13, the following apply:

1. The communication routines from AICCOM.C are invoked with the include statement.

2. The values (in AICSEC) for secondary communication are specified to obtain a 10-kHz sampling rate. From the value 0x1428 in AICSEC,

$$0001 \quad 0100 \quad 0010 \quad 1000$$

RA and TA, each five bits wide, can be obtained because

$$\begin{array}{ll} \text{bits 2–6} & \text{TA} = (01010)_b = 10 \\ \text{bits 7, 8} & \text{don't care} \\ \text{bits 9–13} & \text{TA} = (01010)_b = 10 \end{array}$$

3.4 Programming Examples

```
/*SINE4AIC.C-SINE PROGRAM WITH 4 POINTS USING INTERRUPTS     */
#include "aiccom.c"                    /*AIC comm routines            */
#define VEC_ADDR (volatile int *) 0x00; /*addr of vectors             */
int AICSEC[4] = {0x1428,0x1,0x4A96,0x67}; /*config data for AIC SP0  */
int data_out, loop = 0;                /*declare global variables     */
int sin_table[4] = {0,1000,0,-1000};   /*values for 4-point sinewave  */

void c_int05()                         /*TINT0 interrupt routine      */
{
  PBASE[0x48] = sin_table[loop] << 2;  /*output value from sine table*/
  if (loop < 3) ++loop;                /*increment loop counter < 3   */
  else loop = 0;                       /*reset loop counter           */
}

main()
{
  volatile int *INTVEC = VEC_ADDR;     /*pointer to vectors           */
  INTVEC[5] = (volatile int) c_int05;  /*install interrupt 5 Handler */
  AICSET_I();                          /*function to configure AIC    */
  for (;;);                            /*wait for interrupt           */
}
```

FIGURE 3.13. Real-time sine-generation program with four points, using C code (SINE4AIC.C)

Having "chosen" TA = 10, the switched capacitor filter (SCF) frequency is

$$\text{SCF} = \frac{(\text{AIC master clock frequency})}{2 \times \text{TA}} = \frac{7.5 \text{ MHz}}{20} = 375 \text{ kHz}$$

From the SCF frequency, the **B** value can be found, or

$$\text{TB} = \frac{\text{SCF}}{F_s} = \frac{375 \text{ kHz}}{10 \text{ kHz}} \approx 37 = (100101)_b$$

Using

bits 2–7	for RB
bit 8	for don't care
bits 9–14	for TB
bit 15	for don't care

the third value in the AICSEC array, 0x4A96, can be obtained. This value is found using TB = RB in

x |←TB→| x |←RB→| b1 b0

```c
/*AICCOM.C-COMUNICATION ROUTINES FOR SERIAL PORT 0 OF AIC */
#define TWAIT while (!(PBASE[0x40] & 0x2)) /*wait till XMIT buffer clear*/
extern int AICSEC[4];                      /*array defined in main prog */
volatile int *PBASE = (volatile int *) 0x808000; /*peripherals base addr*/

void AICSET()                       /*function to initialize AIC */
{
  volatile int loop;                /*declare local variables   */
  PBASE[0x28] = 0x00000001;         /*set timer period          */
  PBASE[0x20] = 0x000002C1;         /*set timer control register*/
  asm("    LDI       00000002h,IOF"); /*set IOF low to reset AIC */
  for (loop = 0; loop < 50; loop++); /*keep IOF low for a while  */
  PBASE[0x42] = 0x00000111;         /*set xmit port control     */
  PBASE[0x43] = 0x00000111;         /*set receive port control  */
  PBASE[0x40] = 0x0E970300;         /*set serial port global reg*/
  PBASE[0x48] = 0x00000000;         /*clear xmit register       */
  asm("    OR        00000006h,IOF"); /*set IOF high to enable AIC*/
  for (loop = 0; loop < 4; loop++)  /*loop to configure AIC     */
  {
    TWAIT;                          /*wait till XMIT buffer clear*/
    PBASE[0x48] = 0x3;              /*enable secondary comm     */
    TWAIT;                          /*wait till XMIT buffer clear*/
    PBASE[0x48] = AICSEC[loop];     /*secondary command for SP0 */
  }
}

void AICSET_I()                     /*configure AIC, enable TINT0*/
{
  AICSET();                         /*function to configure AIC */
  asm("    LDI       00000000h,IF"); /*clear IF Register         */
  asm("    OR        00000010h,IE"); /*enable EXINT0 CPU interrupt*/
  asm("    OR        00002000h,ST"); /*global interrupt enable   */
}

int UPDATE_SAMPLE(int output)       /*function to update sample */
{
  int input;                        /*declare local variables   */
  TWAIT;                            /*wait till XMIT buffer clear*/
  PBASE[0x48] = output << 2;        /*left shift and output sample*/
  input = PBASE[0x4C] << 16 >> 18;  /*input sample and sign extend*/
  return(input);                    /*return new sample         */
}
```

FIGURE 3.14. TMS320C30–AIC communication program in C (`AICCOM.C`)

or

0100 1010 1001 0110

which corresponds to 4A96 in hex, using x = 0 as "don't care" and the first two bits b0 = 0 and b1 = 1. Note that the actual F_s is 10.135 kHz because TB was approximated to 37.

The corner frequency (or bandwidth) can be found using

$$\frac{\text{New SCF}}{\text{Set SCF}} = \frac{375 \text{ kHz}}{288 \text{ kHz}} = 1.3$$

Then, the corner frequency (3,600 Hz) is scaled by this ratio, or

$$f_c = (1.3)(3,600) = 4,688 \text{ Hz}$$

using a bandwidth value of 3.6 kHz, corresponding to an SCF frequency of 288 kHz. This cutoff frequency can be verified with the loop program of Example 1.3, by observing that the amplitude of the output sinusoid reduces when the input frequency reaches 4,688 Hz.

The last value in AICSEC, 0x67, specifies the primary input (IN) to the AIC. Change this value to 0x77 to select the second AIC input (AUX IN). Furthermore, the third bit specifies whether the AIC input filter is inserted or deleted. Use a value of 0x63 in order to bypass the input filter (see d2 in Figure 3.4). Although this second input of the AIC on-board the EVM is connected to ground, we will see how an external AIC module, with two available inputs, can be connected through the serial port 1 on the TMS320C30.

3. Interrupt XINT0 is chosen using the interrupt vector c_int05 at memory location 5. Because the timer 0 clock is set at 7.5 MHz for a real-time implementation, the AIC can be configured as illustrated in the previous step to provide the desired rate. With c_int05, the serial port 0 transmit interrupt is used. The previous program example (SINE4C.C) used TINT0, or int09.

4. Within the main program, the function AICSET_I is called to initialize the AIC so that it becomes interrupt-driven. Then, execution proceeds to wait for the interrupt (within an endless loop).

5. On interrupt, execution proceeds to the interrupt vector function c_int05. The data transmit register is at the peripheral address 808048h, where each output value is stored. The peripheral base address for serial port 0 is 808000h. The four values stored in the array (sine table) are sent to the data transmit register, one at a time. The loop count is reinitialized each time it reaches the last value set in the table.

6. Because the AIC has a 14-bit ADC and DAC, the output value is left-shifted by two and the least significant bits are cleared to zero to enable primary communications of the AIC.

TMS320C30 – AIC Communication Routines (AICCOM.C)

The TMS320C30–AIC communication routines, listed in Figure 3.14 can handle both interrupt and polling, using port 0. Later, we will use a similar program to handle serial port 1 as well. The serial port 1 registers correspond to the memory locations of the serial port 0 registers, with an offset by 10. Hence, an offset value within PBASE can be used with this program to select the serial port 1 registers. Note the following (see also Table 3.1):

1. The timer period register 0x808028 is set to 1, the timer global register is set with the control value 0x2C1. Bit 2 of the IOF flag register (OUTXF0) is set to 1. The wait loop is used to keep the IOF low for some period of time (recommended by Texas Instruments, Inc.).
2. The transmit, receive, and serial port global control registers at 0x808042, 0x808043, and 0x808040, respectively, are initialized.
3. The serial port data transmit register at 0x808048 is initialized to zero. The AIC is enabled by setting the IOF register high.
4. Without interrupt (with polling), the AICSET function is accessed to initialize the AIC. If the AIC is interrupt-driven, the AICSET_I function is accessed. This interrupt-driven function initializes first the AIC, as with polling, then enables the AIC for interrupt. This is accomplished with the three asm statements (not needed for polling).
5. The TMS320C30 assembly code within the asm statements is ignored by the C compiler and is executed as specified. The interrupt flag register IF is initialized to zero, to clear any pending interrupts. The EXINT0 CPU interrupt is enabled by setting the fifth bit of the interrupt enable register IE. Then the global interrupt enable register GIE, bit 13 within the status register, is set to 1. Interrupt is then enabled.
6. For input and output, the data receive register at 0x80804C and the data transmit register at 0x808048 are used.

Run this program and observe a sinusoidal waveform on an oscilloscope. Using a sampling frequency of 10 kHz, the frequency of the sinusoidal waveform is

$$f = \frac{F_s}{\text{number of points}} = 2.5 \text{ kHz}$$

Change the third value in AICSEC from 0x4A96 to 0x5EBE, in order to use a sampling rate of 8 kHz, and generate a sinusoidal waveform of frequency 2 kHz. This new sampling frequency would correspond to a **B** value of $\simeq 47$. The second value 0x1, set in AICSEC in the main program, is used for fine-tuning the sampling frequency through the registers TA' and RA' available on the AIC [1, 4].

Example 3.7 *Real-Time Loop Program without Interrupt, Using C Code.* The program LOOPALL.C, introduced in Example 1.3, is modified to obtain a

3.4 Programming Examples

main program LOOPC.C, listed in Figure 3.15. This main program calls the TMS320C30-AIC communication routines AICCOM.C (used in the previous sine-generation example), which is listed in Figure 3.14. A command file similar to that of Example 3.6 can be used for linking. To use EVMLOAD, it is necessary to use VECSIR.OBJ when linking. When the UPDATE_SAMPLE function (in AICCOM) is called, an output first occurs from the data transmit register at location specified by PBASE[0x48], or 808048h. Note that this output is first shifted left by two, to enable primary AIC communication. Before each output and input, the transmit buffer is first cleared. An input sample is obtained from the data receive register at location PBASE[0x4C], or 80804Ch. The input sample is sign-extended by first shifting the data left by 16 bits and then right by 18 bits (not 16).

Run this program and observe an output sinusoidal waveform of the same frequency as that of the input sinusoid, but amplified by approximately 8 (amplified to drive an external speaker). The amplification circuitry is on the EVM.

The two previous examples illustrate how the AIC can be used: with interrupt as in the sine-generation program, or without interrupt as in the loop program. We will now discuss a more general communication program that can be used with either port 0 or 1 on the TMS320C30.

Testing the AIC through Serial Port 1

Alternative input and output devices can be connected through the TMS320C30 serial port 1. This port, available through a 10-pin connector on the EVM, can be used to interface the TMS320C30 to another (external) AIC with two inputs.

```
/*LOOPC.C-LOOP PROGRAM USING PORT 0 WITH POLLING */
#include "aiccom.c" /*AIC Communication routines */
int AICSEC[4] = {0x1428,0x1,0x4A96,0x67}; /*Config data for AIC*/
main()
{
   int  data_in, data_out;            /*Initialize variables   */
   AICSET();                          /*Function to config AIC*/
   while (1)                          /*Create endless loop    */
   {
     data_in = UPDATE_SAMPLE(data_out); /*Call function to update sample*/
     data_out = data_in;              /*Loop input to output            */
   }
}
```

FIGURE 3.15. Main loop program using C code (LOOPC.C)

74 Alternative Input/Output and Extended Development System

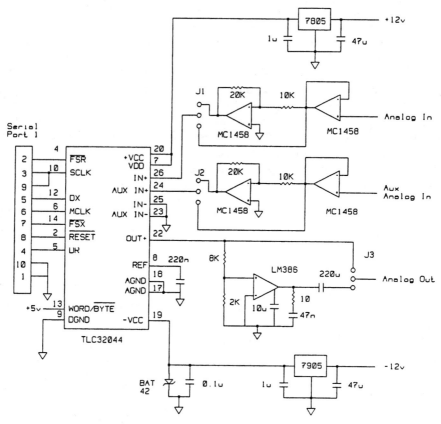

FIGURE 3.16. TMS320C30–AIC interfacing diagram

Figure 3.16 shows the TMS320C30–AIC interfacing diagram. An inexpensive external AIC module can be readily built on a board and connected to the serial port 1 on the EVM. The interfacing circuit diagram includes buffers and circuitry for noise reduction and is used to interface the AIC to the TMS320C30. It provides a similar type of amplification and phase inversion as the interfacing circuitry on the EVM. Jumper connectors are available on the board with the external AIC so that the input and output can be processed with or without amplification. Jumper connectors J1 and J2 can be used to bypass or to select the input amplification available (gain of 2) for each of the two AIC inputs (IN and AUX IN). Connector J3 provides the option for selecting an output with an amplification factor of 4. A ribbon cable connects this board and the 10-pin serial port 1 connector on the EVM. A similar AIC module has been used for several years [2], initially with the

3.4 Programming Examples

```
/*LOOPCI.C-LOOP PROGRAM TO TEST AIC USING INTERRUPTS        */
#include "aiccom01.c"                  /*AIC comm routines         */
#define VEC_ADDR (volatile int *) 0x00 /*addr of vectors           */
int AICSEC[4] = {0x1428,0x1,0x4A96,0x67}; /*config data for SP0    */
int AICSEC1[4];                        /*SP1 not used              */
int data_in, data_out;                 /*declare global variables*/

void c_int05()                         /*TINT0 interrupt routine */
{
  data_in = UPDATE_SAMPLE(SP0, data_out); /*update sample to SP0 AIC*/
  data_out = data_in;                  /*loop input to output      */
}

main()
{
  volatile int *INTVEC = VEC_ADDR;     /*pointer to vectors        */
  INTVEC[5] = (volatile int) c_int05;  /*install interrupt 5 Handler*/
  AICSET_I(SP0);                       /*configure SP0 of AIC      */
  for (;;);                            /*wait for interrupt        */
}
```

FIGURE 3.17. Alternative loop program, which calls AICCOM01.C with interrupt, using C code and serial port 0 (LOOPCI.C)

first-generation TLC32041 (which has no internal reference voltage) and, more recently, with the TLC32044 (the generation on-board the EVM). A newer AIC member, the TLC32046, has characteristics similar to those of the TLC32044, but with a higher bandwidth and a maximum sampling rate of 25 kHz, as compared to the 19.2 kHz available with the TLC32044.

Example 3.8 Alternative Loop Program Suitable for Either Port 0 or 1, Using C Code. The loop program (LOOPCI.C) listed in Figure 3.17 is more general than the one discussed in the previous example because it can be readily modified for use with either port 0 or 1. This program calls the AIC communication program AICCOM01.C, listed in Figure 3.18. This example is tested using port 0. The pertinent registers for the serial port 1 are located at addresses that are 10-higher than those used for serial port 0. Set SP (which designates serial port, not stack pointer) to 0 or to 10 in order to use either the serial port 0 registers or the serial port 1 registers, respectively.

The TMS320C30–AIC communication program listed in Figure 3.18 is more general than the previous AICCOM.C (for serial port 0 only), because it can be used with either serial port 0 or serial port 1, and with either polling or interrupt. The two functions GET_SAMPLE and PUT_SAMPLE in AIC-COM01 are optional. The program in Figure 3.17 is set for serial port 0, using interrupt. The configuration data for the AIC port 0 is defined in AICSEC.

```c
/*AICCOM01.C-COMMUNICATION ROUTINES FOR BOTH SERIAL PORTS 0 AND 1 OF AIC  */
#define TWAIT while (!(PBASE[0x40+SP] & 0x2))  /*wait till XMIT buffer clear*/
#define SP0 0x00                               /*serial port 0 offset       */
#define SP1 0x10                               /*serial port 1 offset       */
extern int AICSEC[4], AICSEC1[4];              /*array defined in main prog */
volatile int *PBASE = (volatile int *) 0x808000; /*peripherals base addr    */

void AICSET(int SP)                            /*function to initialize AIC */
{
  volatile int loop;                           /*declare local variables    */
  PBASE[0x28+SP] = 0x00000001;                 /*set timer period           */
  PBASE[0x20+SP] = 0x000002C1;                 /*set timer control register */
  if (!SP) asm("    LDI     00000062h,IOF");   /*set IOF low to reset PRI AIC*/
  else asm("    LDI     00000026h,IOF");       /*set IOF low to reset AUX AIC*/
  for (loop = 0; loop < 50; loop++);           /*keep IOF low for a while   */
  PBASE[0x42+SP] = 0x00000111;                 /*set xmit port control      */
  PBASE[0x43+SP] = 0x00000111;                 /*set receive port control   */
  PBASE[0x40+SP] = 0x0E970300;                 /*set serial port global reg */
  PBASE[0x48+SP] = 0x00000000;                 /*clear xmit register        */
  if (!SP) asm("    OR      00000006h,IOF");   /*set IOF high enable AIC PRI*/
  else asm("    OR      00000060h,IOF");       /*set IOF high enable AIC AUX*/
  for (loop = 0; loop < 4; loop++)             /*loop to configure AIC      */
  {
    TWAIT(SP);                                 /*wait till XMIT buffer clear*/
    PBASE[0x48+SP] = 0x3;                      /*enable secondary comm      */
    TWAIT(SP);                                 /*wait till XMIT buffer clear*/
    if (!SP)  PBASE[0x48] = AICSEC[loop];      /*secondary command for SP0  */
    else PBASE[0x58] = AICSEC1[loop];          /*secondary command for SP1  */
  }
}

void AICSET_I(int SP)                          /*configure AIC, enable TINT0*/
{
  AICSET(SP);                                  /*function to configure AIC  */
  asm("    LDI     00000000h,IF");             /*clear IF Register          */
  if (!SP) asm("    OR      00000010h,IE");    /*enable EXINT0 CPU interrupt*/
  else asm("    OR      00000040h,IE");        /*enable EXINT1 CPU interrupt*/
  asm("    OR      00002000h,ST");             /*global interrupt enable    */
}

int UPDATE_SAMPLE(int SP, int output)          /*function to update sample  */
{
  int input;                                   /*declare local variable     */
  TWAIT(SP);                                   /*wait till XMIT buffer clear*/
  PBASE[0x48+SP] = output << 2;                /*left shift and output sample*/
  input = PBASE[0x4C+SP] << 16 >> 18;          /*input sample and sign extend*/
  return(input);                               /*return new sample          */
}
int GET_SAMPLE(int SP)                         /*function to input sample   */
{
  int input;                                   /*declare local variable     */
  input = PBASE[0x4C+SP] << 16 >> 18;          /*input and sign extend sample*/
  return(input);                               /*return new sample          */
}
void PUT_SAMPLE(int SP, int output)            /*function to output sample  */
{
  TWAIT(SP);                                   /*wait till XMIT buffer clear*/
  PBASE[0x48+SP] = output << 2;                /*left shift and output sample*/
}
```

FIGURE 3.18. AIC communication routines for serial port 0 or 1 using C code (`AICCOM01.C`)

3.4 Programming Examples

```
/*LOOPC1.C-LOOP PROGRAM TO TEST AIC PORT 1              */
#include "aiccom01.c"                 /*AIC COMM ROUTINES      */
int AICSEC[4];                        /*PORT 0 NOT USED FOR AIC*/
int AICSEC1[4] = {0x1428,0x1,0x4A96,0x67}; /*CONFIG DATA FOR AIC  */
main()
{
  int data_in, data_out;              /*INITIALIZE VARABLES    */
  AICSET(SP1);                        /*FUNCTION TO CONFIGURE AIC*/
  while (1)                           /*CREATE ENDLESS LOOP    */
  {
    data_in = UPDATE_SAMPLE(SP1, data_out); /*CALL FUNCT TO UPDATE SAMPLE*/
    data_out = data_in;               /*LOOP INPUT TO OUTPUT   */
  }
}
```

FIGURE 3.19. Loop program for serial port 1 without interrupt (LOOPC1.C)

To use serial port 1, make the following changes in LOOPCI.C:

1. Define the AIC configuration data in AICSEC1 in lieu of AICSEC.
2. Use TINT1 interrupt routine c_int07 in lieu of TINT0 interrupt routine c_int05. This includes the use of INTVEC[7] in lieu of INTVEC[5].
3. Use SP1 for serial port 1, when calling the functions UPDATE_SAMPLE and AICSET_I.

Note that VECSIR.OBJ should always be included in the command file used with the C code.

Figure 3.19 shows another loop program, LOOPC1.C, which calls AIC-COM01.C for serial port 1. It is similar to LOOPC.C listed in Figure 3.15. This program is not interrupt-driven.

Test these programs (on accompanying disk) in a fashion similar to that in the previous example.

The AIC module, with two inputs and one output, can be very useful for adaptive filtering, as discussed in Chapter 7. One application in adaptive filtering is for noise reduction. In such an application, the primary input would contain a signal with added noise, and the secondary input would contain a reference noise. Note that it is further possible to obtain three inputs, using the on-board AIC and the external AIC module.

Example 3.9 Loop Program Using TMS320C30 Code. This example discusses a loop program LOOPS.ASM in TMS320C30 code, using serial port 0. This main program, listed in Figure 3.20, calls the AIC communication program

```
;LOOPS.ASM-LOOP PROGRAM USING THE AIC-CALLS AICCOMA.ASM
        .TITLE   "LOOPS.ASM"           ;OUTPUT=INPUT(AMPLIFIED)
        .OPTION  X                     ;FOR SYMBOL XREF
        .GLOBAL  RESET,BEGIN,AICSET,AICSEC,AICIO_P,SPSET
        .SECT    "VECTORS"             ;VECTOR SECTION
RESET   .WORD    BEGIN                 ;RESET VECTOR
        .SPACE   63                    ;REMAINDER OF VECTOR SECTION
        .DATA                          ;ASSEMBLE INTO DATA SECTION
AICSEC  .WORD    1428h,1h,4A96h,67h    ;SP0 AIC CONFIG DATA
        .TEXT                          ;ASSEMBLE INTO TEXT SECTION
BEGIN   LDP      SPSET                 ;INIT DATA PAGE
        CALL     AICSET                ;INIT PRIMARY AIC/TMS320C30
        LDI      0,R7                  ;INIT R7 = OUTPUT 0
LOOP    CALL     AICIO_P               ;I/O ROUTINE,INPUT=>R6,OUT=>R7
        LDI      R6,R7                 ;OUTPUT R7=NEW INPUT SAMPLE
        BR       LOOP                  ;LOOP CONTINUOUSLY
        .END                           ;END
```

FIGURE 3.20. Loop program using TMS320C30 code (LOOPS.ASM)

(AICCOMA.ASM) listed in Figure 3.21, also in TMS320C30 code. The main program can be slightly modified to handle interrupt, or access serial port 1. The AIC communication I/O routine expects the input to be in register R6 and the output to be in register R7. A sampling rate of 10 kHz is set in AICSEC.

```
*AICCOMA.ASM-AIC COMMUNICATION ROUTINES IN TMS320C30-POLLING OR INTERRUPT
        .title   "AICCOMA"             ;TMS320C30/AIC COM WITH POLLING OR INTERRUPT
        .global  AICSEC,AICSET,AICSET_I,AICIO_I,AICIO_P,TWAIT,SPSET,IOPRI,IOAUX
        .data                          ;ASSEMBLE INTO DATA SECTION
PBASE   .word    808000h               ;PERIPHERAL BASE ADDRESS
SETSP   .word    0E970300h             ;SERIAL PORT SET-UP DATA
ATABLE  .word    AICSEC                ;SP0 AIC INIT TABLE ADDR
SPSET   .word    0h                    ;SERIAL PORT 0 OFFSET
        .text                          ;ASSEMBLE INTO TEXT
AICSET  PUSH     AR0                   ;SAVE AR0
        PUSH     AR1                   ;SAVE AR1
        PUSH     AR7                   ;SAVE AR7
        PUSH     R0                    ;SAVE R0
        PUSH     R1                    ;SAVE R1
        LDI      @SPSET,AR7            ;SERIAL PORT OFFSET
        LDI      @PBASE,AR0            ;AR0 -> 808000h
        ADDI     AR7,AR0               ;SERIAL PORT OFFSET + PBASE
        LDI      1h,R0                 ;TIMER CLK = H1/2 *(AIC MASTER CLK)
        STI      R0,*+AR0(28h)         ;INIT TIMER PERIOD REG(TCLK0=7.5 MHZ)
        LDI      02C1h,R0              ;INIT TIMER GLOBAL REG
        STI      R0,*+AR0(20h)         ;RESET TIMER
```

FIGURE 3.21. AIC communication program using TMS320C30 code (AICCOMA.ASM)

3.4 Programming Examples

```
               CMPI     0,AR7                ;DETERMINE REQUESTED SERIAL PORT
               BNZ      SP1A                 ;BRANCH TO CONFIGURE SERIAL PORT 1
               LDI      62h,IOF              ;PRI AIC RESET = 0
               B        SP0A                 ;BRANCH TO CONFIGURE SERIAL PORT 0
SP1A           LDI      26h,IOF              ;AUX AIC RESET = 0
SP0A           LDI      @ATABLE,AR1          ;AR1 -> AIC INIT DATA
               RPTS     99                   ;REPEAT NEXT INSTRUCTION 100 TIMES
               NOP                           ;KEEP IOF LOW FOR A WHILE
               LDI      111h,R0              ;X & R PORT CONTROL REGS DATA
               STI      R0,*+AR0(42h)        ;FSX/DX/CLKX = SP OPERATIONAL PINS
               STI      R0,*+AR0(43h)        ;FSR/DR/CLKR = SP OPERATIONAL PINS
               LDI      @SETSP,R0            ;RESET -> SP: 16 BITS, EXT CLKS, STD MODE
               STI      R0,*+AR0(40h)        ;FSX = OUTPUT, & INT ENABL (SP GLOBAL REG)
               LDI      0,R0                 ;CLEAR R0
               STI      R0,*+AR0(48h)        ;CLEAR SERIAL PORT XMIT REGISTER
               CMPI     0,AR7                ;DETERMINE REQUESTED SERIAL PORT
               BNZ      SP1B                 ;BRANCH TO CONFIGURE SERIAL PORT 1
               OR       06h,IOF              ;BRING PRI AIC OUT OF RESET
               B        SP0B                 ;BRANCH TO CONFIGURE SERIAL PORT 0
SP1B           OR       60h,IOF              ;BRING AUX AIC OUT OF RESET
SP0B           LDI      03h,RC               ;RC = 3 TO TRANSMIT 4 VALUES
               RPTB     SECEND               ;REPEAT 4 DATA TRANSMITS OF SECONDARY COMM
               CALL     TWAIT                ;WAIT FOR DATA TRANSMIT
               LDI      03h,R0               ;VALUE FOR SEC XMIT REQUEST FOR AIC
               STI      R0,*+AR0(48h)        ;SECONDARY XMIT REQUEST FOR AIC
               CALL     TWAIT
               LDI      *AR1++(1),R0         ;AR1 -> NEXT AIC INIT DATA
SECEND         STI      R0,*+AR0(48h)        ;DTR =  CURRENT AIC DATA
               POP      R1                   ;RESTORE R1
               POP      R0                   ;RESTORE R0
               POP      AR7                  ;RESTORE AR7
               POP      AR1                  ;RESTORE AR1
               POP      AR0                  ;RESTORE AR0
               RETS                          ;RETURN FROM SUBROUTINE
AICSET_I                                     ;-----CONFIG FOR INTERRUPT-----
               PUSH     AR7                  ;SAVE AR7
               CALL     AICSET               ;CALL AICSET ROUTINE
               LDI      @SPSET,AR7           ;SERIAL PORT OFFSET
               LDI      0h,IF                ;CLEAR IF REGISTER
               CMPI     0,AR7                ;DETERMINE REQUESTED SERIAL PORT
               BNZ      SP1C                 ;BRANCH TO CONFIGURE SERIAL PORT 1
               OR       10h,IE               ;ENABLE EXINT0 CPU INTERRUPT
               B        SP0C                 ;BRANCH TO LABEL A2
SP1C           OR       40h,IE               ;ENABLE EXINT1 CPU INTERRUPT
SP0C           OR       2000h,ST             ;GLOBAL INTERRUPT ENABLE
               POP      AR7                  ;RESTORE AR7
               RETS                          ;RETURN FROM SUBROUTINE
;--------------------TRANSMIT WAIT ROUTINE------------------------
TWAIT          PUSH     AR0                  ;SAVE AR0
               PUSH     AR7                  ;SAVE AR7
               PUSH     R0                   ;SAVE R0
               LDI      @PBASE,AR0           ;AR0 -> 0808000h
               LDI      @SPSET,AR7           ;SERIAL PORT OFFSET
               ADDI     AR7,AR0              ;SERIAL PORT OFFSET + PBASE
TW1            LDI      *+AR0(40h),R0        ;R0 = CONTENT OF SP GLOBAL CONTROL REG
               AND      02h,R0               ;SEE IF TRANSMIT BUFFER IS READY
```

FIGURE 3.21. (*Continued*)

```
                BZ      TW1                 ;IF NOT READY, TRY AGAIN
                POP     R0                  ;RESTORE R0
                POP     AR7                 ;RESTORE AR7
                POP     AR0                 ;RESTORE AR0
                RETS                        ;RETURN FROM SUBROUTINE
;--------------------------AIC TRANSFER ROUTINE--------------------------
AICIO_I LSH     2,R7                ;TWO LSB MUST = 0 FOR PRIMARY AIC COM
IO      PUSH    AR0                 ;SAVE AR0
        PUSH    AR7                 ;SAVE AR7
        LDI     @SPSET,AR7          ;SERIAL PORT OFFSET
        LDI     @PBASE,AR0          ;AR0 -> 0808000h
        ADDI    AR7,AR0             ;SERIAL PORT OFFSET + PBASE
        STI     R7,*+AR0(48h)       ;DTR = NEXT DATA FOR AIC D/A
        LDI     *+AR0(4Ch),R6       ;R6 = DRR DATA FROM AIC A/D
        LSH     16,R6               ;LEFT SHIFT FOR SIGN EXTENSION
        ASH     -18,R6              ;RIGHT SHIFT KEEPING SIGN
        POP     AR7                 ;RESTORE AR7
        POP     AR0                 ;RESTORE AR0
        RETS
;--------------------------AIC POLLING ROUTINE--------------------------
AICIO_P CALL    TWAIT               ;WAIT FOR DATA TO BE TRANSFERRED
        CALL    AICIO_I             ;CALL AIC TRANSFER ROUTINE
        RETS                        ;RETURN FROM SUBROUTINE
SW_IO   LSH     2,R7                ;PREPARE FOR SECONDARY AIC COM
        OR      03h,R7              ;SET TWO LSB FOR SECONDARY COM
        CALL    TWAIT               ;WAIT FOR DATA TO BE TRANSFERRED
        CALL    IO                  ;CALL AIC TRANSFER ROUTINE
        LDI     R1,R7               ;LOAD SECONDARY COM DATA INTO R7
        CALL    TWAIT               ;WAIT FOR DATA TO BE TRANSFERRED
        CALL    IO                  ;CALL AIC TRANSFER ROUTINE
        RETS
;SUBROUTINES FOR PRIMARY OR AUXILIARY INPUT
IOPRI   PUSH    R1                  ;SAVE R1
        LDI     067h,R1             ;LOAD SECONDARY COM DATA INTO R1
        CALL    SW_IO               ;CALL IO ROUTINE TO SWITCH INPUTS
        POP     R1                  ;RESTORE R1
        RETS
IOAUX   PUSH    R1                  ;SAVE R1
        LDI     077h,R1             ;LOAD SECONDARY COM DATA INTO R1
        CALL    SW_IO               ;CALL IO ROUTINE TO SWITCH INPUTS
        POP     R1                  ;RESTORE R1
        RETS
```

FIGURE 3.21. *(Continued)*

To use serial port 1, add the following two instructions in `LOOPS.ASM` (before `CALL AICSET`):

```
LDI  10h,R1
STI  R1,@SPSET
```

An interrupt-driven version of the loop program in TMS320C30 code, `LOOPSI.ASM`, is listed in Figure 3.22. Note that `AICSEC_I` (not `AICSEC`) is used for the AIC configuration data and that `AICIO_I` (not `AICCIO_P`) is called for input and output. The same AIC communication program AIC-

3.4 Programming Examples

```
;LOOPSI.ASM-LOOP PROGRAM USING INTERRUPT.CALLS AICCOMS.ASM
        .title   "LOOPSI"          ;TESTS AIC
        .global  RESET,BEGIN,AICSEC,AICSET,AICSET_I,AICIO_I,SPSET
        .sect    "vectors"         ;VECTOR SECTION
RESET   .word    BEGIN             ;RESET VECTOR
        .space   4                 ;SKIP 4 WORDS
        .word    ISR               ;SP 0 TRANSMIT INTERRUPT SERVICE ROUTINE
        .space   58                ;REMAINDER OF VECTOR SECTION
        .data                      ;ASSEMBLE INTO DATA SECTION
STACKS  .word    809F00h           ;INIT STACK POINTER DATA
AICSEC  .word    1428h,1h,4A96h,67h    ;SP0 AIC CONFIG DATA
        .text                      ;ASSEMBLE INTO TEXT SECTION
BEGIN   LDP      STACKS            ;INIT DATA PAGE
        LDI      @STACKS,SP        ;SP -> 809F00h
        CALL     AICSET_I          ;INIT AIC
        LDI      0,R7              ;R7 = 0h   (OUTPUT)
LOOP    IDLE                       ;WAIT FOR TRANSMIT INTERRUPT
        LDI      R6,R7             ;R7 = NEW INPUT SAMPLE (OUTPUT)
        BR       LOOP              ;BRANCH BACK AGAIN
ISR     CALL     AICIO_I           ;OUTPUT R7 AND PLACE INPUT IN R6
        RETI                       ;RETURN FROM INTERRUPT
        .END                       ;END
```

FIGURE 3.22. Loop program with interrupt, using TMS320C30 code (LOOPSI.ASM)

COMA.ASM is called from both programs LOOPS.ASM (polling) and LOOPSI.ASM (interrupt-driven).

Example 3.10 Loop Program to Test both AIC Inputs, Using TMS320C30 Code. This example tests both the primary and the auxiliary inputs of the external AIC through serial port 1, using TMS320C30 code. The AIC communication routines in the program AICCOMA.ASM are used. The main program LOOPSW1.ASM is listed in Figure 3.23. The subroutine IOPRI (from AICCOMA.ASM) is called to input a new sample from the primary input (IN) of the AIC. This input, placed in R6 within the AIC transfer subroutine in AICCOMA.ASM, is loaded into R7 in the main program for output. When the IOAUX subroutine is called from the main program, the primary input sample obtained previously and loaded in R7 is output, then a new input is placed into R6 from the auxiliary input (AUX IN) of the AIC. This new sample is then loaded into R7 for output when the IOPRI subroutine is called again.

Run this program. Connect a sinusoidal signal to both the primary and auxiliary inputs of the AIC (the same signal through both inputs). Observe the output signal. Disconnect one of the AIC inputs and observe the output waveform.

```
;LOOPSW1.ASM-TO TEST BOTH AIC INPUTS USING SERIAL PORT 1.CALLS AICCOMA
         .TITLE   "LOOPSW1"           ;OUTPUT=INPUT
         .GLOBAL  RESET,BEGIN,AICSET,AICSEC,AICIO_P,SPSET,IOPRI,IOAUX
         .SECT    "VECTORS"           ;VECTOR SECTION
RESET    .WORD    BEGIN               ;RESET VECTOR
         .SPACE   63                  ;REMAINDER OF VECTOR SECTION
         .DATA                        ;ASSEMBLE INTO DATA SECTION
AICSEC   .WORD    1428h,1h,4A96h,67h  ;SP AIC CONFIG DATA
         .TEXT                        ;ASSEMBLE INTO TEXT SECTION
BEGIN    LDP      SPSET               ;INIT DATA PAGE
         LDI      10h,R1              ;SERIAL PORT 1 OFFSET
         STI      R1,@SPSET           ;ENABLE SERIAL PORT 1
         CALL     AICSET              ;INIT PRIMARY AIC/TMS320C30
         LDI      0,R7                ;INIT R7 = OUTPUT 0
LOOP     CALL     IOPRI               ;USE PRIMARY INPUT IN
         LDI      R6,R7               ;OUTPUT R7=PRIMARY INPUT
         CALL     IOAUX               ;USE AUXILIARY INPUT AUXIN
         LDI      R6,R7               ;OUTPUT R7=AUXILIARY INPUT
         BR       LOOP                ;LOOP CONTINUOUSLY
         .END                         ;END
```

FIGURE 3.23. Loop program to test both inputs of external AIC, using serial port 1 (LOOPSW1.ASM)

In Chapter 7, we discuss a real-time adaptive notch filter project where we use both AIC inputs.

3.5 PC-HOST – TMS320C30 COMMUNICATION

In Chapter 1 we saw that it was possible, with the loop program LOOPCTRL.C, to control the output amplitude of a resulting sinusoidal waveform with the function keys of the PC. This control program (with graphics routines) is on the accompanying disk. A simpler version of this program without the extended graphics features, is discussed here. First, it is instructive to see how a variable can be passed from the PC host to the TMS320C30 on the EVM.

Communication between the PC host and the EVM can be achieved with either an 8-bit "command passing" or 16-bit DMA data transfer. This is accomplished through a bidirectional 16-bit register located on the EVM. This 16-bit register consists of two 8-bit registers that interface the lower 8 data bits on the I/O bus of the host computer to the lower 16 bits of the TMS320C30's expansion bus.

Because the communication through the 16-bit bidirectional register is reduced to 8 bits on the PC bus, a control register with an offset of 808h from the I/O base address controls 16-bit transfers. Likewise, a control register located with an offset of 800h from the I/O base address controls 8-bit transfers.

3.5 PC-Host – TMS320C30 Communication

Command passing is accomplished with a double polling protocol using the following steps:

1. The PC host tests a register value at 800h, offset from the I/O base address, to determine if it is clear. If so, this indicates that the TMS320C30 is available to receive a command.
2. The PC host writes the eight-bit command with an offset of 800h from the I/O base address.
3. The TMS320C30 acknowledges receiving the command by transferring the same value back to the PC host.
4. After the PC host receives acknowledgment, it withdraws its request by clearing the register value at 800h. The PC can then initiate a new command request.

This double polling method is a simple protocol for a single eight-bit command request from the PC host. This procedure can be repeated for multiple command requests.

The input/output base address in the PC host is set at 240h. Your PC may have another piece of hardware at this address. If so, this base address can be changed. Other settings are possible with the DIP switches on the EVM, such as a base address of 280h (instead of the default value of 240h). The PC host is capable of controlling the EVM through a register in the I/O space of the PC. The TMS320C30 and the PC can communicate with each other through a specific register, which is located at the base address offset by 800h. This address is mapped to the TMS320C30's expansion bus (804000h–805FFFh). The TMS320C30 can read what is transferred from the PC; likewise, the PC can read what is transferred from the TMS320C30. Neither the PC nor the TMS320C30 can read back what it wrote. The communication between the PC and the EVM, through the base address offset by 800h, represents an eight-bit read/write on the port address of the PC. The expansion bus on the TMS320C30 is connected to this PC port address through registers on board the EVM.

A DMA data transfer can be initiated to transfer 16-bit data between the PC host and the EVM. The procedures for transferring data between the EVM and the PC host, along with the appropriate macros, can be found in references [3] and [4].

The following examples will further illustrate the communication between the PC host and the EVM.

Example 3.11 Loop Program with Amplitude Control, Using C Code. Create the following batch file HOSTLOOP.BAT:

```
PCLOOP.EXE
EVMLOAD HOSTLOOP.OUT
```

84 Alternative Input/Output and Extended Development System

```c
/*PCLOOP.C COMMAND PROGRAM FOR COMMUNICATION BETWEEN PC/EVM */
#include <stdio.h>
#include <dos.h>

#define IOBASE      0x240              // base address on PC
#define COM_CMD     IOBASE + 0x0800    // addr of 8 bit R/W port

void main()
{
  int i;                               // declare variable
  clrscr();                            // clear screen
  outp(COM_CMD,0x00);                  // clear 8 bit R/W port to zero
  for (;;)                             // create endless loop
  {
    do
    {
      printf("\nPRESS CTRL BREAK TO QUIT");
      printf("\nEnter attenuation value (1-10) : ");
      scanf("%i", &i);                 // get attenuation value
    }
    while ((i<1) || (i>10));           // repeat if value not valid
    outp(COM_CMD,i);                   // output value to R/W port
    clrscr();                          // clear screen for next value
  }
}
```

FIGURE 3.24. Interfacing program for communication between PC and TMS320C30 using C code (`PCLOOP.C`)

This batch file contains the instructions to (a) execute the C program, `PCLOOP.C` (shown in Figure 3.24) and (b) download and run `HOSTLOOP.OUT`. The source file `HOSTLOOP.C` is shown in Figure 3.25.

Run the batch file by typing `HOSTLOOP.BAT`.

1. Consider the program `PCLOOP.C` in Figure 3.24. When executed, it creates the following messages on the PC screen:

```
PRESS CTRL BREAK TO QUIT
Enter attenuation value (1- 10):
```

Enter an attenuation value of 1. Connect a sinusoidal source with an amplitude of less than 3 V peak-to-peak to the EVM input. Observe the EVM output on an oscilloscope; the output amplitude is approximately 8 times the input amplitude. Enter an attenuation value of 4, and verify that the output amplitude is reduced by 4.

Initially, a value of zero is output to the PC port address (240h + 800h), defined by `COM_CMD` in the program `PCLOOP.C`, listed in Figure 3.24. When a valid value for i between 1 and 10 is entered, it is sent to the TMS320C30,

3.5 PC-Host – TMS320C30 Communication

```
/*HOSTLOOP.C-LOOP PROGRAM WITH AMPLITUDE CONTROL*/
#include "aiccom.c"                        /* AIC communications routines*/
int AICSEC[4] = {0x1428,0x1,0x4A96,0x67};  /* AIC setup data             */
volatile int *host = (volatile int *) 0x804000; /* address of host       */

void main(void)
{
  int data_IN, data_OUT, ampt = 1, i;      /* declare variables          */
  AICSET();                                /* initialize AIC             */
  do
  {
    i = *host & 0xFF;                      /* i = 8 bit word from host*/
    if (i > 0 && i < 11) ampt = i;         /* i is attenuation value   */
    data_IN = UPDATE_SAMPLE(data_OUT);     /* input output sample      */
    data_OUT = data_IN / ampt;             /* scale input to output    */
  }
  while (1);                               /* endless loop             */
}
```

FIGURE 3.25. Loop program with amplitude control, using C code (HOSTLOOP.C)

by outputting it to the address specified by COM_CMD. The statement outp
() is a macro for an eight-bit I/O space access [3, 4].

   2. Consider the program HOSTLOOP.C in Figure 3.25. The HOSTLOOP.C
program calls the AIC communication routines in AICCOM.C. The
TMS320C30 gets the i value through its I/O address 804000h (shown in
HOSTLOOP.C with a pointer to host specified at 804000h). After the AIC is
initialized, using AICSET, the i value passed from the PC (the content of
804000h) to the TMS320C30 is masked to reduce it to an eight-bit word (by
ANDing the i value with FFh). If this value is between 1 and 10, then
the variable ampt takes on this new i value. Otherwise, it remains set with
the previous value. The output amplitude is the input (obtained through the
UPDATE_SAMPLE function in AICCOM.C) divided by the attenuation variable ampt.

Example 3.12 Resetting and Running the TMS320C30 with PC Host. This
example illustrates how the TMS320C30 can be placed in or out of reset.
Figure 3.26 shows a listing of the program (EVMCTRL.C) to stop and run the
TMS320C30. A control register is at the PC I/O base address (240h), offset
by Ah (0xA). This program example shows how data (0x800) can be passed
from the PC to the TMS320C30 using

```
outport  (0x240 + 0xA, 0x800)
```

where outport () is a function for a 16-bit I/O space access. To test this

```c
/*EVMCTRL.C-TO PLACE TMS320C30 IN OR OUT OF RESET*/
#include "stdio.h"
main(argc,argv)
int argc;
char *argv[];
{
  if (argc!=2)
  {
    printf("\n\n\tUsage: evmctrl [go] or [stop]\n\n");
    exit(1);
  }
  if (strncmp(argv[1],"go")==0)
  {
    printf("\n\n\tStarting C30\n\n");
    outport(0x0240 + 0x0A, 0x800);
    exit(0);
  }
  if (strncmp(argv[1],"stop")==0)
  {
    printf("\n\n\tStopping C30\n\n");
    outport(0x0240 + 0x0A, 0x808);
    exit(0);
  }
  printf("\n\n\tUsage: evmctrl [go] or [stop]\n\n");
  exit(1);
}
```

FIGURE 3.26. Program listing to place TMS320C30 in or out of reset (EVMCTRL.C)

program, run the HOSTLOOP program, by typing

```
EVMLOAD HOSTLOOP.OUT
```

Verify that this program is running by observing an output sinusoidal waveform (with a sinusoid as input), as in the previous example. Type

```
evmctrl stop
```

to place the TMS320C30 in reset, stopping the execution of the HOSTLOOP program. Type

```
evmctrl go
```

to run the `HOSTLOOP` program, taking the TMS320C30 out of reset. Offsetting the PC I/O base address by 0xA and transferring the data value 0x808 resets the TMS320C30 (through circuitry on the EVM).

Example 3.13 DMA Communication Between the PC Host and the TMS320C30. This examples illustrates DMA communication between the PC host and the TMS320C30. Create a batch file `DMAHOST.BAT` to execute the program `DMAHOST.EXE` and download the program `DMAEVM.OUT` (the source programs are on the accompanying disk, and supporting files are available with the EVM package). On running `DMAHOST.BAT` a set of 10 data values [1000, 2000, 3000, ..., 10000] initialized as an array in the `PCHOST.C` program is printed on the PC monitor in reversed order. This illustrates the following:

1. The data is transferred to the TMS320C30 through DMA with the program `DMAHOST.C`.
2. The TMS320C30 program `DMAEVM.C` reverses the order of the data and then, through DMA, transfers the set of data back to the PC host.

The EVM package [3, 4] includes two program examples:

1. a linear predictive coding (LPC) as an application in speech processing
2. an oscilloscope

These programs contain many useful macros and routines in C and TMS320C30 code, supporting much interaction between the PC host and the TMS320C30 on the EVM. Several of the projects discussed in Chapter 8 illustrate the transfer of data between the PC host and the TMS320C30.

3.6 BURR–BROWN TWO-CHANNEL ANALOG EVALUATION FIXTURE

A two-channel analog evaluation fixture, DEM1163, is available from Burr–Brown. This reasonably priced ($200) fixture is a complete analog I/O system. It includes the DSP102 analog-to-digital converter (ADC), the DSP202 digital-to-analog converter (DAC), antialiasing input and reconstruction output filters, and an on-board conversion rate generator. The ADC has a maximum conversion rate of 200 kHz, and the DAC 500 kHz, both with 18-bit conversion capability. Figure 3.27 shows the DEM1163 block diagram. Loop program examples are included in Appendix D to demonstrate the interfacing of the Burr–Brown analog board through the serial port 1 connector on the EVM. The interfacing diagram and the communication routines are also included in Appendix D.

88 Alternative Input/Output and Extended Development System

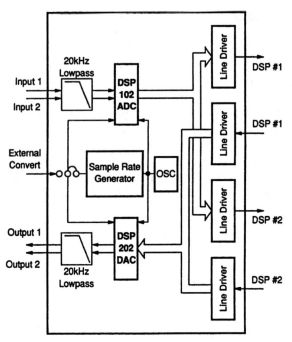

FIGURE 3.27. Burr–Brown analog evaluation fixture block diagram (© 1991 Burr–Brown Corporation. Reprinted, in whole or in part, with the permission of Burr–Brown Corporation)

3.7 TEXAS INSTRUMENTS' ANALOG INTERFACE BOARD

The analog interface board (AIB) is available from Texas Instruments, Inc. This board connects to an alternative DSP development system, the XDS1000 emulator, which will be discussed in the next section. The AIB contains the (Burr–Brown) 16-bit PCM76 ADC and 16-bit PCM53 DAC and can also be used to interface to the second-generation TMS320C25 digital signal processor. The AIB includes selectable input and output filters, mounted on a header, to allow ease of changing the filters' bandwidth. It has a maximum sampling rate capability of 58 kHz and an analog input and output range of ±10 V. An interrupt-driven program run on the simulator, with output selected at port address 804002h, can run on the XDS1000 connected to the AIB without changing the linked COFF object file. Examples using the AIB will be presented in conjunction with the XDS1000 in the next section and in Appendix E.

3.8 EXTENDED DEVELOPMENT SYSTEM XDS1000 EMULATOR

A powerful, but rather expensive, DSP development system is the XDS1000, available from Texas Instruments, Inc. The XDS1000 includes two boards and a cable to connect to a target system through a 12-pin connector. All the software tools used previously are also included with the XDS1000 package. The windows in the emulator-based XDS1000 are identical to those in the simulator, allowing for easy migration from simulation to a real-time environment.

SPOX, a real-time DSP operating system developed by Spectron Microsystems, Inc., can be used with the XDS1000. C-callable functions, such as the fast Fourier transform, are included in SPOX [6].

Target System

One of the two boards included with the XDS1000 package is an emulator. The other board, an application board that includes a TMS320C30, can be used for program development. A "homemade" target system was designed and built to interface to the XDS1000 emulator. This target module, used as an application board, includes the TMS320C30, 16K × 32 of static memory (SRAM). Three connectors are on the module to allow interfacing between the XDS1000 and either an external AIC module, or the AIB. Appropriate connectors for ± 5 V, -12 V, and ground, are also on the target module. A photo of the target module, supporting diagrams, and examples are included in Appendix E.

Exercise 3.1 Pseudorandom Noise Generation with the AIC. Implement a real-time noise generator (see Examples 3.2 and 3.3) using the AIC. Choose a sampling frequency of 10 kHz. The output to a spectrum analyzer should be relatively "flat" up to the cutoff frequency of the AIC.

REFERENCES

[1] *TMS320C25 User's Guide*, Texas Instruments, Inc., Dallas, Tex., 1989.

[2] R. Chassaing and D. W. Horning, *Digital Signal Processing with the TMS320C25*, Wiley, New York, 1990.

[3] *TMS320C30 Evaluation Module Technical Reference*, Texas Instruments, Inc., Dallas, Tex., 1990.

[4] *Digital Signal Processing Applications with the TMS320C30 Evaluation Module—Selected Application Notes*, Texas Instruments, Inc., Dallas, Tex., 1991.

[5] *Linear Circuits Data Book Volume 2: Data Acquisition and Conversion*, Texas Instruments, Inc., Dallas, Tex., 1989.

[6] *SPOX*, Spectron Microsystems, Santa Barbara, Calif., 1990.

4
Finite Impulse Response Filters

CONCEPTS AND PROCEDURES
- *Introduction to the Z-transform*
- *Design and implementation of finite impulse response (FIR) filters*
- *Use of filter design packages*

In this chapter, discrete-time and time-invariant systems are discussed. The Z-transform (ZT), used for discrete-time signals, is introduced. Mapping from the s-plane, associated with the Laplace transform, to the z-plane, associated with the Z-transform is illustrated.

Finite impulse response (FIR) filters are introduced from the convolution equation,

$$y(n) = \sum_{k=0}^{N} h(k)x(n-k)$$

The design of FIR filters is discussed using the Fourier series method. Window functions to improve the characteristics of such filters are covered. Two filter design packages are discussed:

1. *One from Hyperception, Inc., to design FIR filters (also IIR filters, discussed in the next chapter). Both C and TMS320C30 code can be generated with this filter package. Similar filter design packages are available from several other companies, such as Atlanta Signal Processors, Inc.*
2. *A "homemade" filter design package, included on the accompanying disk. An FIR filter using the rectangular, Hanning, Hamming, Blackman, or Kaiser window can be designed using this package.*

4.1 INTRODUCTION TO THE Z-TRANSFORM

Whereas the Laplace transform is used in conjunction with continuous-time signals, the Z-transform (ZT) is used for the analysis of discrete-time signals. In recent years, discrete-time signals have become more and more important. In a basic circuit course, we use the Laplace transform to solve a differential equation representing an analog filter. Similarly, we can use the Z-transform to solve a difference equation representing a digital filter.

Consider an analog signal $x(t)$, ideally sampled to yield

$$x_s(t) = \sum_{k=0}^{\infty} x(t)\delta(t - kT) \quad (4.1)$$

where $T = 1/F_s$ is the sampling period, and $\delta(t - kT)$ is the impulse function delayed by kT. This function is zero everywhere except at $t = kT$. The Laplace transform of the sampled signal $x_s(t)$ is

$$X_s(s) = \int_0^{\infty} x_s(t)e^{-st}\, dt$$

$$= \int_0^{\infty} \{x(t)\delta(t) + x(t)\delta(t - T) + \cdots\}e^{-st}\, dt \quad (4.2)$$

From the property of the impulse function,

$$\int_0^{\infty} f(t)\delta(t - kT)\, dt = f(kT).$$

Therefore, (4.2) becomes

$$X_s(s) = x(0) + x(T)e^{-sT} + x(2T)e^{-2sT} + \cdots = \sum_{n=0}^{\infty} x(nT)e^{-nsT} \quad (4.3)$$

Let $z = e^{sT}$ in (4.3), which becomes

$$X(z) = \sum_{n=0}^{\infty} x(nT)z^{-n} \quad (4.4)$$

92   Finite Impulse Response Filters

Let the sampling period T be implied, then $x(nT)$ can be written as $x(n)$. Equation (4.4) becomes

$$X(z) = \text{ZT}\{x(n)\} = \sum_{n=0}^{\infty} x(n) z^{-n} \tag{4.5}$$

which represents the Z-transform of $x(n)$. There is a one-to-one correspondence between $x(n)$ and $X(z)$, making the ZT a unique transformation.

Example 4.1 *ZT of Exponential Sequence.* Let $x(n) = e^{na}$, $n \geq 0$, and a is a constant. The ZT of $x(n)$ is

$$X(z) = \sum_{n=0}^{\infty} e^{na} z^{-n} = \sum_{n=0}^{\infty} (e^a z^{-1})^n \tag{4.6}$$

The geometric series

$$\sum_{n=0}^{\infty} u^n = \frac{1}{1-u}, \quad |u| < 1$$

can be obtained from a Taylor series approximation and used in (4.6) to yield

$$X(z) = \frac{1}{1 - e^a z^{-1}} = \frac{z}{z - e^a} \tag{4.7}$$

for $|e^a z^{-1}| < 1$, or $|z| > |e^a|$. Note that if $a = 0$, then the ZT of $x(n) = 1$ is $X(z) = z/(z-1)$.

Example 4.2 *ZT of Sinusoidal Sequence.* Let $x(n) = \sin n\omega T$. The sinusoidal function can be written in terms of complex exponentials, using Euler's formula. From $e^{+ju} = \cos u + j \sin u$, one can obtain

$$\cos n\omega T = \frac{e^{jn\omega T} + e^{-jn\omega T}}{2}$$

and

$$\sin n\omega T = \frac{e^{jn\omega T} - e^{-jn\omega T}}{2j}$$

Then

$$X(z) = \frac{1}{2j} \sum_{n=0}^{\infty} \{e^{jn\omega T}z^{-n} - e^{-jn\omega T}z^{-n}\} \quad (4.8)$$

A procedure similar to the previous example follows, applying the geometric series, or the results in (4.7) can be used with $a = j\omega T$ in the first summation and $a = -j\omega T$ in the second, to yield

$$X(z) = \frac{1}{2j}\left\{\frac{z}{z - e^{j\omega T}} - \frac{z}{z - e^{-j\omega T}}\right\}$$

$$= \frac{1}{2j}\left\{\frac{z^2 - ze^{-j\omega T} - z^2 + ze^{j\omega T}}{z^2 - z(e^{-j\omega T} + e^{+j\omega T}) + 1}\right\}$$

$$= \frac{z \sin \omega T}{z^2 - 2z \cos \omega T + 1}$$

$$= \frac{Cz}{z^2 - Az - B}, \quad |z| > 1 \quad (4.9)$$

where

$$A = 2\cos \omega T$$
$$B = -1$$
$$C = \sin \omega T$$

In Chapter 5, we generate a sinusoid using a difference equation based on this result.

It can be shown in a similar fashion that the ZT of $x(n) = \cos n\omega T$ is

$$X(z) = \frac{z^2 - z \cos \omega T}{z^2 - 2z \cos \omega T + 1}, \quad |z| > 1$$

Mapping from s-plane to z-plane

The s-plane associated with the Laplace transform can be used to determine whether a system is stable. For example, if the poles of a system are on the left side of the $j\omega$ axis, the system will have a time-decaying response, hence stable. Poles on the right side of the $j\omega$ axis will yield a response that will grow in time. A sinusoidal response can be obtained if the poles are purely imaginary (located on the $j\omega$ axis). It is then useful to find what the corresponding regions are in the z-plane. Note that the $j\omega$ axis represents the sinusoidal frequency ω, and $j\omega = 0$ represents DC. Because $z = e^{sT}$ can be expressed in terms of $s = \sigma + j\omega$,

$$z = e^{\sigma T} e^{j\omega T} \tag{4.10}$$

The magnitude of z is $|z| = e^{\sigma T}$ with a phase of $\theta = \omega T = 2\pi f/F_s$, where F_s is the sampling frequency. Consider the following regions in the s-plane.

1. $\sigma = 0$ Equation (4.10) becomes $z = e^{j\omega T}$, yielding a magnitude of $|z| = 1$. This implies that the $j\omega$ axis in the s-plane ($\sigma = 0$) corresponds to a unit circle in the z-plane. Hence, an oscillator with poles on the imaginary axis in the s-plane would have its poles *on* the unit circle in the z-plane. In the next chapter, we implement a digital oscillator by programming a difference equation with poles on the unit circle. From the previous example, the poles of $X(z) = ZT\{\sin n\omega T\}$ are the roots of $z^2 - 2z \cos \omega T + 1$, or

$$p_{1,2} = \frac{2\cos \omega T \pm \sqrt{4\cos^2 \omega T - 4}}{2}$$
$$= \cos \omega T \pm \sqrt{-\sin^2 \omega T}$$
$$= \cos \omega T \pm j \sin \omega T$$

The magnitude of each pole is $\sqrt{\cos^2 \omega T + \sin^2 \omega T} = 1$. We will see in the next chapter how we can generate sinusoidal waveforms of different frequencies by simply changing the value of ω in (4.9).

2. $\sigma < 0$ Equation (4.10) yields a magnitude of $|z| < 1$, because $e^{\sigma T} < 1$. As σ varies from 0^- to $-\infty$, $|z|$ will vary from 1^- to 0. Because a system that has poles on the left side of the $j\omega$ axis in the s-plane is stable, corresponding poles *inside* the unit circle in the z-plane would yield a stable system. The response of such a stable system would be either a decaying exponential, if the poles are real, or a decaying sinusoid, if the poles are complex.

3. $\sigma > 0$ Equation (4.10) yields a magnitude of $|z| > 1$, because $e^{\sigma T} > 1$. As σ varies from 0^+ to ∞, $|z|$ will vary from 1^+ to ∞. Because a system that has poles on the right side of the $j\omega$ axis in the s-plane is unstable, corresponding poles *outside* the unit circle would yield an unstable system. The response of such an unstable system would be either an increasing exponential, if the poles are real, or a growing sinusoid, if the poles are complex.

The phase of z is $\theta = \omega T = 2\pi f/F_s$. As f varies from 0 (DC) to $\pm F_s/2$, the phase θ will vary from 0 to π. The mapping from the s-plane to the z-plane is shown in Figure 4.1.

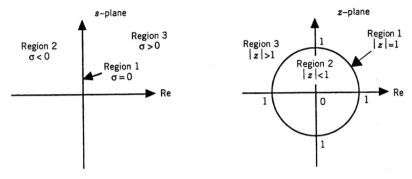

FIGURE 4.1. Mapping from *s*-plane to *z*-plane

Solution of Difference Equations

An analog filter can be described by a differential equation. Similarly, a digital filter can be represented by a difference equation. To solve a difference equation, we need to find the Z-transform of expressions such as $x(n-k)$, which corresponds to the kth derivative of an analog signal $x(t)$, or $d^k x(t)/dt^k$. The order of the difference equation is determined from the largest value of k in the expression $x(n-k)$. From (4.5), the Z-transform of $x(n)$ is

$$X(z) = \text{ZT}\{x(n)\}$$
$$= \sum_{n=0}^{\infty} x(n) z^{-n}$$
$$= x(0) + x(1)z^{-1} + x(2)z^{-2} + x(3)z^{-3} + \cdots \quad (4.11)$$

Therefore, the ZT of $x(n-1)$, equivalent to a first derivative dx/dt, is

$$\text{ZT}\{x(n-1)\} = \sum_{n=0}^{\infty} x(n-1) z^{-n}$$
$$= x(-1) + x(0)z^{-1} + x(1)z^{-2} + x(2)z^{-3} + \cdots$$
$$= x(-1) + z^{-1}\{x(0) + x(1)z^{-1} + x(2)z^{-2} + \cdots\}$$
$$= x(-1) + z^{-1} X(z) \quad (4.12)$$

where $x(-1)$ represents the initial condition associated with a first-order difference equation. To find the ZT of $x(n-2)$, equivalent to a second derivative $d^2x(t)/dt^2$,

$$\begin{aligned}
\text{ZT}\{x(n-2)\} &= \sum_{n=0}^{\infty} x(n-2)z^{-n} \\
&= x(-2) + x(-1)z^{-1} + x(0)z^{-2} + x(1)z^{-3} + \cdots \\
&= x(-2) + x(-1)z^{-1} + z^{-2}\{x(0) + x(1)z^{-1} + \cdots\} \\
&= x(-2) + x(-1)z^{-1} + z^{-2}X(z)
\end{aligned} \quad (4.13)$$

where $x(-2)$ and $x(-1)$ represent the two initial conditions imposed on a second-order difference equation. In general,

$$\text{ZT}\{x(n-k)\} = z^{-k} \sum_{m=1}^{k} x(-m)z^{m} + z^{-k}X(z) \quad (4.14)$$

If the initial conditions are all zero, then all the terms $x(-m)$, $m = 1, 2, \ldots, k$, are zero and (4.14) reduces to

$$\text{ZT}\{x(n-k)\} = z^{-k}X(z) \quad (4.15)$$

4.2 DISCRETE SIGNALS

A discrete signal $x(n)$ consists of a sequence of values $x(1), x(2), x(3), \ldots$, where n is the time, and each sample $x(1), x(2), \ldots$ is taken one sample-time apart, determined by the sampling period or sampling interval T. The signal $x(n)$ can be written as

$$x(n) = \sum_{m=-\infty}^{\infty} x(m)\delta(n-m) \quad (4.16)$$

where $\delta(n-m)$, the impulse sequence $\delta(n)$ delayed by m, is equal to 1 for $n = m$, and is 0 otherwise.

The signals and systems that we will be dealing with in this text are linear and time-invariant. Both superposition and shift invariance apply in those systems. Let an input $x(n)$ yield an output response $y(n)$, or $x(n) \rightarrow y(n)$. If $a_1 x_1(n) \rightarrow a_1 y_1(n)$ and $a_2 x_2(n) \rightarrow a_2 y_2(n)$, then $a_1 x_1(n) + a_2 x_2(n) \rightarrow a_1 y_1(n) + a_2 y_2(n)$, where a_1 and a_2 are constants. This is the *superposition property*, where by the overall output response is the sum of the individual responses to each input. *Shift invariance* implies that $x(n-m) \rightarrow y(n-m)$, or if the input is delayed by m samples, the output will also be delayed by m samples. If the input is a unit impulse $\delta(n)$, the resulting output impulse response is designated as $h(n)$, or $\delta(n) \rightarrow h(n)$. A delayed impulse $\delta(n-m)$ yields $h(n-m)$, using the shift-invariance property. Furthermore, if this impulse is multiplied by a constant $x(m)$, then

$$x(m)\delta(n - m) \rightarrow x(m)h(n - m) \qquad (4.17)$$

Using the signal $x(n)$ represented by (4.16), the property of superposition, and (4.17), the response becomes

$$y(n) = \sum_{m=-\infty}^{\infty} x(m)h(n - m) \qquad (4.18)$$

Equation (4.18) represents the convolution equation and, for a causal system, reduces to

$$y(n) = \sum_{m=-\infty}^{n} x(m)h(n - m) \qquad (4.19)$$

Letting $k = n - m$ in (4.19),

$$y(n) = \sum_{k=0}^{\infty} h(k)x(n - k) \qquad (4.20)$$

4.3 FINITE IMPULSE RESPONSE FILTERS

One of the most useful signal processing operations is filtering. With advances in very large scale integration (VLSI), digital signal processors are now available to implement digital filters in real time. The instruction set and architecture of these special-purpose microprocessors make them well suited for filtering operations. An analog filter operates on continuous signals and is typically realized with discrete devices such as operational amplifiers, resistors, and capacitors. On the other hand, a digital filter operates on discrete-time signals and can be implemented with a digital signal processor such as the TMS320C30. Such implementations involve using an analog-to-digital converter to capture an external input signal, processing these input samples with a digital signal processor, and sending the resulting output signal through a digital-to-analog converter.

In the last few years, the cost of these digital signal processors has been significantly reduced. This economical factor adds to the numerous advantages that a digital filter has over an analog counterpart, which include higher reliability, accuracy, and less sensitivity to temperature and aging. Very stringent magnitude and phase characteristics can be realized with a digital filter. Furthermore, the characteristics of a digital filter such as center frequency, bandwidth, and filter type can be readily modified. For example, in a matter of a few minutes, a finite impulse response (FIR) filter with some desired characteristics can be implemented in real time. Available filter design packages can be used as a tool to quickly design a filter. The design of a filter consists of the approximation of a transfer function, with a resulting set of coefficients associated with delayed input samples. An actual implementation makes use of both software and hardware. The TMS320C30 has a 32-bit wide memory, with floating-point operations capability. As a result, errors due to the approximation of filter coefficients are significantly reduced.

Different techniques are available for the design of FIR filters. A commonly used technique utilizes the Fourier series, which we will discuss later. Computer-aided design techniques such as that of Parks and McClellan [1, 2] has become popular in recent years.

Equation (4.20) is a very important equation for the design of finite impulse response filters. In designing such filters, we use a finite number of terms, or

$$y(n) = \sum_{k=0}^{N} h(k)x(n-k) \qquad (4.21)$$

If the input is the unit impulse $x(n) = \delta(0)$, the output response will be the impulse response $y(n) = h(n)$. We will see how to design an FIR filter using $N+1$ coefficients $h(0), h(1), \ldots, h(N)$, and $N+1$ input samples $x(n), x(n-1), \ldots, x(n-N)$. The input sample at time n is $x(n)$, and the delayed input samples are $x(n-1), \ldots, x(n-N)$.

The ZT of (4.21) yields (with zero initial conditions)

$$Y(z) = h(0)X(z) + h(1)z^{-1}X(z)$$
$$+ h(2)z^{-2}X(z) + \cdots + h(N)z^{-N}X(z) \qquad (4.22)$$

Equation (4.21) represents a convolution in time between the coefficients and the input samples. This convolution in time is equivalent to a multiplication in the frequency domain, or

$$Y(z) = H(z)X(z)$$

where $H(z)$, the ZT of $h(k)$, is the transfer function

$$H(z) = \sum_{k=0}^{N} h(k)z^{-k} = h(0) + h(1)z^{-1} + h(2)z^{-2} + \cdots + h(N)z^{-N}$$
$$(4.23)$$

$$= \frac{h(0)z^N + h(1)z^{N-1} + h(2)z^{N-2} + \cdots + h(N)}{z^N} \qquad (4.24)$$

with N poles only at the origin. Hence, an FIR filter is inherently stable. Figure 4.2 shows an FIR filter structure, representing either (4.21) or (4.22).

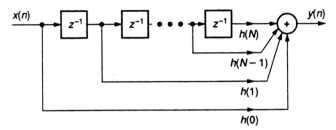

FIGURE 4.2. FIR filter structure showing delays

Equation (4.21) shows that an FIR filter can be implemented with knowledge of the input $x(n)$ at time n and of the delayed inputs $x(n - k)$. No feedback or past outputs are required. Other names used for FIR filters are nonrecursive, transversal, and tapped-delay filters. Filters requiring the knowledge of past outputs (with feedback) are discussed in the next chapter.

An important feature of an FIR filter is that it can guarantee linear phase. With linear phase, all input sinusoidal components are delayed by the same amount. This feature can be very useful in applications such as speech analysis where phase distortion can be very critical. For example, the Fourier transform of a delayed input sample $x(n - k)$ is $e^{-j\omega kT}X(j\omega)$, yielding a phase or $\theta = -\omega kT$. Note that the phase is a linear function in terms of ω. The group delay function, defined as the derivative of the phase, $d\theta/d\omega$, becomes a constant $-kT$.

4.4 FIR LATTICE STRUCTURE

The lattice structure is commonly used for applications in adaptive filtering and speech processing [3, 4]. A linear predictive coding (LPC) application program example, which is included with the EVM package, contains a filtering routine that uses a lattice structure. An Nth-order lattice structure is shown in Figure 4.3. The coefficients $k_1, k_2, \ldots, k_N$ are commonly referred to as reflection coefficients (or k-parameters). An advantage of this structure is that the frequency response is not as sensitive as the previous structure to

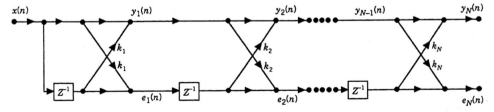

FIGURE 4.3. FIR lattice structure

small changes in the coefficients. From the first section in Figure 4.3, with $N = 1$, we have

$$y_1(n) = x(n) + k_1 x(n - 1) \qquad (4.25)$$
$$e_1(n) = k_1 x(n) + x(n - 1) \qquad (4.26)$$

From the second section (cascaded with the first), using (4.25) and (4.26),

$$\begin{aligned} y_2(n) &= y_1(n) + k_2 e_1(n - 1) \\ &= x(n) + k_1 x(n - 1) + k_2 k_1 x(n - 1) + k_2 x(n - 2) \\ &= x(n) + (k_1 + k_1 k_2) x(n - 1) + k_2 x(n - 2) \end{aligned} \qquad (4.27)$$

and

$$\begin{aligned} e_2(n) &= k_2 y_1(n) + e_1(n - 1) \\ &= k_2 x(n) + k_2 k_1 x(n - 1) + k_1 x(n - 1) + x(n - 2) \\ &= k_2 x(n) + (k_1 + k_1 k_2) x(n - 1) + x(n - 2) \end{aligned} \qquad (4.28)$$

For a specific section i,

$$y_i(n) = y_{i-1}(n) + k_i e_{i-1}(n - 1) \qquad (4.29)$$
$$e_i(n) = k_i y_{i-1}(n) + e_{i-1}(n - 1) \qquad (4.30)$$

It is instructive to see that (4.27) and (4.28) have the same coefficients but in reversed order. It can be shown that this property also holds true for a higher-order structure. In general, for an Nth-order FIR lattice system, (4.27) and (4.28) become

$$y_N(n) = \sum_{i=0}^{N} a_i x(n - i) \qquad (4.31)$$

and

$$e_N(n) = \sum_{i=0}^{N} a_{N-i} x(n - i) \qquad (4.32)$$

with $a_0 = 1$. If we take the ZT of (4.31) and (4.32) and find their impulse responses,

$$Y_N(z) = \sum_{i=0}^{N} a_i z^{-i} \qquad (4.33)$$

$$E_N(z) = \sum_{i=0}^{N} a_{N-i} z^{-i} \qquad (4.34)$$

4.4 FIR Lattice Structure

It is interesting to note that

$$E_N(z) = z^{-N} Y_N(1/z) \tag{4.35}$$

Equations (4.33) and (4.34) are referred to as image polynomials. For two sections, $k_2 = a_2$; in general,

$$k_N = a_N \tag{4.36}$$

For this structure to be useful, it is necessary to find the relationship between the k-parameters and the impulse response coefficients. The lattice network is highly structured, as seen in Figure 4.3 and as demonstrated through the previous difference equations. Starting with k_N in (4.36), we can recursively (with reverse-recursion) compute the preceding k-parameters, $k_{N-1}, \ldots, k_1$.

Consider an intermediate section r and, using (4.33) and (4.34),

$$Y_r(z) = Y_{r-1}(z) + k_r z^{-1} E_{r-1}(z) \tag{4.37}$$

$$E_r(z) = k_r Y_{r-1}(z) + z^{-1} E_{r-1}(z) \tag{4.38}$$

Solving for $E_{r-1}(z)$ in (4.38) and substituting it into (4.37), $Y_r(z)$ becomes

$$Y_r(z) = Y_{r-1}(z) + k_r z^{-1} \frac{E_r(z) - k_r Y_{r-1}(z)}{z^{-1}} \tag{4.39}$$

Equation (4.39) now can be solved for $Y_{r-1}(z)$ in terms of $Y_r(z)$, or

$$Y_{r-1}(z) = \frac{Y_r(z) - k_r E_r(z)}{1 - k_r^2}, \quad |k_r| = 1 \tag{4.40}$$

Using (4.35) with $N = r$, (4.40) becomes

$$Y_{r-1}(z) = \frac{Y_r(z) - k_r z^{-r} Y_r(1/z)}{1 - k_r^2} \tag{4.41}$$

Equation (4.41) is an important relationship which shows that by using a reversed recursion procedure, we can find Y_{r-1} from Y_r, where $1 \le r \le N$. Consequently, we can also find the k-parameters starting with k_r and proceeding to k_1. For r sections, (4.33) can be written as

$$Y_r(z) = \sum_{i=0}^{r} a_{ri} z^{-i} \tag{4.42}$$

Replacing i by $r - i$, and z by $1/z$, (4.42) becomes

$$Y_r\left(\frac{1}{z}\right) = \sum_{i=0}^{r} a_{r(r-i)} z^{r-i} \tag{4.43}$$

102 Finite Impulse Response Filters

Using (4.42) and (4.43), equation (4.41) becomes

$$\sum_{i=0}^{r} a_{(r-1)i} z^{-i} = \frac{\sum_{i=0}^{r} a_{ri} z^{-i} - k_r z^{-r} \sum_{i=0}^{r} a_{r(r-i)} z^{r-i}}{1 - k_r^2} \qquad (4.44)$$

$$= \frac{\sum_{i=0}^{r} a_{ri} z^{-i} - k_r \sum_{i=0}^{r} a_{r(r-i)} z^{-i}}{1 - k_r^2} \qquad (4.45)$$

from which,

$$a_{(r-1)i} = \frac{a_{ri} - k_r a_{r(r-i)}}{1 - k_r^2}, \qquad i = 0, 1, \ldots, r-1 \qquad (4.46)$$

with $r = N, N-1, \ldots, 1$, $|k_r| \neq 1$, $i = 0, 1, \ldots, r-1$, and

$$k_r = a_{rr}, \qquad r = N, N-1, \ldots, 1 \qquad (4.47)$$

Example 4.3 *FIR Lattice Structure.* This example illustrates the use of (4.46) and (4.47) to compute the k-parameters. Given that the impulse response of an FIR filter in the frequency domain is

$$Y_2(z) = 1 + 0.2z^{-1} - 0.5z^{-2}$$

From (4.42), with $r = 2$,

$$Y_2(z) = a_{20} + a_{21}z^{-1} + a_{22}z^{-2}$$

where $a_{20} = 1$, $a_{21} = 0.2$, and $a_{22} = -0.5$. Starting with $r = 2$ in (4.47),

$$k_2 = a_{22} = -0.5$$

Using (4.46), for $i = 0$,

$$a_{10} = \frac{a_{20} - k_2 a_{22}}{1 - k_2^2} = \frac{1 - (-0.5)(-0.5)}{1 - (-0.5)^2} = 1$$

and, for $i = 1$,

$$a_{11} = \frac{a_{21} - k_2 a_{21}}{1 - k_2^2} = \frac{0.2 - (-0.5)(0.2)}{1 - (-0.5)^2} = 0.4$$

From (4.47),

$$k_1 = a_{11} = 0.4$$

Note that the values for the k-parameters $k_2 = -0.5$ and $k_1 = 0.4$ can be verified using (4.27). In the next chapter, we will continue our discussions on lattice structures in conjunction with IIR filters.

4.5 FIR IMPLEMENTATION USING FOURIER SERIES

The basic method which follows is to design an FIR filter such that the magnitude response of its transfer function $H(z)$ approximates a desired magnitude response, using a finite Fourier series. The desired transfer function is

$$Hd(\omega) = \sum_{n=-\infty}^{\infty} C_n e^{jn\omega T}, \qquad |n| < \infty \qquad (4.48)$$

where C_n are the Fourier series coefficients, which will be determined such that the transfer function $H(z)$ approximates this desired transfer function. A normalized frequency variable ν can be introduced, such that $\nu = f/F_N$, where F_N is the Nyquist frequency, or $F_N = F_s/2$, and the sampling period $T = 1/F_s$. The desired transfer function can then be written as

$$Hd(\nu) = \sum_{n=-\infty}^{\infty} C_n e^{jn\pi\nu} \qquad (4.49)$$

because $\omega T = 2\pi f/F_s = \pi\nu$ and $|\nu| < 1$. The Fourier series coefficients are defined as

$$C_n = \tfrac{1}{2} \int_{-1}^{1} Hd(\nu) e^{-jn\pi\nu} \, d\nu$$

$$= \tfrac{1}{2} \int_{-1}^{1} Hd(\nu) \{\cos n\pi\nu - j\sin n\pi\nu\} \, d\nu \qquad (4.50)$$

Assume that $Hd(\nu)$ is an even function, then (4.50) reduces to

$$C_n = \int_0^1 Hd(\nu) \cos n\pi\nu \, d\nu, \qquad n \geq 0 \qquad (4.51)$$

since $Hd(\nu)\sin n\pi\nu$ is an odd function and $\int_{-1}^{1}(\text{odd function}) = 0$. Furthermore, the coefficients $C_{-n} = C_n$. Note that $Hd(\nu)$ in (4.49) requires an infinite number of coefficients C_n. In order to obtain a realizable filter, we must truncate (4.49), which yields the approximated transfer function

$$Ha(\nu) = \sum_{n=-Q}^{Q} C_n e^{jn\pi\nu} \qquad (4.52)$$

where Q is positive and finite. The value of Q will effectively determine the order of the FIR filter. The larger the value of Q, the better the approximation in (4.52), because a larger number of terms are used to approximate (4.49). The truncation of the infinite series results in ignoring the terms outside of a rectangular window function between $-Q$ and $+Q$. We will see in the next section that window functions other than rectangular can be used to improve the characteristics of a filter.

Letting $z = e^{j\pi \nu}$, (4.52) becomes

$$Ha(z) = \sum_{n=-Q}^{Q} C_n z^n \qquad (4.53)$$

with impulse response coefficients $C_{-Q}, C_{-Q+1}, \ldots, C_{-1}, C_0, C_1, \ldots, C_{Q-1}, C_Q$. The approximated transfer function in (4.53), with positive powers of z, implies a noncausal filter that would produce an output before an input is applied, hence not realizable. To remedy this situation, a delay of Q samples is introduced in (4.53) to give

$$H(z) = z^{-Q} Ha(z) = \sum_{n=-Q}^{Q} C_n z^{n-Q} \qquad (4.54)$$

Let $n - Q = -i$, then (4.54) becomes

$$H(z) = \sum_{i=0}^{2Q} C_{Q-i} z^{-i} \qquad (4.55)$$

If we let $h_i = C_{Q-i}$ and $N = 2Q$, $H(z)$ becomes

$$H(z) = \sum_{i=0}^{N} h_i z^{-i} \qquad (4.56)$$

The impulse response coefficients are h_i, where $h_0 = C_Q$, $h_1 = C_{Q-1}, \ldots, h_{Q-1} = C_1$, $h_Q = C_0$, $h_{Q+1} = C_1, \ldots, h_{2Q-1} = C_{-Q+1}$, and $h_{2Q} = C_{-Q}$. The impulse response is symmetric about h_Q, with $C_n = C_{-n}$. The order of the filter is $N + 1$, or $2Q + 1$. For example, if $Q = 22$, the filter will have 45 coefficients $h_0, h_1, \ldots, h_{44}$ with symmetry about h_{22}, or

$$h_0 = h_{44} = C_{22}$$
$$h_1 = h_{43} = C_{21}$$
$$h_2 = h_{42} = C_{20}$$
$$\vdots$$
$$h_{21} = h_{23} = C_1$$
$$h_{22} = C_0$$

4.5 FIR Implementation Using Fourier Series

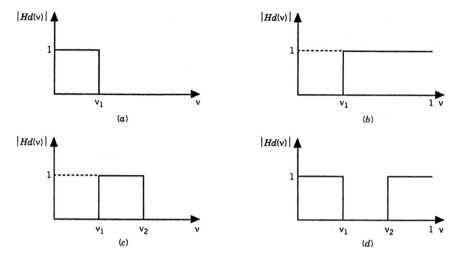

FIGURE 4.4. Desired magnitude transfer function: (*a*) lowpass; (*b*) highpass; (*c*) bandpass; (*d*) bandstop

The coefficients C_n can be found for the following frequency-selective FIR filters using Figure 4.4, with the desired transfer functions $Hd(\nu)$ ideally represented:

1. *Lowpass*

$$C_n = \int_0^{\nu_1} Hd(\nu)\cos n\pi\nu \, d\nu = \frac{\sin n\pi\nu_1}{n\pi} \quad (4.57)$$

2. *Highpass*

$$C_n = \int_{\nu_1}^{1} Hd(\nu)\cos n\pi\nu \, d\nu = -\frac{\sin n\pi\nu_1}{n\pi} \quad (4.58)$$

3. *Bandpass*

$$C_n = \int_{\nu_1}^{\nu_2} Hd(\nu)\cos n\pi\nu \, d\nu = \frac{\sin n\pi\nu_2 - \sin n\pi\nu_1}{n\pi} \quad (4.59)$$

4. *Bandstop*

$$C_n = \int_0^{\nu_1} Hd(\nu)\cos n\pi\nu \, d\nu + \int_{\nu_2}^{1} Hd(\nu)\cos n\pi\nu \, d\nu$$

$$= \frac{\sin n\pi\nu_1 - \sin n\pi\nu_2}{n\pi} \quad (4.60)$$

where ν_1 and ν_2 are the normalized cutoff frequencies, as shown in Figure 4.4.

106 Finite Impulse Response Filters

Example 4.4 *FIR Lowpass Filter.* This example is to implement an FIR lowpass filter with 11 coefficients, with a cutoff frequency f_c of 1 kHz and a sampling frequency of 10 kHz.

From (4.57),

$$C_n = \frac{\sin 0.2n\pi}{n\pi}, \quad n = 1, 2, \ldots, 10 \quad (4.61)$$

because $\nu_1 = f_c/F_N$, where $f_c = 1$ kHz, and the Nyquist frequency is $F_N = 5$ kHz.

For $n = 0$,

$$C_0 = \int_0^{\nu_1} Hd(\nu)\, d\nu = \nu_1 = 0.2 \quad (4.62)$$

Using (4.61), (4.62), and $h_i = C_{Q-i}$,

$$\begin{aligned}
h_0 &= C_5 = 0 \\
h_1 &= C_4 = 0.0468 \\
h_2 &= C_3 = 0.1009 \\
h_3 &= C_2 = 0.1514 \\
h_4 &= C_1 = 0.1872 \\
h_5 &= C_0 = 0.2 \\
h_6 &= C_{-1} = C_1 = h_4 \\
h_7 &= C_{-2} = C_2 = h_3 \\
h_8 &= C_{-3} = C_3 = h_2 \\
h_9 &= C_{-4} = C_4 = h_1 \\
h_{10} &= C_{-5} = C_5 = h_0
\end{aligned} \quad (4.63)$$

Note the symmetry of the coefficients about $h_Q = h_5$. These coefficients can also be obtained readily using a filter design package as shown later in this chapter. These coefficients will be used in a lowpass filter programming example later in this chapter.

4.6 WINDOW FUNCTIONS

In order to obtain a realizable FIR filter, the infinite series in the transfer function equation (4.49) was truncated, yielding a finite number of coefficients. In essence, this is equivalent to multiplying the filter's coefficients by a rectangular window function with an amplitude of 1 between $-Q$ and $+Q$, and 0 elsewhere. The rectangular window function is defined as

$$w_R(n) = \begin{cases} 1, & |n| \leq Q \\ 0, & \text{otherwise} \end{cases} \quad (4.64)$$

The transform of the rectangular window function $w_R(n)$ yields a sinc function in the frequency domain that has a mainlobe and high sidelobes. It can be shown that

$$W_R(\nu) = \sum_{n=-Q}^{Q} e^{jn\pi\nu} = e^{-jQ\pi\nu}\left[\sum_{n=0}^{2Q} e^{jn\pi\nu}\right]$$

$$= \frac{\sin\left[\left(\frac{2Q+1}{2}\right)\pi\nu\right]}{\sin(\pi\nu/2)}$$

An FIR filter with the rectangular window function exhibits high sidelobes or oscillations caused by the abrupt truncation, specifically, near discontinuities. The amplitude of the oscillations decreases as you move away from the discontinuity. Other window functions have been used to reduce these oscillations. A measure of the filter is a ripple factor that compares (takes the ratio of) the peak of the first sidelobe to the peak of the mainlobe. A number of window functions, which provide a more gradual truncation to the infinite series expansion, have lower sidelobes, although at the same time they have a wider mainlobe (lower selectivity). Hence, a compromise or trade-off is to select a window function that can reduce the sidelobes while approaching the selectivity of the rectangular window function. The plot of the magnitude response of an FIR filter, implemented in real time later in this chapter, shows the undesirable sidelobes.

In general, the Fourier series coefficients are

$$C'_n = C_n w(n) \qquad (4.65)$$

where $w(n)$ is the window function. In the case of the rectangular window function, $C'_n = C_n$. The transfer function in (4.56) can then be written as

$$H'(z) = \sum_{i=0}^{N} h'_i z^{-i} \qquad (4.66)$$

where

$$h'_i = C'_{Q-i}, \qquad 0 \le i \le 2Q \qquad (4.67)$$

The following window functions are commonly used in the design of FIR filters. A thorough discussion of the trade-offs of a number of window functions can be found in [5].

Hanning Window

The Hanning or raised cosine window function is

$$w_{HA}(n) = \begin{cases} 0.5 + 0.5\cos(n\pi/Q), & \text{for } |n| \le Q \\ 0, & \text{otherwise} \end{cases} \qquad (4.68)$$

with the highest or first sidelobe level at approximately -31 dB from the peak of the mainlobe.

Hamming Window

The Hamming window function is

$$w_H(n) = \begin{cases} 0.54 + 0.46\cos(n\pi/Q), & \text{for } |n| \le Q \\ 0, & \text{otherwise} \end{cases} \quad (4.69)$$

with the highest sidelobe level at approximately -43 dB from the peak of the mainlobe.

Blackman Window

The Blackman window function is

$$w_B(n) = \begin{cases} 0.42 + 0.5\cos(n\pi/Q) + 0.08\cos(2n\pi/Q), & |n| \le Q \\ 0, & \text{otherwise} \end{cases} \quad (4.70)$$

with the highest sidelobe level down to approximately -58 dB from the peak of the mainlobe. Although the Blackman window provides the largest reduction in sidelobe (the smallest peak sidelobe) compared to the previous window functions, it has the widest mainlobe. The width of the mainlobe can be decreased by increasing the width of the window (the number of coefficients).

The rectangular window has its highest sidelobe level down by only -13 dB from the peak of its mainlobe, resulting in oscillations of considerable size. On the other hand, it has the narrowest mainlobe, with sharp transitions at a discontinuity, providing high selectivity.

Kaiser Window

The Kaiser window has a variable parameter to control the size of the sidelobe with respect to the mainlobe. As with the previous windows, the width of the mainlobe can be decreased by increasing the length of the window. The Kaiser window function is

$$w_K(n) = \begin{cases} I_0(\mathbf{b})/I_0(\mathbf{a}), & |n| \le Q \\ 0, & \text{otherwise} \end{cases} \quad (4.71)$$

where $\mathbf{a}$ is an empirically determined variable, and $\mathbf{b} = \mathbf{a}[1 - (n/Q)^2]^{1/2}$.

$I_0(x)$ is the modified Bessel function of the first kind defined by,

$$I_0(x) = 1 + \frac{0.25x^2}{(1!)^2} + \frac{(0.25x^2)^2}{(2!)^2} + \cdots$$

$$= 1 + \sum_{n=1}^{\infty} \left[\frac{(x/2)^n}{(n!)} \right]^2 \qquad (4.72)$$

This series expansion converges rapidly. By changing the length of the window and the parameter **a**, a trade-off between the size of the sidelobe and the width of the mainlobe can be achieved. The design of FIR filters with the Kaiser window has become very popular in recent years.

Parks – McClellan Optimum Approximation

The straightforward window design technique is such that the passband ripple and the stopband ripple are required to be the same. A more efficient technique is the computer-aided iterative design developed by Parks and McClellan [1, 2], based on the Remez exchange algorithm.

The Parks–McClellan algorithm produces equiripple approximation of FIR filters. The order of the filter and the edges of both passbands and stopbands are fixed, and the coefficients are varied to provide this equiripple approximation. This efficient algorithm is designed to minimize the ripple in both the passbands and the stopbands. The transition regions are left unconstrained and are considered as "don't care" regions where the solution may fail. An FIR filter using the Parks–McClellan algorithm can be designed with the Hypersignal-Plus DSP [6] Software (Appendix C) or with the digital filter design package (DFDP) from Atlanta Signal Processors, Inc. [7].

4.7 FILTER DESIGN PACKAGES

Several commercially available filter design packages can be used to design FIR filters, such as the Hypersignal-Plus DSP Software from Hyperception, Inc., and the digital filter design package (DFDP) from Atlanta Signal Processors, Inc. These packages also include a code generator program, specific to the TMS320C30 processor.

We discuss here two filter design packages:

1. The Hypersignal-Plus DSP Software package. Appendix C contains examples showing the various features of the Hypersignal package, which includes utilities for plotting, for spectral analysis, and so forth.
2. A "homemade" filter design package (FDP), on the accompanying disk calculates the coefficients to implement lowpass, highpass, bandpass,

and bandstop FIR filters using the rectangular, Hanning, Hamming, Blackman, and Kaiser windows. These coefficients can be obtained readily in a format compatible with TMS320C30 code. The use of this package will be demonstrated at the end of this chapter.

4.8 PROGRAMMING EXAMPLES USING C AND TMS320C30 CODE

Several examples will illustrate the implementation of FIR filters using TMS320C30 assembly code, mixed code with a C program calling an assembly function, and C code. Simulation and real-time realizations are covered. Appendix C illustrates the use of the Hypersignal-Plus DSP Software to design a specific filter, and the utilities available for time and frequency analysis. A walk-through example is included in Appendix C for the design of a bandpass filter in C or TMS320C30 code.

In programming these examples, we will use N to represent the filter's length, not $N + 1$ as previously done. We will arrange the coefficients in memory such that the starting (lower) memory address contains the last coefficient $h(N - 1)$, and the ending (higher) memory address contains the first coefficient $h(0)$. Table 4.1 shows the memory organization for the coefficients and samples. Initially, all the samples $x(n), x(n - 1), \ldots$ are set to zero. When the newest sample $x(n)$ is acquired, it is placed at the bottom (higher) memory address. Then AR1 is incremented in a circular fashion to point at the top memory address. The output at time n is

$$y(n) = h(N - 1)x(n - (N - 1)) + h(N - 2)x(n - (N - 2))$$
$$+ \cdots + h(1)x(n - 1) + h(0)x(n)$$
$$= \sum_{k=0}^{N-1} h(k)x(n - k)$$

which is the convolution equation (4.21) with N (instead of $N + 1$) coefficients. After the last multiply operation $h(0)x(n)$, AR1 is incremented to point at the top or starting memory address of the samples, where a newly

TABLE 4.1 TMS320C30 Memory Organization for Convolution

	Input Samples		
Coefficients	Time n	Time $n + 1$	Time $n + 2$
AR0 → $h(N - 1)$	AR1 → $x(n - (N - 1))$	newest → $x(n + 1)$	$x(n + 1)$
$h(N - 2)$	$x(n - (N - 2))$	AR1 → $x(n - (N - 2))$	newest → $x(n + 2)$
$h(N - 3)$	$x(n - (N - 3))$	$x(n - (N - 3))$	AR1 → $x(n - (N - 3))$
$\vdots$	$\vdots$	$\vdots$	$\vdots$
$h(1)$	$x(n - 1)$	$x(n - 1)$	$x(n - 1)$
$h(0)$	newest → $x(n)$	$x(n)$	$x(n)$

4.8 Programming Examples Using C and TMS320C30 Code

acquired sample $x(n + 1)$ can be placed. AR1 is then postincremented to point at $x(n - (N - 2))$. This process is illustrated in Table 4.1. The output at time $n + 1$ is

$$y(n + 1) = h(N - 1)x(n - (N - 2)) + h(N - 2)x(n - (N - 3)) \\ + \cdots + h(1)x(n) + h(0)x(n + 1)$$

and the output at time $n + 2$ is

$$y(n + 2) = h(N - 1)x(n - (N - 3)) + h(N - 2)x(n - (N - 4)) \\ + \cdots + h(1)x(n + 1) + h(0)x(n + 2)$$

Note that for each time $n, n + 1, n + 2, \ldots$, the last multiply operation involves $h(0)$ and the newest sample.

Example 4.5 *FIR Lowpass Filter Using TMS320C30 Code.* This programming example discusses the implementation of a lowpass FIR filter, with 11 coefficients and a bandwidth of 1 kHz. The impulse response coefficients were found in the previous example, equation (4.63). The program LP11.ASM, listed in Figure 4.5, implements this filter. The command file is listed in Figure 4.6. The program example PREFIR.ASM in Chapter 2, provides a good background for this example.

A description of the program LP11.ASM follows:

1. An impulse value of 10,000, used as input to this filter, is set in the program. We will show later how the file IMPULSE.DAT (on the accompanying disk) can be used as input, because this file contains the value 2710h (equivalent to 10,000), followed by zeros. The output port address is set to 804001h.
2. XN_ADDR specifies the last "bottom" (higher-memory) address, where the first sample value 10,000 is to be stored. HN_ADDR specifies the starting address of the coefficient table.
3. .USECT "XN_BUFF", LENGTH specifies the buffer size in a user-defined, named section. This section is named in the command file as XN_BUFF and is listed under quotes in the main program. This assembler directive reserves spaces (specified by length) for the data samples. The BEGIN address is to specify where code begins on reset.
4. AR5, AR6, and AR1 are loaded with the impulse address, the output address, and the "bottom" sample address, respectively. BK contains the size of the circular buffer. The content in memory, pointed by AR5 (memory location that contains the impulse value) is loaded into R3, then stored in memory pointed by AR1 (the bottom memory location used for the samples). AR1 is then incremented to point at the "top" address of the samples buffer, because a circular mode of addressing is used in conjunction with AR1.

112 Finite Impulse Response Filters

```
;LP11.ASM-LOWPASS FIR WITH 11 COEFFICIENTS
            .TITLE    "LP11.ASM"              ;LP @ 1 kHz
            .GLOBAL   MAIN,BEGIN,FILT,COEFF  ;REF/DEF SYMBOLS
            .DATA                             ;ASSEMBLE -> DATA SECT
IMP_ADDR    .WORD     IMPULSE                 ;INIT VALUE IMPULSE
IO_OUT      .WORD     804001H                 ;OUTPUT ADDRESS
XN_ADDR     .WORD     XN+LENGTH-1             ;(LAST) SAMPLE ADDR
HN_ADDR     .WORD     COEFF                   ;COEFF TABLE ADDR
XN          .USECT    "XN_BUFF",LENGTH        ;BUFFER SIZE OF SAMPLES
            .SECT     "VECTORS"               ;ASSEMBLE -> VECT SECT
MAIN        .WORD     BEGIN                   ;BEGIN @ RESET (0H)
            .TEXT                             ;ASSEMBLE -> TEXT SECT
BEGIN       LDI       @IMP_ADDR,AR5           ;IMPULSE ADDR -> AR5
            LDI       @IO_OUT,AR6             ;OUTPUT ADDR  -> AR6
            LDI       @XN_ADDR,AR1            ;"LAST"SAMPLE ADDR->AR1
            LDI       LENGTH,BK               ;BK=SIZE OF CIRC BUFFER
            LDF       0,R0                    ;INIT R0=0
            LDF       0,R7                    ;INIT R7=0
            LDI       LENGTH,R4               ;COUNTER FOR FILT SUB
            LDF       *AR5,R3                 ;INPUT IMPULSE VALUE
            STF       R3,*AR1++(1)%           ;STORE IMPULSE
LOOP        LDI       @HN_ADDR,AR0            ;COEFF H(N-1) ADDR->AR0
            CALL      FILT                    ;GO TO SUBROUTINE FILT
            FIX       R2,R1                   ;R1=INTEGER(R2)
            STI       R1,*AR6                 ;OUTPUT INTEGER VALUE
            SUBI      1,R4                    ;DECREMENT R4
            STF       R7,*AR1++(1)%           ;ALL OTHER SAMPLES=0
            BNZ       LOOP                    ;BRANCH UNTIL R4 < 0
WAIT        BR        WAIT                    ;WAIT
           ;SUBROUTINE FILT
FILT        LDF       0,R0                    ;INIT R0=0
            LDF       0,R2                    ;INIT R2=0
            RPTS      LENGTH-1                ;N MULTIPLY
            MPYF      *AR0++,*AR1++%,R0       ; HN*XN -> R0
||          ADDF      R0,R2                   ;// WITH ACC -> R2
            ADDF      R0,R2                   ;LAST ACC -> R2
            RETS                              ;RETURN FROM SUBROUTINE
            .DATA                             ;ASSEMBLE -> DATA SECT
COEFF       .FLOAT    0.0                     ;H10
            .FLOAT    0.0468                  ;H9
            .FLOAT    0.1009                  ;H8
            .FLOAT    0.1514                  ;H7
            .FLOAT    0.1872                  ;H6
            .FLOAT    0.2                     ;H5
            .FLOAT    0.1872                  ;H4
```

FIGURE 4.5. Lowpass FIR filter program using TMS320C30 code (LP11.ASM)

```
              .FLOAT    0.1514              ;H3
              .FLOAT    0.1009              ;H2
              .FLOAT    0.0468              ;H1
H0            .FLOAT    0.0                 ;H0
LENGTH        .SET      H0-COEFF+1          ;LENGTH=11
IMPULSE       .FLOAT    10000               ;IMPULSE VALUE
              .END                          ;END
```

FIGURE 4.5. *(Continued)*

5. The top (lower-memory) address of the coefficient table (address of memory containing H10) is loaded into AR0. The filter subroutine FILT is called. This routine was discussed previously in the program PREFIR.ASM, in Chapter 2.
6. The resulting filter output for a specific time is stored at the memory address specified by AR6 (the output port address). R4, the loop counter register, loaded initially with the length of the filter, is decremented.
7. Steps 5 and 6 are repeated (length − 1) times. AR0 is reinitialized each time to point at the top (lower-memory) address of the coefficient table,

```
/*LP11.CMD       COMMAND FILE                              */
LP11.OBJ         /*FIR PROGRAM                             */
-E BEGIN         /*SPECIFIES ENTRY POINT FOR OUTPUT*/
-O LP11.OUT      /*LINKED COFF OUTPUT FILE                 */
MEMORY
{
  VECS: org = 0          len = 0x40    /*VECTOR LOCATIONS*/
  SRAM: org = 0x40       len = 0x3FC0  /*16K EXTERNAL RAM*/
  RAM : org = 0x809800   len = 0x800   /*2K INTERNAL RAM */
}
SECTIONS
{
  .data:  {} > SRAM       /*DATA SECTION              */
  .text:  {} > SRAM       /*CODE                      */
  .cinit: {} > SRAM       /*INITIALIZATION TABLES     */
  .stack: {} > RAM        /*SYSTEM STACK              */
  .bss:   {} > RAM        /*BSS SECTION IN RAM        */
  VECTORS:{} > VECS       /*RESET & INTERRUPT VECTORS */
  XN_BUFF  ALIGN(64): {} > RAM  /*CIRCULAR BUFFER     */
}
```

FIGURE 4.6. Command file for FIR lowpass filter (LP11.CMD)

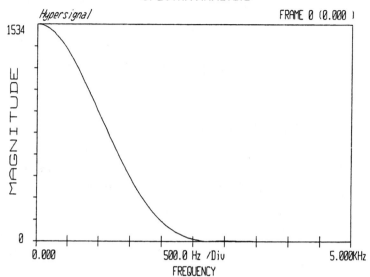

FIGURE 4.7. Plot of frequency response of lowpass filter

which contains $h(N - 1)$. We define here the filter's length to be $N = 11$ coefficients.

The output of the filter at time n is

$$y(n) = h(10)x(n - 10) + h(9)x(n - 9) + \cdots + h(1)x(n - 1) + h(0)x(n)$$

The last multiply operation consists of the coefficient $h(0)$ times the newest sample. The first time through, it is $h(0)x(n)$. In this example, there is only one newest nonzero value. However, note that at time $n + 1$,

$$y(n + 1) = h(10)x(n - 9) + h(9)x(n - 8) + \cdots + h(1)x(n) + h(0)x(n + 1)$$

where the newest sample (at time $n + 1$) is zero. A nonzero value is contributed by $h(1)x(n)$.

Assemble, link, and run this program. The output file `LP11.DAT` is used to plot Figure 4.7, with the Hypersignal utilities described in Appendix C.

Example 4.6 *FIR Lowpass Filter with Two Circular Buffers, Using TMS320C30 Code.* This example implements the same lowpass filter as in the previous example. However, the "noise" sequence (`NOISEINT.DAT`) generated in

4.8 Programming Examples Using C and TMS320C30 Code

```
LP11BF2P-PARTIAL PROGRAM LOWPASS FIR WITH 2 BUFFERS
IO_IN      .WORD    804000H              ;INPUT ADDRESS
HN1_ADDR   .WORD    COEFF                ;COEFF TABLE ADDR IN SRAM
HN2_ADDR   .WORD    HN                   ;COEFF TABLE ADDR IN RAM
HN         .USECT   "HN_BUFF",LENGTH     ;COEFF BUFFER SIZE IN RAM
BEGIN      LDI      @IO_IN,AR5           ;INPUT ADDR -> AR5
           LDF      0,R0                 ;INIT R0=0
           RPTS     LENGTH-1             ;INIT ALL SAMPLES TO ZERO
           STF      R0,*AR1--%           ;START/END AT LAST ADDR
           LDI      LENGTH-1,RC          ;REPEAT COUNTER=LENGTH
           LDI      @HN1_ADDR,AR0        ;1ST COEFF ADDR IN SRAM
           LDI      @HN2_ADDR,AR2        ;1ST COEFF ADDR IN RAM
           RPTB     EXCH                 ;TRANSFER BLOCK SRAM->RAM
           LDF      *AR0++,R5            ;COEFF IN SRAM->R5
EXCH       STF      R5,*AR2++            ;STORE IN RAM
           LDI      @HN2_ADDR,AR0        ;1ST COEFF ADDR->AR0
LOOP       FLOAT    *AR5,R3              ;INPUT NEW SAMPLE
           STF      R3,*AR1++%           ;STORE TO MODEL DELAY
FILT       LDF      0,R0                 ;INIT R0=0
           MPYF     *AR0++%,*AR1++%,R0   ; HN*XN -> R0
           :
```

FIGURE 4.8. Partial program listing of lowpass filter using two circular buffers (LP11BF2P)

Example 3.2 is used as the input to the filter in lieu of an impulse value (as in example 4.5). Note that NOISEINT.DAT is the output of the pseudorandom noise generator. Two circular buffers are used: one for the samples XN as before, and one for the coefficients HN. This eliminates the need to reinitialize AR0 after each filter output sample to point at the top coefficient table address. A partial program LP11BF2P, listed in Figure 4.8, shows the changes from the previous program example. The complete program is on the accompanying disk as LP11BF2.ASM. This program also illustrates the transfer of the coefficients from a table (memory section) in external SRAM to a table (a different memory section) in internal RAM. A description of this program follows:

1. An input port address is specified for the input noise NOISEINT.DAT. H1_ADDR specifies the starting address of the coefficient table in external SRAM, and H2_ADDR specifies the starting address in internal RAM [where the coefficients $h(N-1), \ldots, h(0)$ will be transferred].
2. In conjunction with the single-repeat instruction RPTS, the memory locations for the samples XN are initialized to zero, starting with the bottom (higher-memory) address and finishing at the top (lower-memory) address. AR1 will point back to the bottom memory address when

116   Finite Impulse Response Filters

postdecremented from the top memory address in a circular mode of addressing.

3. In order to use the block repeat instruction RPTB, the repeat counter register RC must first be loaded in order to specify the number of times the block of code is to be repeated. Two special registers, which contain the starting address (RS) and the ending address (RE), are used in conjunction with the RPTB instruction. RE contains the address specified by the label EXCH, and RS contains the address of the instruction (LDF *AR0++, AR5) in the block of code to be repeated. Note that before this transfer, the starting address of the coefficients in external SRAM is first loaded into AR0, and the starting address where the coefficients are to be transferred in internal RAM is loaded into AR2.

4. AR0 is reinitialized to the starting address of the coefficients in RAM, before the LOOP section. An input sample value (from the noise sequence) is obtained and loaded in R3 (after being converted into a floating-point format), then stored. From the filter subroutine, in the multiply operation, both AR0 and AR1 are incremented using a circular mode of addressing.

Run this program (LP11BF2.ASM, on accompanying disk), and verify that the magnitude of the lowpass filter is similar to that obtained in the previous impulse response example. Use NOISEINT.DAT as the input file at port 804000h, and the simulator output specified at 804001h. A file IMPULSE.DAT containing the value 10,000 (2710h) followed by zeros represents an impulse and can be used in lieu of NOISEINT.DAT (as shown in the next example). Note that there are two circular buffers and that both XN_BUFF and HN_BUFF are ALIGNed, as specified in the command file.

Example 4.7 *FIR Bandpass Filter with 45 Coefficients, Using TMS320C30 Code.* This program example discusses a bandpass FIR filter with 45 coefficients. This filter has a center frequency of 1 kHz and a sampling frequency of 10 kHz. The filter's coefficients, using the Kaiser window function, were obtained with the Hypersignal package in Example C.3 of Appendix C. We can implement this filter by replacing the coefficients set in the lowpass filter of Example 4.5 with those found in Example C.3. For example, the filter's length can be calculated in the program using the label COEFF to specify the last coefficient H44, and the label H0 to specify the first coefficient H0. The label COEFF represents the starting address of the coefficient table, and the label H0 the ending address. The filter's length, minus 1, is the difference between the addresses of the coefficient H0 and the coefficient HN-1 specified by COEFF.

We will discuss instead a slightly different (and more efficient) version of the lowpass filter program example. This will permit us later to implement a more efficient filter in mixed mode (main program in C calling an assembly

4.8 Programming Examples Using C and TMS320C30 Code

```
;BP45A.ASM-FIR BANDPASS FILTER WITH 45 COEFF,USING TMS320C30 CODE
         .TITLE    "BP45A.ASM"      ;BP FIR,Fc=1 kHz,Fs=10 kHz
         .GLOBAL   FILT,MAIN,BEGIN  ;GLOBAL REF/DEF
         .DATA                      ;ASSEMBLE INTO DATA SECTION
XN_ADDR  .WORD     XN+LENGTH-1      ;(LAST) SAMPLE ADDRESS
HN_ADDR  .WORD     COEFF            ;ADDR OF COEFF H(N-1)
IO_IN    .WORD     804000H          ;INPUT PORT ADDRESS
IO_OUT   .WORD     804001H          ;OUTPUT PORT ADDRESS
XN       .USECT    "XN_BUFF",LENGTH ;BUFFER SIZE OF SAMPLES
         .SECT     "VECTORS"        ;VECTOR SECTION
MAIN     .WORD     BEGIN            ;BEGIN @ RESET (0H)
         .TEXT                      ;ASSEMBLE INTO TEXT
BEGIN    LDI       @IO_IN,AR5       ;POINTER TO INPUT PORT ADDR
         LDI       @IO_OUT,AR6      ;POINTER TO OUTPUT PORT ADDR
         LDI       LENGTH,BK        ;SIZE OF CIRCULAR BUFFER
         LDI       @XN_ADDR,AR1     ;LAST SAMPLE ADDR -> AR1
         LDI       @COEFF,AR0       ;COEFF H(N-1) ADDR-> AR0
         LDF       0,R0             ;INITIALIZE R0
         RPTS      LENGTH-1         ;INITIALIZE ALL XN TO 0
         STF       R0,*AR1--%       ;START/END AT LAST SAMPLE
FILT     LDI       LENGTH,AR4       ;LENGTH IN AR4
         SUBI      1,AR4            ;AR4=LOOP COUNTER
LOOP     FLOAT     *AR5,R3          ;INPUT NEW SAMPLE
         STF       R3,*AR1++        ;STORE NEWEST SAMPLE
         LDI       @HN_ADDR,AR0     ;AR0 POINTS TO COEFF H(N-1)
         LDF       0,R0             ;INITIALIZE R0
         LDF       0.0,R2           ;INITIALIZE R2
         RPTS      LENGTH-1         ;REPEAT LENGTH-1 TIMES
         MPYF      *AR0++,*AR1++%,R0   ;R0 = HN*XN
||       ADDF      R0,R2            ;R2 IS THE ACCUMULATOR
         DBNZD     AR4,LOOP         ;DELAYED BRANCH UNTIL AR4<0
         ADDF      R0,R2            ;LAST VALUE ACCUMULATED
         FIX       R2,R0            ;FLOAT(R2) TO INTEGER(R0)
         STI       R0,*AR6          ;OUTPUT R0 TO IO_OUT
WAIT     BR        WAIT             ;WAIT
         .DATA                      ;ASSEMBLE INTO DATA SECTION
COEFF    .FLOAT    -1.839E-3        ;H44
         .FLOAT    -2.657E-3        ;H43
         .FLOAT    -1.437E-7        ;H42
         .FLOAT    3.154E-3         ; .
         .FLOAT    2.595E-3         ; .
         .FLOAT    -4.159E-3
         .FLOAT    -1.540E-2
         .FLOAT    -2.507E-2
         .FLOAT    -2.547E-2
```

FIGURE 4.9. Bandpass FIR filter program with 45 coefficients using TMS320C30 code (BP45A.ASM)

```
         .FLOAT   -1.179E-2
         .FLOAT    1.392E-2
         .FLOAT    4.206E-2
         .FLOAT    5.888E-2
         .FLOAT    5.307E-2
         .FLOAT    2.225E-2
         .FLOAT   -2.410E-2
         .FLOAT   -6.754E-2
         .FLOAT   -8.831E-2
         .FLOAT   -7.475E-2
         .FLOAT   -2.956E-2
         .FLOAT    3.030E-2
         .FLOAT    8.050E-2
         .FLOAT    1.000E-1
         .FLOAT    8.050E-2
         .FLOAT    3.030E-2
         .FLOAT   -2.956E-2
         .FLOAT   -7.475E-2
         .FLOAT   -8.831E-2
         .FLOAT   -6.754E-2
         .FLOAT   -2.410E-2
         .FLOAT    2.225E-2
         .FLOAT    5.307E-2
         .FLOAT    5.888E-2
         .FLOAT    4.206E-2
         .FLOAT    1.392E-2
         .FLOAT   -1.179E-2
         .FLOAT   -2.547E-2
         .FLOAT   -2.507E-2
         .FLOAT   -1.540E-2
         .FLOAT   -4.159E-3
         .FLOAT    2.595E-3
         .FLOAT    3.154E-3
         .FLOAT   -1.437E-7
         .FLOAT   -2.657E-3         ;H1
H0       .FLOAT   -1.839E-3         ;H0
LENGTH   .SET     H0-COEFF+1        ;LENGTH = 45
         .END                       ;END
```

FIGURE 4.9. (*Continued*)

function). Figure 4.9 shows the FIR bandpass filter program BP45A.ASM. A description of this program follows (see also the previous two examples).

1. XN_ADDR specifies the bottom (last) address of the samples, reserved for $x(n)$. HN_ADDR specifies the starting address of the coefficient table. IO_IN and IO_OUT point at the input (804000h) and output

(804001h) port addresses, respectively. XN_BUFF is the name used in a user-defined section to reserve 45 (length) memory locations for the samples. The reset vector is always at 0h.

2. The input and output port addresses are loaded into AR5 and AR6, respectively. The filter's length (45) is loaded in AR4 and also into the special register BK, which specifies the size of the circular buffer for the samples. AR4 is used as a loop counter.

3. AR1 and AR0 are loaded with the bottom, or higher-memory, address of the sample $x(n)$ and the starting address of the coefficient table, COEFF, respectively. All the memory locations reserved for the samples are initialized to zero, starting and ending with the bottom address.

4. The instruction FLOAT *AR5,R3 is to input a sample. This sample is in memory address pointed by AR5 (initialized previously with the input port address 804000h). This newest sample is converted to its floating-point equivalent in R3, then stored at the last address in the sample table, reserved for $x(n)$. AR1 is then postincremented to point at the top or first address in the sample table. AR0 is loaded with the starting address of the coefficient table.

5. The multiply operation is executed in parallel with the next ADDF addition operation, a total number of length (45) times. The result of each multiplication is stored in R0. Then R0 is accumulated into R2. However, as the first multiplication is being performed, the next addition to accumulate R0 into R2 does not yet find the first product result, because this addition is being executed in parallel with the multiply operation. Therefore, after the first parallel addition, R2 contains zero (the initialized value of R0). The second time the parallel addition instruction is executed, R2 accumulates the result of the first multiplication. Hence we need a last addition instruction to accumulate the last product. This is done with the second addition instruction ADDF R0,R2 (without the parallel symbol).

6. The instruction DBNZD AR4,LOOP (to decrement AR4 and to branch based on the condition that AR4 is not zero) causes a branch with delay. This delayed-branch instruction makes more efficient use of the four levels of pipelining in the architecture of the TMS320C30. It allows the next three instructions to be executed before branching occurs. This effectively produces a one-cycle instruction branching. The non-delayed-branching instruction takes four instruction cycle times. Note that AR4 is decremented, then tested for a nonzero condition before the delayed-branching occurs. This instruction is more effective than using a register as a loop counter and decrementing this register before a conditional branch occurs, as in Example 4.5.

7. The floating-point value R0 is converted to integer format into R2 for output. This integer value, represents one output sample point, $y(n)$. It is stored at the output port address 804001h pointed by AR6.

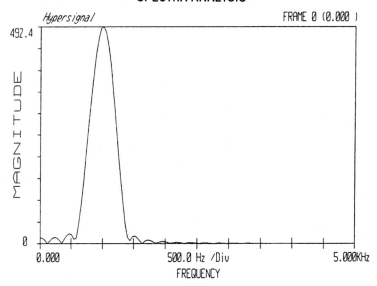

FIGURE 4.10. Plot of frequency response of 45-coefficient bandpass filter

8. Steps 4–7 are repeated to access a new sample and calculate a new output value $y(n + 1)$, and so on. This loop is executed `AR4 + 1` times (45 times), producing 45 output values: $y(n), y(n + 1), y(n + 2), \ldots, y(n + 44)$. Note that the second input sample is stored at the top, or lower-memory, address location reserved for the sample $x(n - (N - 1)) = x(n - 44)$. The third input sample is stored at $x(n - 43)$, and so on. This technique is used to model the delay samples $x(n), x(n - 1), \ldots, x(n - (N - 1))$.

Assemble and link this program using a command file similar to the one used with the lowpass filter in Figure 4.6 (change only the program names). Download this program into the simulator. Specify the input file `IMPULSE.DAT`, the impulse sequence (on accompanying disk) that contains a first value of 10,000 (2710h) followed by zeros, as the input to port 804000h. Run this program to create the output file `BP45A.DAT`. This filter's output response is plotted in Figure 4.10.

A different filter with a different number of coefficients can be implemented with this program by replacing the 45 coefficients with the new set of coefficients. Again use the labels `COEFF` and `H0` to represent, respectively, the starting and ending addresses of the coefficient table. Also, you could use the "noise" input sequence `NOISEINT.DAT` in lieu of the impulse sequence.

Example 4.8 *FIR Bandpass Filter Using C and C-Called Assembly Function.* This example implements the bandpass filter of the previous example, using a

4.8 Programming Examples Using C and TMS320C30 Code

main program in C (BP45MC.C, listed in Figure 4.11) and a C-called function in TMS320C30 code (BP45MCF.ASM, listed in Figure 4.12). The mixed-mode Examples 2.8 and 2.9 in Chapter 2 provide additional background necessary for this example. The coefficients of this filter are the same as in the previous example.

The buffer size used for the samples is twice the filter's length because it is difficult to align the circular buffer in a data section defined in C. This will ensure that any adjacent data will not be accidentally overwritten. The input and output port addresses are defined, and the function filt is called. This function is a C-identifier and hence must be referenced with an underscore (_filt) in the assembly function BP45MCF.ASM.

The assembly function BP45MCF.ASM is very similar to the one in the previous programming example, except for the following:

1. The frame FP set in auxiliary register AR3 is used for passing the addresses of arguments from the main C program BP45MC.C to the assembly function. Note that the old frame pointer FP is at the first location on the stack.
2. The registers AR4, AR5, and AR6 are dedicated and, because they are used, they must be saved using the PUSH instructions and later restored using the POP instructions. The frame pointer also must be saved and

```
/*BP45MC.C-FIR WITH C AND C-CALLED ASSEMBLY FUNCTION BP45F.ASM*/
#define N 45                     /*length of impulse response*/
float DLY[2*N];                  /*delay samples             */
const float H[N] = {/* filter coefficients*/
-1.839E-03,-2.657E-03,-4.312E-10, 3.154E-03, 2.595E-03,-4.159E-03,
-1.540E-02,-2.507E-02,-2.547E-02,-1.179E-02, 1.392E-02, 4.206E-02,
 5.888E-02, 5.307E-02, 2.225E-02,-2.410E-02,-6.754E-02,-8.831E-02,
-7.475E-02,-2.956E-02, 3.030E-02, 8.050E-02, 1.000E-01, 8.050E-02,
 3.030E-02,-2.956E-02,-7.475E-02,-8.831E-02,-6.754E-02,-2.410E-02,
 2.225E-02, 5.307E-02, 5.888E-02, 4.206E-02, 1.392E-02,-1.179E-02,
-2.547E-02,-2.507E-02,-1.540E-02,-4.159E-03, 2.595E-03, 3.154E-03,
-4.312E-10,-2.657E-03,-1.839E-03};
extern void filt(float *, float *, int *, int *, int);

main ()
{
 int loop;
 volatile int *IO_INPUT = (volatile int *) 0x804000; /*in port addr */
 volatile int *IO_OUTPUT= (volatile int *) 0x804001; /*out port addr*/
 for (loop = 0; loop < 2*N; loop++) DLY[loop] = 0.0; /*init samples */
 filt((float *)H, (float *)DLY, (int *)IO_INPUT, (int *)IO_OUTPUT, N);
}
```

FIGURE 4.11. Main FIR bandpass filter program in C (BP45MC.C)

122 Finite Impulse Response Filters

```
*BP45MCF.ASM-FIR FUNCTION IN ASSEMBLY CALLED FROM (BP45MC.C) C
FP       .SET      AR3                ;FRAME POINTER IN AR3
         .GLOBAL   _filt              ;GLOBAL REF/DEF FILTER ROUTINE
_filt    PUSH      FP                 ;SAVE FRAME POINTER
         LDI       SP,FP              ;LOAD STACK POINTER INTO FP
         PUSH      AR4                ;SAVE AR4
         PUSH      AR5                ;SAVE AR5
         PUSH      AR6                ;SAVE AR6
         LDI       *-FP(2),AR0        ;ADDR OF HN POINTER->AR0
         LDI       *-FP(3),AR1        ;ADDR OF XN POINTER->AR1
         LDI       *-FP(4),AR5        ;ADDR OF IO_INPUT POINTER->AR5
         LDI       *-FP(5),AR6        ;ADDR OF IO_OUTPUT POINTER->AR6
         LDI       *-FP(6),AR2        ;FILTER LENGTH->AR2
         LDI       AR2,BK             ;SIZE OF CIRCULAR BUFFER->BK
         SUBI      1,AR2              ;DECREMENT AR2
         ADDI      AR2,AR1            ;AR1=XN ADDR+LENGTH-1(BOTTOM)
         LDI       AR2,AR4            ;AR4 IS LOOP COUNTER
LOOP     FLOAT     *AR5,R3            ;INPUT NEW SAMPLE
         STF       R3,*AR1++%         ;STORE NEWEST SAMPLE
         LDI       *-FP(2),AR0        ;AR0 POINTS TO COEFF H(N-1)
         LDF       0,R0               ;INITIALIZE R0
         LDF       0,R2               ;INITIALIZE R2
         RPTS      AR2                ;REPEAT LENGTH-1 TIMES
         MPYF      *AR0++,*AR1++%,R0  ;R0 = HN*XN
||       ADDF      R0,R2              ;R2 = ACCUMULATOR
         DBNZD     AR4,LOOP           ;DELAYED BRANCH UNTIL AR4<0
         ADDF      R0,R2              ;LAST VALUE ACCUMULATED
         FIX       R2,R0              ;FLOAT(R2) TO INTEGER(R0)
         STI       R0,*AR6            ;OUTPUT R0 TO IO_OUTPUT
         POP       AR6                ;RESTORE THE CONTENTS OF AR6
         POP       AR5                ;RESTORE THE CONTENTS OF AR5
         POP       AR4                ;RESTORE THE CONTENTS OF AR4
         POP       FP                 ;RESTORE THE FRAME POINTER
         RETS                         ;RETURN TO C PROGRAM
```

FIGURE 4.12. C-called FIR bandpass filter function in TMS320C30 code (BP45MCF.ASM)

restored. The stack pointer SP is loaded into the frame pointer FP (the old frame pointer is at the first location on the stack).

3. The frame pointer, offset appropriately (starting with an offset of -2) is used to point at the starting addresses of the coefficients and the samples, the input and output port addresses, and the memory address containing the filter's length. These arguments are specified in the main C program where the filter function filt is called. The main part of the assembly function is very much the same as in the previous example.

4.8 Programming Examples Using C and TMS320C30 Code

Assemble the function program BP45MCF.ASM. Compile and assemble the main program BP45MC.C. Link, using a command file similar to LOOPALL.CMD, listed in Chapter 1. Note that both BP45MC.OBJ and BP45MCF.OBJ are to be specified in the command file for linking. Use the same input file IMPULSE.DAT as in the previous example. Run this program and verify that results identical to those of the previous example are obtained.

Example 4.9 *FIR Bandpass Filter Using C Code with the Modulo Operator.* This example implements the FIR bandpass filter of the previous two examples, using C code only. In Appendix C, the filter's coefficients were generated and incorporated within either a C or an assembly program. The generated C code (without input or output specification) is listed in Appendix C and uses the modulo operator % for circular buffering. This C code in Appendix C is relatively easy to follow because it uses the same approach as with assembly. However, this technique executes much slower (depending on the C compiler version used) than an implementation without the modulo operation, as seen in the next example. The filter program BP45H.C (on accompanying disk) is obtained by modifying the C program generated by the Hypersignal package to incorporate input and output, and it yields results identical to those of Example 4.8. A partial version of this program (BP45HPT) is shown in Figure 4.13.

Example 4.10 *FIR Bandpass Filter with Data Move Using C Code.* The program BP45NMD.C, listed in Figure 4.14, implements the same bandpass filter, centered at 1 kHz. To update the delay samples, a "data move" type of instruction is used, or

```
dly[i] = dly[i-1];
```

Run this program and verify that this filter executes faster than the one utilizing the modulo operator, with results identical to those previously obtained. Still another version, which follows, executes even faster (although it is more difficult to follow).

Example 4.11 *FIR Bandpass Filter without Modulo Operation, Using C Code.* This program example is another version of the bandpass filter, without the use of the modulo operator %. The program BP45NM.C, listed in Figure 4.15 executes 2.07 times faster than the previous implementation with the data move type of instruction. With an execution time of approximately 0.25 ms, it is still 1.5 times slower than the mixed-mode version using C and assembly code. Note that this execution time depends on the C compiler version. Specifying the appropriate memory locations in the command file slightly reduces the execution time. The execution time can be found by setting

124   Finite Impulse Response Filters

```
/*BP45HPT-PARTIAL FIR USING HYPERSIGNAL MODIFIED FOR I/O*/
#define N 45            /*length of impulse response*/
int start_index = 0;    /*circ buffer start position*/
double H[N];            /* filter coefficients*/
double DLY[N];          /* delay samples        */
double filt(stage_input)    /*filter routine*/
double stage_input;
{
  double acc;
  int    i, j;
  DLY[start_index] = stage_input;
  j = --start_index;
  acc = 0.0;
  for (i=0; i<N; i++)
  {
    j = ++j % N;  /*circular buffer requires modulo*/
    acc += H[i] * DLY[j];
  }
  start_index = j;
  return acc;
}

main ()
{
  #define IMPULSE_LENGTH 45    /* length of impulse response */
  volatile int *IO_OUT = (volatile int *) 0x804001; /*added for i/o*/
  int   n;
  *IO_OUT = filt(10000.0);  /*the "impulse"         */
  for (n=1; n<IMPULSE_LENGTH; n++)
  {
    *IO_OUT = filt(0.0);        /*other values are zero*/
  }
}
```

FIGURE 4.13. Partial FIR bandpass filter program generated with the Hypersignal package and modified for I/O (BP45HPT)

breakpoints at the beginning and at the end of the main processing section code. The first breakpoint is set at the instruction PUSH AR3 (after _filt) and the second breakpoint is set at the instruction SUBI 5,SP (after CALL_filt). The clock register CLK in the simulator window indicates the number of cycles, and can be converted to execution time by multiplying the CLK value by 60 ns, the cycle time of the TMS320C30.

The coefficients are listed in a separate file, BP45COEF.H, in Figure 4.16. This coefficient file is included (called) from the main program using the #include statement. For a different filter, a different coefficient file can be readily generated and "included" in the C program. The filter's length N is specified in the coefficients header file. No changes are necessary in the C

4.8 Programming Examples Using C and TMS320C30 Code

```c
/*BP45NMD.C-FIR BANDPASS FILTER IN C WITHOUT USING MODULO*/
#include "bp45coef.h"      /*include coefficient file   */
float DLY[N];              /*delay samples              */
void filt(float *, float *, int *, int *, int);

main ()
{
  int i;
  volatile int *IO_INPUT = (volatile int *) 0x804000;
  volatile int *IO_OUTPUT = (volatile int *) 0x804001;
  for (i = 0; i < N; i++) DLY[i] = 0.0;
  filt((float *)H, (float *)DLY, (int *)IO_INPUT, (int *)IO_OUTPUT, N);
}

void filt(float *h, float *dly, int *IO_input, int *IO_output, int n)
{
  int i, t;
  float acc;
  for (t = 0; t < n; t++)
  {
    acc = 0.0;
    dly[0] = *IO_input;
    for (i = 0; i < n; i++)
      acc += h[i] * dly[i];
    for (i = n-1; i > 0; i--)
      dly[i] = dly[i-1];
    *IO_output = acc;
  }
}
```

FIGURE 4.14. FIR bandpass filter program with data move to update the delay samples (BP45NMD.C)

program for a different set of coefficients. Operations such as n- m occurring within the processing loop are defined before the loop as n_m in order to increase the execution speed.

The two loops, with *i* and *j*, are used to achieve the effects of a circular buffer. Table 4.2 shows how the convolution equation is implemented.

Run this FIR bandpass filter program and obtain results identical to those in the previous examples. The next example illustrates an alternative approach.

Example 4.12 *FIR Bandpass Filter with Samples Shifted, Using C Code.* This program example implements the FIR bandpass filter of the previous example using a different method. It uses an array size of $2N - 1$ for the sample delays. A slightly shorter execution time is achieved with this method. Figure 4.17 shows the program listing BP45ERIC.C for this filter. The same coefficient file used in the previous example is included in this program. A

```
/*BP45NM.C-FIR BANDPASS FILTER IN C WITHOUT USING MODULO*/
#include "bp45coef.h"    /*include coefficient file      */
float DLY[N];            /*delay samples                 */
void filt(float *, float *, int *, int *, int);

main ()
{
  int i;
  volatile int *IO_INPUT = (volatile int *) 0x804000;
  volatile int *IO_OUTPUT = (volatile int *) 0x804001;
  for (i = 0; i<N; i++) DLY[i] = 0.0;
  filt((float *)H, (float *)DLY, (int *)IO_INPUT, (int *)IO_OUTPUT, N);
}

void filt(float *h, float *dly, int *IO_input, int *IO_output, int n)
{
  int i, j, m, N1, N1_m, n_m;
  float acc = 0;
  N1 = n-1;
  dly[0] = *IO_input;
  for (m = 0; m < n; m++)
  {
    N1_m = N1-m;
    n_m = n-m;
    for (i = 0; i < n_m; i++)
      acc += h[i] * dly[N1_m-i];
    for (j = m; j > 0; j--)
      acc += h[n-j] * dly[N1_m+j];
    *IO_output = acc;
    acc = 0.0;
    dly[N1_m] = *IO_input;
  }
}
```

FIGURE 4.15. FIR bandpass filter program without modulo operator (BP45NM.C)

```
/*BP45COEF.H-HEADER FILE COEFF FOR BANDPASS FILTER USED BY FILT*/
#define N 45                    /*length of impulse response*/
const float H[N] = {/* filter coefficients*/
-1.839E-03,-2.657E-03,-4.312E-10, 3.154E-03, 2.595E-03,-4.159E-03,
-1.540E-02,-2.507E-02,-2.547E-02,-1.179E-02, 1.392E-02, 4.206E-02,
 5.888E-02, 5.307E-02, 2.225E-02,-2.410E-02,-6.754E-02,-8.831E-02,
-7.475E-02,-2.956E-02, 3.030E-02, 8.050E-02, 1.000E-01, 8.050E-02,
 3.030E-02,-2.956E-02,-7.475E-02,-8.831E-02,-6.754E-02,-2.410E-02,
 2.225E-02, 5.307E-02, 5.888E-02, 4.206E-02, 1.392E-02,-1.179E-02,
-2.547E-02,-2.507E-02,-1.540E-02,-4.159E-03, 2.595E-03, 3.154E-03,
-4.312E-10,-2.657E-03,-1.839E-03};
```

FIGURE 4.16. Header coefficients file for bandpass filter (BP45COEF.H)

4.8 Programming Examples Using C and TMS320C30 Code

TABLE 4.2 Convolution for Bandpass Filter Example

	Time $n = 0$	Time $n = 1$	Time $n = 2$
	$h(0)x(44)$	$h(0)x(43)$	$h(0)x(42)$
	$h(1)x(43)$	$h(1)x(42)$	$h(1)x(41)$
	$h(2)x(42)$	$h(2)x(41)$	$h(2)x(40)$
loop i	$\vdots$	$\vdots$	$\vdots$
	$h(44)x(0)$	$h(43)x(0)$	$h(42)x(0)$
loop j	—	$h(44)x(44)$	$h(43)x(44)$
			$h(44)x(43)$

```
/*BP45ERIC.C-FIR BANDPASS WITH 45 COEFFICIENTS. SAMPLES SHIFTED*/
#include "bp45coef.h"        /*include coefficient file      */
#define N 45                 /*length of impulse response    */
float DLY[N*2-1];            /*init for 2*N-1 samples        */
void filt(float *,float *,int *,int *,int); /*filter routine */

main()
{
  volatile int *in_addr =(volatile int *)0x804000; /*input addr */
  volatile int *out_addr=(volatile int *)0x804001; /*output addr*/
  filt((float *)H,(float *)DLY,(int *)in_addr,(int *)out_addr,N);
}

void filt(float *h,float *dly,int *in_port,int *out_port,int NS)
{
  float acc=0.0;                /*init accumulator            */
  int   i,j,k=NS-1;             /*index variables             */
  for (i = 0; i<N*2-1; i++) DLY[i] = 0.0; /*init samples*/
  for (i=0;i<NS;i++)
  {
    dly[i+k] =*in_port;         /*get new sample              */
    for (j=0;j<NS;j++)
      acc += h[j]*dly[i+j];     /*perform convolution         */
    *out_port=acc;              /*output new value            */
    acc=0.0;                    /*reset accumulator           */
  }
  for (i=0;i<k;i++)             /*shift values from           */
    dly[i]=dly[i+NS];           /*lower half to upper half    */
}
```

FIGURE 4.17. Alternative FIR bandpass filter program (BP45ERIC.C)

TABLE 4.3 Initial Assignment of Delay Samples

dly[0]	→ $x(n - 43)$
dly[1]	→ $x(n - 42)$
dly[2]	→ $x(n - 41)$
⋮	
dly[42]	→ $x(n - 1)$
dly[43]	→ $x(n)$
dly[44]	→ newest sample

description of this program follows.

1. At the start of the filter function, the memory locations with the samples are shown in Table 4.3. The last memory location is reserved for the newest sample $x(n + 1)$, or dly[44], obtained with the instruction dly[i + k] = *in_port.

2. The convolution equation is performed next to obtain an output value representing $y(n + 1)$, or

$$y(n + 1) = h(0)\mathtt{dly[0]} + h(1)\mathtt{dly[1]} + \cdots + h(44)\mathtt{dly[44]}$$

Note that dly[0] specifies the sample $x(n - 43)$, and dly[44] is the newest sample $x(n + 1)$.

3. A new sample $x(n + 2)$ is acquired and placed at the bottom memory location following $x(n + 1)$. This newest sample value $x(n + 2)$ is specified with dly[45]. The convolution equation is again performed to yield $y(n + 2)$, or

$$y(n + 2) = h(0)\mathtt{dly[1]} + h(1)\mathtt{dly[2]} + \cdots + h(44)\mathtt{dly[45]}$$

with the sample delays updated.

4. For each time n, a new sample is acquired and placed at the "bottom" of the memory array. After 45 samples, the memory locations with the samples are as shown in Table 4.4. (Note that the first half of the table values are as in Table 4.3.) Note that the size of the array for the samples is $2N - 1 = 89$.

5. The last loop

```
for (i = 0;i<k;i + + )
    dly[i] = dly[i + NS];
```

is used to shift the samples in the lower-half memory locations to the upper-half locations, as shown in Table 4.5. This is equivalent to the

4.8 Programming Examples Using C and TMS320C30 Code

TABLE 4.4 Sample Delays after $n = 45$

dly[0] → $x(n - 43)$
dly[1] → $x(n - 42)$
dly[2] → $x(n - 41)$
$\vdots$
dly[42] → $x(n - 1)$
dly[43] → $x(n)$
dly[44] → $x(n + 1)$
dly[45] → $x(n + 2)$
dly[46] → $x(n + 3)$
dly[47] → $x(n + 4)$
$\vdots$
dly[86] → $x(n + 43)$
dly[87] → $x(n + 44)$
dly[88] → $x(n + 45)$

way the samples were displayed the first time, except that here the "older" samples are not set to zero.

This shift is useful (it can be deleted for this example) if a real-time implementation is desired, where the process is continually repeated. The filter function can be called repeatedly from the `main` program with

```
while (1)
{
   filt (...)
}
```

Run this program and verify that identical results for this bandpass filter are obtained as in the pervious implementations. The resulting output sequence can be plotted to obtain the same magnitude response of the bandpass filter as in the previous example.

TABLE 4.5 Sample Delays if Process is Repeated

dly[0] → $x(n + 2)$
dly[1] → $x(n + 3)$
dly[2] → $x(n + 4)$
$\vdots$
dly[42] → $x(n + 44)$
dly[43] → $x(n + 45)$
dly[44] → $x(n + 1)$

130 Finite Impulse Response Filters

Breakpoints can be set appropriately in order to measure the execution time of the filter routine. Verify that this filter routine executes slightly faster than the previous implementation.

Example 4.13 *Real-Time FIR Bandpass Filter Using C Code.* A real-time version of the filter program in Example 4.11 is listed in Figure 4.18 as BP45NMR.C. It calls the AIC communication routines in the program AICCOM.C from Chapter 3. Observe the following from BP45NMR.C:

1. The idle instruction is used to wait for an interrupt. When an interrupt occurs, execution proceeds to the interrupt vector function c_int05.
2. From the interrupt vector function c_int05, the UPDATE_SAMPLE (from AICCOM.C) function is called. Because the C function UPDATE_SAMPLE is called from another C function (c_int05), the compiler creates a number of PUSH / PUSHF and POP / POPF, which slows down the execution time of the filter. For the filter to execute faster, the following instructions from the UPDATE_SAMPLE function in AICCOM.C can be incorporated directly in the c_int05 function:

```
PBASE[0x48] = data_out << 2;
data_in = PBASE[0x4C] << 16 >> 18;
```

as shown in the next example, using a mixed-mode implementation Note that the filter function is called within an "infinite" loop.
3. The magnitude response of the real-time bandpass filter is plotted in Figure 4.19. This plot is obtained from a Hewlett Packard (HP) HP3561A signal analyzer. Random noise, available as a source of noise from the HP signal analyzer, provides the input to the filter. The filter coefficients included in the file BP45COEF.H were obtained with the Hypersignal-Plus DSP Software package, using the Kaiser window function. This filter can be tested in the following fashion. Connect a 1-kHz sinusoidal signal as input to the EVM in lieu of the random noise. Observe, from an oscilloscope, that the EVM output is also a 1-kHz signal. Slightly increase or decrease the frequency of the input signal, and observe the effect of the bandpass filter on the output signal.

Example 4.14 *Real-Time FIR Bandpass Filter Using Mixed C and TMS320C30 Code.* The main program in C for this example (BP45MCR.C) is listed in Figure 4.20 and calls the function BP45MCFR.ASM. This implementation is interrupt-driven. The function BP45MCFR.ASM is similar to the function BP45MCF.ASM (used in the simulated version of this example), listed in Figure 4.12, with the exception of an IDLE instruction added where the label

4.8 Programming Examples Using C and TMS320C30 Code

```
/*BP45NMR.C-REAL-TIME FIR BANDPASS FILTER IN C.CALLS AICCOM */
#include "aiccom.c"              /*include AIC com routines */
#include "bp45coef.h"             /*include coefficients file*/
#define VEC_ADDR (volatile int *) 0x00
float DLY[N];                    /*delay samples            */
int data_in, data_out;
int AICSEC[4] = {0x1428,0x1,0x4A96,0x67};  /*AIC config data*/

void filt(float *h, float *dly, int *IO_input, int *IO_output, int n)
{
  int i, j, m, N1, N1_m, n_m, index = 0;
  float acc = 0;
  N1 = n-1;
  dly[0] = *IO_input;
  for (m = 0; m < n; m++)
  {
    asm("    idle");              /*wait for interrupt        */
    N1_m = N1-m;
    n_m = n-m;
    for (i = 0; i < n_m; i++)     /*addr below new sample to 0*/
      acc += h[i] * dly[N1_m-i];
    for (j = m; j > 0; j--)       /*from n to latest sample   */
      acc += h[n-j] * dly[N1_m+j]; /*latest sample last        */
    *IO_output = acc;             /*output result             */
    acc = 0.0;                    /*clear accumulator         */
    dly[N1_m] = *IO_input;        /*get new sample            */
  }
}

void c_int05()
{
  data_in = UPDATE_SAMPLE(data_out);
}

main ()
{
  volatile int *INTVEC = VEC_ADDR;
  int *IO_INPUT, *IO_OUTPUT;
  IO_INPUT = &data_in;
  IO_OUTPUT = &data_out;
  INTVEC[5] = (volatile int) c_int05;
  AICSET_I();
  for(;;)
    filt((float *)H, (float *)DLY, (int *)IO_INPUT, (int *)IO_OUTPUT, N);
}
```

FIGURE 4.18. Real-time FIR bandpass filter program in C (BP45NMR.C)

132 Finite Impulse Response Filters

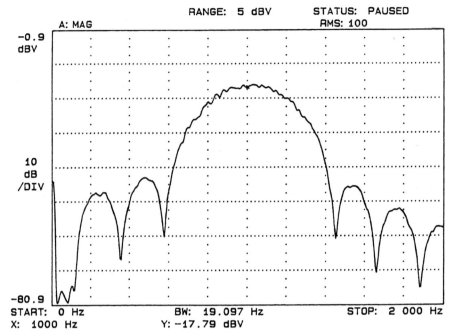

FIGURE 4.19. Frequency response of FIR bandpass filter using the Kaiser window function

address LOOP is, or

```
            .
            .
            .
         LDI      AR2,AR4
LOOP     IDLE     (added instruction)
         FLOAT    *AR5,R3
            .
            .
```

This idle instruction is to wait for an interrupt and accomplishes the same effect as the previous example in C code. The magnitude response of this filter is identical to the one in the previous example.

Example 4.15 *Real-Time FIR Bandpass Filter Using TMS320C30 Code.* This program example implements in real time with TMS320C30 code the same FIR bandpass filter as in the previous example. The program BP45AR.ASM

4.8 Programming Examples Using C and TMS320C30 Code

```
/*BP45MCR.C-REAL-TIME FIR BANDPASS FILTER.CALLS ASSEMBLY FUNCTION   */
#include "aiccom.c"                     /*include AIC com routines     */
#include "bp45coef.h"                   /*include coefficients file    */
#define VEC_ADDR (volatile int *) 0x00
float DLY[2*N];                         /*delay samples                */
int AICSEC[4] =  {0x1428,0x1,0x4A96,0x67}; /*AIC config data           */
int data_in, data_out;
extern void filt(float *, float *, int *, int *, int);

void c_int05()
{
  PBASE[0x48] = data_out << 2;
  data_in = PBASE[0x4C] << 16 >> 18;
}

main ()
{
  volatile int *INTVEC = VEC_ADDR;
  int *IO_INPUT, *IO_OUTPUT;
  IO_INPUT = &data_in;
  IO_OUTPUT = &data_out;
  INTVEC[5] = (volatile int) c_int05;
  AICSET_I();
  for (;;)
    filt((float *)H, (float *)DLY, (int *)IO_INPUT, (int *)IO_OUTPUT, N);
}
```

FIGURE 4.20. Real-time FIR bandpass filter program in C calling an assembly function (BP45MCR.C)

listed in Figure 4.21 calls the AIC communication program AICCOMA.ASM listed in Chapter 3 (Figure 3.21). In the next section, we will see how a similar program can be used as a generic filter program and linked with an appropriate coefficient file. The output rate is obtained from polling, as opposed to the previous interrupt-driven program examples. The AIC communication routines in TMS320C30 code can be used for either port 0 or 1 and can handle either a polling or an interrupt-driven calling program. Consider the main program (see also the simulated program version BP45A.ASM listed in Figure 4.9) and note that the data page is loaded with SPSET to ensure consistency in the page accessed. SPSET, which is set to zero in AICCOMA.ASM, is used to access the desired port. To access serial port 1, the content of SPSET must be set to 10h, because the default value of SPSET is 0, selecting serial port 0. The AIC configuration data with AICSEC is set for a sampling rate of 10 kHz.

All supporting files are on the accompanying disk. Run this program and observe the same frequency response as in the previous example. To control the rate with interrupt, make the following changes in BP45AR.ASM to

```
;BP45AR.ASM-FIR BANDPASS FILTER WITH 45 COEFF,USING TMS320C30 CODE
        .TITLE   "BP45AR.ASM"       ;BP FIR,Fc=1 kHz,Fs=10 kHz
        .GLOBAL  FILT,MAIN,BEGIN,AICSET,AICSEC,SPSET,AICIO_P
        .DATA                       ;ASSEMBLE INTO DATA SECTION
XN_ADDR .WORD    XN+LENGTH-1        ;(LAST) SAMPLE ADDRESS
HN_ADDR .WORD    COEFF              ;ADDR OF COEFF H(N-1)
AICSEC  .WORD    1428h,1h,4A96h,67h ;AIC CONFIG DATA
XN      .USECT   "XN_BUFF",LENGTH   ;BUFFER SIZE OF SAMPLES
        .SECT    "VECTORS"          ;VECTOR SECTION
MAIN    .WORD    BEGIN              ;BEGIN @ RESET (0H)
        .TEXT                       ;ASSEMBLE INTO TEXT
BEGIN   LDP      SPSET              ;INIT DATA PAGE
        CALL     AICSET             ;INIT AIC
        LDI      LENGTH,BK          ;SIZE OF CIRCULAR BUFFER
        LDI      @XN_ADDR,AR1       ;LAST SAMPLE ADDR ->AR1
FILT    LDI      LENGTH,AR4         ;LENGTH IN AR4
        SUBI     1,AR4              ;AR4 = LENGTH - 1
LOOP    CALL     AICIO_P            ;AICIO ROUTINE,IN->R6 OUT->R7
        FLOAT    R6,R3              ;INPUT NEW SAMPLE
        STF      R3,*AR1++          ;STORE NEWEST SAMPLE
        LDI      @HN_ADDR,AR0       ;AR0 POINTS TO COEFF H(N-1)
        LDF      0,R0               ;INITIALIZE R0
        LDF      0,R2               ;INITIALIZE R2
        RPTS     LENGTH-1           ;REPEAT LENGTH-1 TIMES
        MPYF     *AR0++,*AR1++%,R0  ;R0 = HN*XN
||      ADDF     R0,R2              ;R2 IS THE ACCUMULATOR
        DBNZD    AR4,LOOP           ;DELAYED BRANCH UNTIL AR4<0
        ADDF     R0,R2              ;LAST VALUE ACCUMULATED
        FIX      R2,R7              ;FLOAT(R2) TO INTEGER(R7)
        NOP                         ;ADDED DUE TO DELAY BRANCH
        BR       FILT               ;BRANCH BACK TO FILTER ROUTINE
        .DATA                       ;ASSEMBLE INTO DATA SECTION
COEFF   .FLOAT   -1.839E-3          ;H44
   .FLOAT  -2.657E-3,-1.437E-7, 3.154E-3, 2.595E-3,-4.159E-3,-1.540E-2
   .FLOAT  -2.507E-2,-2.547E-2,-1.179E-2, 1.392E-2, 4.206E-2, 5.888E-2
   .FLOAT   5.307E-2, 2.225E-2,-2.410E-2,-6.754E-2,-8.831E-2,-7.475E-2
   .FLOAT  -2.956E-2, 3.030E-2, 8.050E-2, 1.000E-1, 8.050E-2, 3.030E-2
   .FLOAT  -2.956E-2,-7.475E-2,-8.831E-2,-6.754E-2,-2.410E-2, 2.225E-2
   .FLOAT   5.307E-2, 5.888E-2, 4.206E-2, 1.392E-2,-1.179E-2,-2.547E-2
   .FLOAT  -2.507E-2,-1.540E-2,-4.159E-3, 2.595E-3, 3.154E-3,-1.437E-7
        .FLOAT   -2.657E-3          ;H1
H0      .FLOAT   -1.839E-3          ;H0
LENGTH  .SET     H0-COEFF+1         ;LENGTH = 45
        .END                        ;END
```

FIGURE 4.21. Real-time FIR bandpass filter program using TMS320C30 code (BP45AR.ASM)

4.8 Programming Examples Using C and TMS320C30 Code

obtain `BP45ARI.ASM` (on accompanying disk):

1. In the `.GLOBAL` assembler directive, change `AICIO_P` and `AICSET` to `AICIO_I` and `AICSET_I`, respectively
2. After the assembler directive `MAIN .WORD BEGIN`, insert

   ```
   .SPACE   4
   .WORD    ISR
   .SPACE   58
   ```

   This places the interrupt service routine address in location 5h for `XINTO` and skips the remaining memory locations used for interrupt and traps (see Chapter 2 and Appendix B).
3. Change `CALL AICSET` to `CALL AICSET_I`

```
;BP45ARM.ASM-FIR BANDPASS FILTER WITH MACRO, USING TMS320C30 CODE
         .TITLE    "BP45ARM.ASM"      ;BP FIR,Fc=1 kHz,Fs=10 kHz
         .GLOBAL   MAIN,BEGIN,AICSET,AICSEC,SPSET,AICIO_P,MFILT
         .DATA                        ;ASSEMBLE INTO DATA SECTION
XN_ADDR  .WORD     XN+LENGTH-1        ;(LAST) SAMPLE ADDRESS
HN_ADDR  .WORD     COEFF              ;ADDR OF COEFF H(N-1)
AICSEC   .WORD     1428h,1h,4A96h,67h ;AIC CONFIG DATA
XN       .USECT    "XN_BUFF",LENGTH   ;BUFFER SIZE OF SAMPLES
         .SECT     "VECTORS"          ;VECTOR SECTION
MAIN     .WORD     BEGIN              ;BEGIN @ RESET (0H)
         .TEXT                        ;ASSEMBLE INTO TEXT
         .MLIB     "MACRO.LIB"        ;MACRO LIBRARY
BEGIN    LDP       SPSET              ;INIT DATA PAGE
         CALL      AICSET             ;INIT AIC
         LDI       LENGTH,BK          ;SIZE OF CIRCULAR BUFFER
         LDI       @XN_ADDR,AR1       ;LAST SAMPLE ADDR ->AR1
         MFILT                        ;MACRO FILTER
         .DATA                        ;ASSEMBLE INTO DATA SECTION
COEFF    .FLOAT    -1.839E-3          ;H44
   .FLOAT  -2.657E-3,-1.437E-7, 3.154E-3, 2.595E-3,-4.159E-3,-1.540E-2
   .FLOAT  -2.507E-2,-2.547E-2,-1.179E-2, 1.392E-2, 4.206E-2, 5.888E-2
   .FLOAT   5.307E-2, 2.225E-2,-2.410E-2,-6.754E-2,-8.831E-2,-7.475E-2
   .FLOAT  -2.956E-2, 3.030E-2, 8.050E-2, 1.000E-1, 8.050E-2, 3.030E-2
   .FLOAT  -2.956E-2,-7.475E-2,-8.831E-2,-6.754E-2,-2.410E-2, 2.225E-2
   .FLOAT   5.307E-2, 5.888E-2, 4.206E-2, 1.392E-2,-1.179E-2,-2.547E-2
   .FLOAT  -2.507E-2,-1.540E-2,-4.159E-3, 2.595E-3, 3.154E-3,-1.437E-7
         .FLOAT    -2.657E-3          ;H1
H0       .FLOAT    -1.839E-3          ;H0
LENGTH   .SET      H0-COEFF+1         ;LENGTH = 45
         .END                         ;END
```

FIGURE 4.22. Real-time FIR bandpass filter using TMS320C30 code with macro (`BP45ARM.ASM`)

136 Finite Impulse Response Filters

4. Change `LOOP CALL AICIO_P` to `LOOP IDLE`
5. After the `BR FILT` instruction, insert the interrupt service routine (at the end of the program)

```
ISR   CALL  AICIO_I
      RETI
```

Example 4.16 *Real-Time FIR Bandpass Filter Using TMS320C30 Code with Macro.* This example implements in real time the FIR bandpass filter discussed in the previous example, using macro. The program `BP45ARM.ASM` listed in Figure 4.22 (on page 135) calls the macro `MFILT` shown in Figure 4.23. A macro library (for example, `MACRO.LIB`) is provided to the assembler with the `.mlib` assembler directive. This macro library may contain a number of files including `MFILT`. All the files in the macro library are collected into one single file by the archiver [10]. When a specific macro is called, the assembler extracts only that macro from the macro library. The following instruction creates a macro library using the archiver program `AR30.EXE`, included with the EVM package:

```
AR30 -A MACRO MFILT.ASM
```

```
*MFILT.ASM-MACRO FOR FIR FILTER
MFILT   $MACRO                        ;FILTER MACRO
FILT    LDI     LENGTH,AR4            ;LENGTH IN AR4
        SUBI    1,AR4                 ;AR4 = LENGTH - 1
LOOP    CALL    AICIO_P               ;AICIO ROUTINE,IN->R6 OUT->R7
        FLOAT   R6,R3                 ;INPUT NEW SAMPLE
        STF     R3,*AR1++             ;STORE NEWEST SAMPLE
        LDI     @HN_ADDR,AR0          ;AR0 POINTS TO COEFF H(N-1)
        LDF     0,R0                  ;INITIALIZE R0
        LDF     0,R2                  ;INITIALIZE R2
        RPTS    LENGTH-1              ;REPEAT LENGTH-1 TIMES
        MPYF    *AR0++,*AR1++%,R0     ;R0 = HN*XN
        ADDF    R0,R2                 ;R2 IS THE ACCUMULATOR
        DBNZD   AR4,LOOP              ;DELAYED BRANCH UNTIL AR4<0
        ADDF    R0,R2                 ;LAST VALUE ACCUMULATED
        FIX     R2,R7                 ;FLOAT(R2) TO INTEGER(R7)
        NOP                           ;ADDED DUE TO DELAY BRANCH
        BR      FILT                  ;BRANCH BACK TO FILTER ROUTINE
        $ENDM                         ;END OF FILTER MACRO
```

FIGURE 4.23. Macro definition for real-time bandpass filter implementation (`MFILT.ASM`)

4.8 Programming Examples Using C and TMS320C30 Code

```
;BP45ARPN.ASM-FIR BANDPASS FILTER WITH NOISE GENERATOR
         .TITLE   "BP45ARPN.ASM"   ;BP FIR, Fc=Fs/10, Fs=8 kHz
         .GLOBAL  MAIN,BEGIN,AICSET_I,AICSEC,SPSET,AICIO_I,FILT
         .DATA                     ;ASSEMBLE INTO DATA SECTION
XN_ADDR  .WORD    XN+LENGTH-1      ;(LAST) SAMPLE ADDRESS
HN_ADDR  .WORD    COEFF            ;ADDR OF COEFF H(N-1)
AICSEC   .WORD    1428h,1h,5EBEh,67h   ;AIC CONFIG DATA
PLUS     .WORD    1000H            ;POSITIVE NOISE LEVEL
MINUS    .WORD    0FFFFF000H       ;NEGATIVE NOISE LEVEL
SEED     .WORD    7E521603H        ;INITIAL SEED VALUE
SCALER   .FLOAT   0.25             ;OUTPUT SCALE FACTOR
XN       .USECT   "XN_BUFF",LENGTH ;BUFFER SIZE OF SAMPLES
         .SECT    "VECTORS"        ;VECTOR SECTION
MAIN     .WORD    BEGIN            ;BEGIN @ RESET (0H)
         .SPACE   4                ;SKIP 4 SPACES
         .WORD    ISR              ;INTERRUPT @ 5H
         .SPACE   58               ;REMAINDER OF VECTOR/TRAP
         .TEXT                     ;ASSEMBLE INTO TEXT
BEGIN    LDP      SPSET            ;INIT DATA PAGE
         CALL     AICSET_I         ;INIT AIC
         LDI      LENGTH,BK        ;SIZE OF CIRCULAR BUFFER
         LDI      @XN_ADDR,AR1     ;LAST SAMPLE ADDR ->AR1
         LDI      @SEED,R0         ;R0=INITIAL SEED VALUE
WAIT     IDLE                      ;WAIT FOR INTERRUPT
         BR       WAIT             ;BRANCH BACK TO WAIT
;INTERRUPT VECTOR, START OF NOISE GENERATOR
ISR      LDI      0,R4             ;INIT R4=0
         LDI      R0,R2            ;PUT SEED IN R2
         LSH      -17,R2           ;MOVE BIT 17 TO LSB    =>R2
         ADDI     R2,R4            ;ADD BIT (17)          =>R4
         LSH      -11,R2           ;MOVE BIT 28 TO LSB    =>R2
         ADDI     R2,R4            ;ADD BITS (28+17)      =>R4
         LSH      -2,R2            ;MOVE BIT 30 TO LSB    =>R2
         ADDI     R2,R4            ;ADD BITS (30+28+17)   =>R4
         LSH      -1,R2            ;MOVE BIT 31 TO LSB    =>R2
         ADDI     R2,R4            ;ADD BITS (31+30+28+17)=>R4
         AND      1,R4             ;MASK LSB OF R4
         LDIZ     @MINUS,R7        ;IF R4 = 0, R7 = @MINUS
         LDINZ    @PLUS,R7         ;IF R4 = 1, R7 = @PLUS
         LSH      1,R0             ;SHIFT SEED LEFT BY 1
         OR       R4,R0            ;PUT R4 INTO LSB OF R0
;MAIN SECTION
         FLOAT    R7,R3            ;INPUT NEW NOISE SAMPLE
         STF      R3,*AR1++%       ;STORE NEWEST SAMPLE
         LDI      @HN_ADDR,AR0     ;AR0 POINTS TO COEFF H(N-1)
```

FIGURE 4.24. Program listing for FIR bandpass filter with noise generation, using TMS320C30 code (BP45ARPN.ASM)

138 Finite Impulse Response Filters

```
              CALL     FILT                  ;CALL FILTER ROUTINE
              MPYF     @SCALER,R3            ;SCALE OUTPUT SAMPLE
              FIX      R3,R7                 ;R7=INTEGER(R3)
              CALL     AICIO_I               ;AIC I/O ROUTINE FOR OUTPUT
              RETI                           ;RETURN FROM INTERRUPT
;FILTER ROUTINE
FILT          LDF      0,R1                  ;INIT R1=0
              LDF      0,R3                  ;INIT R3=0
              RPTS     LENGTH-1              ;NEXT 2 INSTR. LENGTH TIMES
              MPYF     *AR0++,*AR1++%,R1     ;R1=HN*XN
||            ADDF     R1,R3,R3              ;R3 IS ACCUMULATOR
              ADDF     R1,R3,R3              ;LAST VALUE ACCUMULATED
              RETS                           ;RETURN FROM SUBROUTINE
              .DATA                          ;ASSEMBLE INTO DATA SECTION
COEFF         .FLOAT   -1.839E-3             ;H44
       .FLOAT  -2.657E-3,-1.437E-7, 3.154E-3, 2.595E-3,-4.159E-3,-1.540E-2
       .FLOAT  -2.507E-2,-2.547E-2,-1.179E-2, 1.392E-2, 4.206E-2, 5.888E-2
       .FLOAT   5.307E-2, 2.225E-2,-2.410E-2,-6.754E-2,-8.831E-2,-7.475E-2
       .FLOAT  -2.956E-2, 3.030E-2, 8.050E-2, 1.000E-1, 8.050E-2, 3.030E-2
       .FLOAT  -2.956E-2,-7.475E-2,-8.831E-2,-6.754E-2,-2.410E-2, 2.225E-2
       .FLOAT   5.307E-2, 5.888E-2, 4.206E-2, 1.392E-2,-1.179E-2,-2.547E-2
       .FLOAT  -2.507E-2,-1.540E-2,-4.159E-3, 2.595E-3, 3.154E-3,-1.437E-7
              .FLOAT   -2.657E-3             ;H1
H0            .FLOAT   -1.839E-3             ;H0
LENGTH        .SET     H0-COEFF+1            ;LENGTH = 45
              .END                           ;END
```

FIGURE 4.24. *(Continued)*

The -A option adds the macro file MFILT.ASM in Figure 4.23 into a macro library called MACRO.LIB. Note that Figures 4.22 and 4.23 can be readily obtained from the program BP45AR.ASM listed in Figure 4.21. To delete a file from the macro library, use a -D option (instead of the -A option).

The macro MFILT is contained within $MACRO, which identifies the first line, and $ENDM, which designates the end of the macro definition. The main program BP45ARM.ASM contains the assembler directive .MLIB. Note how the macro is called, using MFILT.

Run this program and verify that the same frequency response as in Figure 4.19 is obtained.

Example 4.17 Real-Time FIR Bandpass Filter with Noise Generation, Using TMS320C30 code. This programming example illustrates the implementation of the 45-coefficient bandpass filter incorporating the noise generator in Example 3.2. The output of the noise generator is either 1000h or its negative equivalent. The output pseudorandom noise sequence (NOISEINT.DAT) is the input to the filter, hence, no external input is required. Figure 4.24 shows the program listing BP45ARPN.ASM for this example. The sampling fre-

4.9 Filter Development Package and Digital Filter Design Package

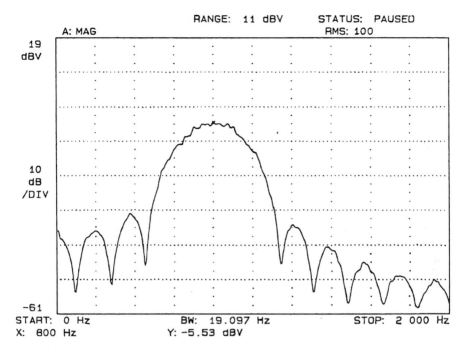

FIGURE 4.25. Frequency response of FIR bandpass filter with internal noise generation

quency is set at 8 kHz by changing 4A96h to 5EBEH in the AIC configuration data (AICSEC). The same 45 coefficients as in the previous examples are used to yield a center frequency of $\frac{1}{10}F_s$.

On interrupt, program execution continues to the interrupt vector ISR, to generate a noise sample (see Example 3.2). This noise sample (either a 0 or 1) is scaled accordingly (1000h or FFFFF000h) into R7. The equivalent (scaled) floating-point value becomes the input to the FIR filter (see MAIN SECTION of program).

Run this program. The frequency response of this filter, with a center frequency of 800 Hz, is displayed in Figure 4.25. Set the AIC for a 10-kHz sampling rate and verify that the center frequency is at 1 kHz.

Note that the filter subroutine is in a separate section in order to make the program easier to follow. This is done at the expense of execution speed (see the previous two examples).

4.9 FILTER DEVELOPMENT PACKAGE (FDP) AND DIGITAL FILTER DESIGN PACKAGE (DFDP)

A noncommercial filter development package (FDP) is included on the accompanying disk. This filter design package includes a program

FIRPROG.BAS, written in BASIC, that allows for the calculation of the coefficients of an FIR filter. Lowpass, highpass, bandpass, and bandstop FIR filters, using the rectangular, Hanning, Hamming, Blackman, and Kaiser windows can be designed. The FDP does not include the Parks–McClellan design algorithm, which can be found in commercially available packages, such as the Hypersignal package discussed in Appendix C and the digital filter design package (DFDP) from Atlanta Signal Processors, Inc. The coefficients obtained from the filter development package are in the appropriate format with a .FLOAT assembler directive preceding each coefficient value. This coefficient file needs to be only slightly edited and linked with a generic (without coefficients) program. It can also be used to replace the coefficients already set in the FIR assembly programs.

In the previous example, we implemented an FIR bandpass filter with the 45 coefficients obtained with the Hypersignal-Plus DSP Software using the Kaiser window function. We will now walk through an example using the (noncommercial) FDP package.

Example 4.18 Real-Time FIR Filter Using a Noncommercial Filter Development Package. This example illustrates the use of the filter package FDP as well as the effects of various windows. The coefficients of the bandpass filter are obtained and linked with a generic program. The frequency response of the filter is displayed on a signal analyzer. Proceed with the following:

1. Run BASIC, and load and run the program FIRPROG.BAS. Figure 4.26 displays the menu for available windows. Choose option 5 to select the Kaiser window. When this option is invoked, a separate module (FIRPROGA.BAS) for the Kaiser window is called from the program FIRPROG.BAS and executed.
2. Selections for lowpass, highpass, bandpass, and bandstop filters are next displayed. Choose option 3 to select an FIR bandpass filter.
3. Figure 4.27 shows the characteristics of the desired filter, such as a sampling frequency of 10,000 Hz, and so on. Enter 45 for the number of coefficients, even though the default value, or the minimum number of coefficients, is 43 (as is done in Appendix C with the Hypersignal-Plus DSP Software). Next select F (Figure 4.27) to save or send the coefficients to a file. Then choose C30 in order to obtain the coefficients with the .FLOAT assembler directive (TMS320C30 format). Enter BP45K.COF as the coefficients filename to be saved. Exit to DOS.
4. The filter coefficients created in the previous step can now replace those in the program BP45AR.ASM (use an editor to delete a section of a file and to insert another file). This new program can now be assembled and linked with no additional changes (except to save it under a different name).

4.9 Filter Development Package and Digital Filter Design Package

```
                    FIR DEVELOPMENT PACKAGE



                         Main Menu
                         ---------

                         1....RECTANGULAR

                         2....HANNING

                         3....HAMMING

                         4....BLACKMAN

                         5....KAISER

                         6....Exit to DOS

Enter window desired (number only) --> 5




       *** FIR COEFFICIENT GENERATION USING THE KAISER WINDOW ***



                    Selections:

                              1....LOWPASS

                              2....HIGHPASS

                              3....BANDPASS

                              4....BANDSTOP

                              5....Exit back to Main Menu

Enter desired filter type (number only) --> 3
```

FIGURE 4.26. Main menu of FIRPROG.BAS using the filter development package FDP

```
Specifications:
            BANDPASS
            Passband Ripple (AP) = 6 db
            Stopband Attenuation (AS) = 26 db
            Lower Passband Frequency = 900 Hz
            Upper Passband Frequency = 1100 Hz
            Lower Stopband Frequency = 600 Hz
            Upper Stopband Frequency = 1400 Hz
            Sampling Frequency (Fs) = 10000 Hz


The calculated # of coefficients required is: 43

Enter # of coefficients desired ONLY if greater than 43
otherwise, press <Enter> to continue --> 45


Send coefficients to:
                (S)creen
                (P)rinter
                (F)ile: contains TMS320 (C25 or C30) data format
                (R)eturn to Filter Type Menu
                (E)xit to DOS

Enter desired path --> F

Enter DSP type (C25 OR C30):? C30
```

FIGURE 4.27. Filter characteristics with Kaiser window

Alternatively, a generic FIR program without coefficients can be obtained by editing BP45AR.ASM in the following fashion:

1. Add the two assembler directives

.GLOBAL	COEFF,LENGTH	to define globally COEFF and LENGTH
LENGTH	.SET 45	to set the filter's length to 45 within the .DATA section

2. End the program with the BR FILT instruction.
3. Delete the second .DATA section (with the coefficients).

Rename this new program. It is on the accompanying disk as FIRGEN.ASM.

Edit the coefficient file (created with the FDP design package) in the

4.9 Filter Development Package and Digital Filter Design Package

following fashion:

1. Add (before the coefficients) the two assembler directives

   ```
   .GLOBAL   COEFF
   .DATA
   ```

2. delete the assembler directive

   ```
   LENGTH        .SET       H0-COEFF+1
   ```

   because the length of the filter is directly set in the main program.

Rename this coefficient file. It is on the accompanying disk as `COEFGEN.ASM`.

To obtain the linked COFF output file, modify the previously used command file in order to link the generic FIR program (`FIRGEN.OBJ`), the program with the coefficients (`COEFGEN.OBJ`), and the communication routines `AICCOMA.OBJ`. This command file is on the accompanying disk as `FIRGEN.CMD`. In this fashion a different FIR filter with a new coefficient file can be quickly obtained, assembled, and linked with the generic program, to obtain the COFF output file.

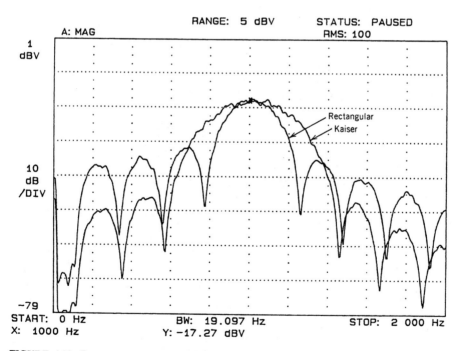

FIGURE 4.28. Frequency responses of FIR bandpass filter using the rectangular and Kaiser windows

Other window functions can be selected, such as the Hanning window (option 2 in the main menu in Figure 4.26). For such a window, with a design similar to the previous one, the lower and upper cutoff frequencies are 900 and 1,100 Hz, respectively. The desired number of coefficients is obtained by specifying the duration of the impulse response. Enter a duration D of 4.4 ms for a 45-coefficient filter, because

$$\text{number of coefficients} = (D \times F_s) + 1$$

Using random noise as input (available from the HP signal analyzer), the output can be sent to an HP signal analyzer. Figure 4.28 shows the frequency responses of the FIR bandpass filters using the rectangular and the Kaiser windows. Figure 4.29 shows the frequency responses using the rectangular, the Hanning, and the Blackman windows. The coefficients used for these filters are obtained with the FDP filter package. Note how the peak of the first sidelobe is relatively high with the rectangular window, compared to the peak of the mainlobe. On the other hand, the rectangular window provides the highest selectivity, or the sharpest transition between passbands and stopbands.

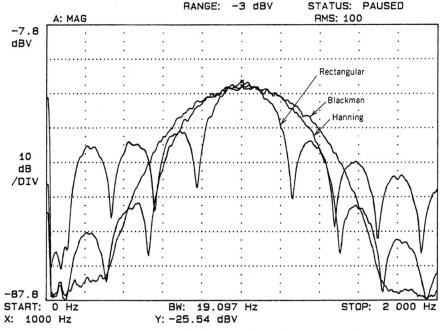

FIGURE 4.29. Frequency responses of FIR bandpass filter using the rectangular, Hanning, and Blackman windows

4.9 Filter Development Package and Digital Filter Design Package 145

Digital Filter Design Package (DFDP)

The noncommercial FDP package can be quite helpful although it does not provide the interactive features usually available with a commercial filter design package. In that respect, the Hypersignal package, discussed in Appendix C can be very useful. The digital filter design package (DFDP) available from Atlanta Signal Processors, Inc., is equally useful. For example, other types of filters such as a differentiator filter or a multiband filter can be readily designed using the DFDP.

The simulated frequency response of a 31-coefficient differentiator filter is shown in Figure 4.30. Figures 4.31 and 4.32 show, respectively, the simulated and realtime frequency responses of a 31-coefficient multiband (two passbands) filter. The multiband filter is used in Chapter 8 for the design of a parametric equalizer. Both the differentiator and the multiband filters were designed with the DFDP package.

Exercise 4.1 Implementation of a 60-Hz Bandstop (Notch) Filter. Design a bandstop filter with the following characteristics.

1. Window: Kaiser
2. Sampling frequency: 600 Hz
3. Center frequency: 60 Hz
4. Bandwidth: 10 Hz

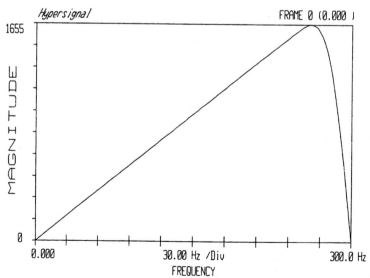

FIGURE 4.30. Frequency response of 31-coefficient differentiator

146 Finite Impulse Response Filters

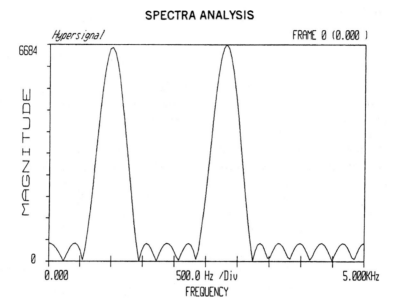

FIGURE 4.31. Frequency response of 31-coefficient multiband filter

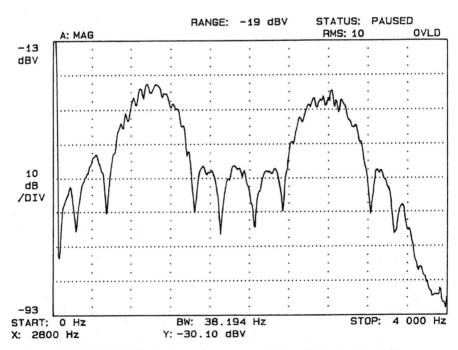

FIGURE 4.32. Real-time frequency response of 31-coefficient multiband filter

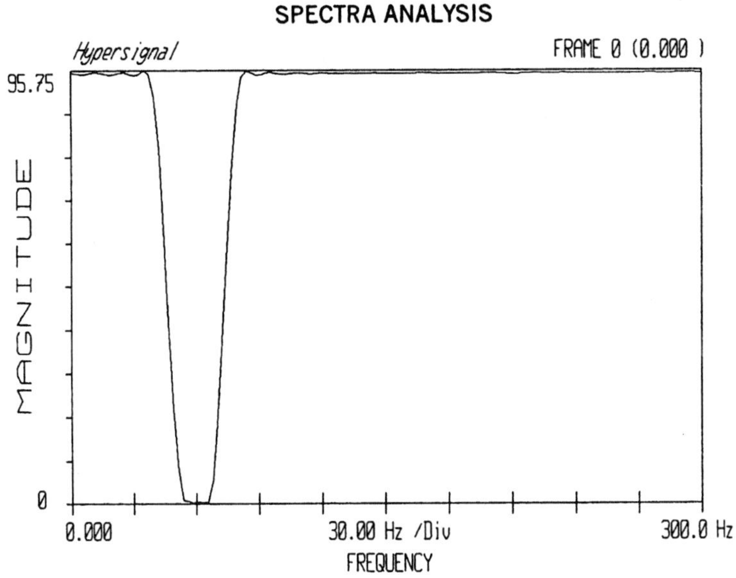

4.9 Filter Development Package and Digital Filter Design Package 147

SPECTRA ANALYSIS

FIGURE 4.33. Frequency response of 95-coefficient bandstop filter, centered at 60 Hz

5. Transition bandwidths 1 and 2: 18 Hz each (see the Hypersignal FIR filter construction menu in Figure C.10, Appendix C)
6. Stopband attenuation: −50 dB (a negative value is used for a stopband attenuation)
7. Filter length: 95 coefficients
8. Passband ripple: 10 dB

The source program BS95.ASM and the coefficient file BS95.COF generated with the Hypersignal package are on the accompanying disk.

Construct and run this filter with the simulator. Use an impulse value of 100,000. Verify that the frequency response is as shown in Figure 4.33.

Exercise 4.2 Real-Time Implementation of Bandstop Filter. Implement with the AIC the filter with the characteristics specified in Exercise 4.1 (coefficient file is on the accompanying disk as BS95.COF). A program such as BP45AR.ASM, listed in Figure 4.21, can be edited and used with a different set of coefficients. The AIC configuration data as set in AICSEC (1428h, 1h, 4A96h, 63h) is for a sampling rate of 10 kHz.

1. Verify that the frequency response is as shown in Figure 4.34. Note that the center frequency is at $F_s/10$, or 1 kHz [even though the coefficients were generated using a sampling frequency of 600 Hz and center frequency of 60 Hz ($F_s/10$)]. Exercise 4.3 illustrates the use of the AIC with a different sampling rate.

148   Finite Impulse Response Filters

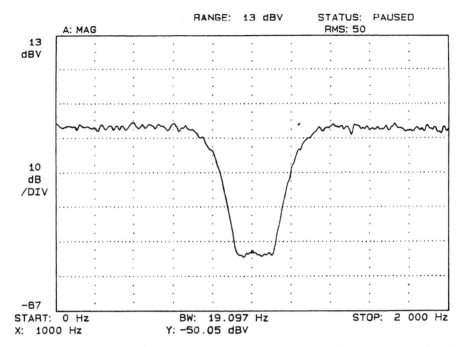

FIGURE 4.34. Real-time frequency response of bandstop filter with center frequency of 1,000 Hz

2. Note the response at low frequencies because the AIC input filter was deleted. Now insert this input filter (change the fourth value in `AICSEC` from 63h to 67h). What is the effect of the AIC input filter on the frequency response at low frequencies?

Exercise 4.3 Use of AIC Master Clock Frequency for Obtaining Different Sampling Rate. The AIC master clock frequency can be further divided in order to obtain different sampling rates. This can be achieved with the instruction

`LDI   1h, R0`

which is the ninth instruction (within the text section) in the AIC communication program `AICCOMA.ASM`, listed in Figure 3.21. This instruction can be seen as

`LDI   K, R0`

where K divides further the master clock. This effectively scales the sampling frequency by K, or F_s/K.

4.9 Filter Development Package and Digital Filter Design Package

For example, set K = 2 in the LDI instruction. Reassemble, link, and run the bandpass filter program BP45AR.ASM centered at 1 kHz (Example 4.15). Is the center frequency now at 500 Hz? Change K to 4, and observe the new center frequency.

Exercise 4.4 Real-Time Implementation of 60-Hz Bandstop Filter with Sample Rate Control. This exercise illustrates the use of the AIC configuration data to set different sampling frequencies. In order to obtain a low sampling rate of 600 Hz, it is necessary to divide the AIC master clock frequency further as in the previous example. Care also must be exercised in order to obtain values for TA and TB that can be represented by 5 or 6 bits, respectively (see also Chapter 3). Proceed with the following:

1. Choose a value for TA such as 16, or

$$TA = 16 = (10000)_b$$

2. Let K = 8 in the instruction LDI K, R0 in the AIC communication program AICCOMA.ASM (see Exercise 4.3). The switch capacitor filter (SCF) frequency is then

$$SCF = \frac{(\text{AIC master clock frequency})/K}{2 \times TA} = \frac{(7.5 \text{ MHz})/8}{2 \times TA}$$
$$= 29{,}297 \text{ Hz}$$

3. $TB = SCF/F_s = 29{,}297/600 = 49 = (110001)_b$.
4. Let TA = RA, and TB = RB.
5. The two values in the AIC configuration for a sampling frequency of 600 Hz follow from TA and RA (see Figure 3.4):

xx |←TA→| xx |←RA→| b1 b0

where xx represents "don't care" and is set to 00, or

0010 0000 0100 0000

which represents a value of 2040h. Similarly,

X |←TB→| X |←RB→| b1 b0

yields

0110 0010 1100 0110

150 Finite Impulse Response Filters

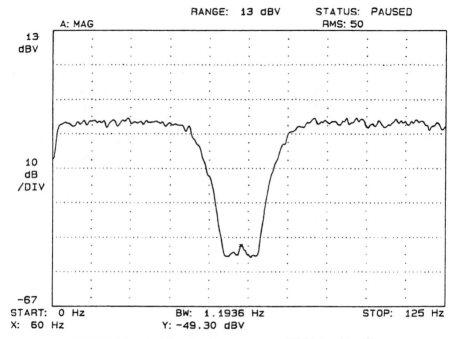

FIGURE 4.35. Real-time frequency response of 60-Hz bandstop filter

which represents a value of 62C6h. The values 2040h and 62C6h are the first and third values, respectively, to be specified in AICSEC (2040h, 1h, 62C6h, 63h). This AIC configuration data sets the AIC for a sampling rate of 600 Hz and deletes the input filter (with 63h in lieu of 67h).

6. The filter coefficients generated in Exercise 4.1 are in the file BS95.COF (on accompanying disk). Verify that the frequency response of the 60-Hz bandstop filter is as shown in Figure 4.35.

7. Make any appropriate changes in order to obtain a center frequency of 30 Hz (only *one* change is necessary).

Exercise 4.5 Real-Time Filter Implementation with Macros and Noise Generator. Implement with the AIC the bandpass filter in Example 4.17 using macros for both the filter subroutine and the noise generator.

REFERENCES

[1] T. W. Parks and J. H. McClellan, Chebychev approximation for nonrecursive digital filter with linear phase, *IEEE Trans. Circuit Theory*, **CT-19**, March 1972, pp. 189–194.

References

[2] J. H. McClellan and T. W. Parks, A unified approach to the design of optimum linear phase digital filters, *IEEE Trans. Circuit Theory*, **CT-20**, November 1973, pp. 697–701.

[3] A. H. Gray and J. D. Markel, Digital lattice and ladder filter synthesis, *IEEE Trans. Acoustics, Speech, and Signal Processing*, **ASSP-21**, December 1973, pp. 491–500.

[4] A. H. Gray and J. D. Markel, A normalized digital filter structure, *IEEE Trans. Acoustics, Speech, and Signal Processing*, **ASSP-23**, June 1975, pp. 258–277.

[5] F. J. Harris, On the use of windows for harmonic analysis with the discrete Fourier transform, *Proceedings of the IEEE*, **66**, January 1978, pp. 51–83.

[6] *Hypersignal-Plus DSP Software*, Hyperception, Inc., Dallas, Tex., 1991.

[7] *Digital Filter Design Package (DFDP)*, Atlanta Signal Processors, Inc., Atlanta, Ga., 1991.

[8] *Proceedings of the 1991 TMS320 Educators Conference*, Texas Instruments, Inc., July 1991.

[9] R. Chassaing, B. Bitler, and D. W. Horning, Real-time digital filters in C, in *Proceedings of the 1991 ASEE Annual Conference*, June 1991.

[10] *TMS320 Floating-Point DSP Assembly Language Tools User's Guide*, Texas Instruments, Inc., Dallas, Tex., 1991.

[11] W. D. Stanley, *Digital Signal Processing*, Reston, Reston, Va., 1975.

[12] A. V. Oppenheim and R. Schafer, *Discrete-Time Signal Processing*, Prentice-Hall, Englewood Cliffs, N.J., 1989.

[13] T. W. Parks and C. S. Burrus, *Digital Filter Design*, Wiley, New York, 1987.

[14] D. J. DeFatta, J. G. Lucas, and W. S. Hodgkiss, *Digital Signal Processing: A System Approach*, Wiley, New York, 1988.

[15] L. B. Jackson, *Digital Filters and Signal Processing*, Kluwer Academic, Norwell, Mass., 1986.

[16] J. G. Proakis and D. G. Manolakis, *Introduction to Signal Processing*, Macmillan, New York, 1988.

[17] L. R. Rabiner and B. Gold, *Theory and Application of Digital Signal Processing*, Prentice-Hall, Englewood Cliffs, N.J., 1975.

[18] J. F. Kaiser, Nonrecursive digital filter design using the I_0–sin h window function, in *Proceedings of the IEEE International Symposium on Circuits and Systems*, 1974.

[19] N. Ahmed and T. Natarajan, *Discrete-Time Signals and Systems*, Reston, Reston, Va., 1983.

[20] R. D. Strum and D. E. Kirk, *First Principles of Discrete Systems and Digital Signal Processing*, Addison Wesley, Reading, Mass., 1988.

[21] L. C. Ludemen, *Fundamentals of Digital Signal Processing*, Harper & Row, New York, 1986.

[22] C. S. Williams, *Designing Digital Filters*, Prentice-Hall, Englewood Cliffs, N.J., 1986.

[23] M. G. Bellanger, *Digital Filters and Signal Analysis*, Prentice-Hall, Englewood Cliffs, N.J., 1986.

[24] T. Kailath, *Modern Signal Processing*, Hemisphere, New York, 1985.

[25] R. W. Hamming, *Digital Filters*, Prentice-Hall, Englewood Cliffs, N.J., 1983.

[26] J. F. Kaiser, Some practical considerations in the realization of linear digital filters, in *Proceedings of the Third Allerton Conference on Circuit System Theory*, October 1965, pp. 621–633.

[27] R. Chassaing and D. W. Horning, *Digital Signal Processing with the TMS320C25*, Wiley, New York, 1990.

[28] P. Papamichalis (editor), *Digital Signal Processing Applications with the TMS320 Family—Theory, Algorithms, and Implementations*, Vol. 3, Texas Instruments, Inc., Dallas, Tex., 1990.

[29] R. Chassaing and P. Martin, Digital filtering with the floating-point TMS320C30 digital signal processor, in *Proceedings of the 21st Annual Pittsburgh Conference on Modeling and Simulation*, May 1990.

[30] *TMS320C30 Evaluation Module Technical Reference*, Texas Instruments, Inc., Dallas, Tex., 1990.

[31] *Digital Signal Processing Applications with the TMS320C30 Evaluation Module—Selected Applications Notes*, Texas Instruments, Inc., Dallas, Tex., 1991.

5
Infinite Impulse Response Filters

CONCEPTS AND PROCEDURES
- *IIR structures: direct form I, direct form II, direct form II transpose, cascade, parallel, and lattice forms*
- *Bilinear transformation for filter design*
- *Sinusoid generation with difference equation*
- *Programming examples using C and TMS320C30 code*
- *Use of filter design packages*

In the previous chapter, we discussed the finite impulse response (FIR) filter. This type of filter has no counterpart in analog form. In this chapter, we discuss the infinite impulse response (IIR) filter, which makes use of the vast knowledge already acquired with analog filters. The design procedure involves the conversion of an analog filter to a discrete filter using the bilinear transformation (BLT). The program BLT.BAS (on the accompanying disk) makes use of the bilinear transformation, which enables us to convert readily the transfer function of an analog filter in the s-domain into a discrete-time transfer function in the z-domain.

5.1 INTRODUCTION

Consider the general input–output equation,

$$y(n) = \sum_{k=0}^{N} a_k x(n-k) - \sum_{j=1}^{M} b_j y(n-j) \qquad (5.1)$$

$$= a_0 x(n) + a_1 x(n-1) + a_2 x(n-2) + \cdots + a_N x(n-N)$$
$$- b_1 y(n-1) - b_2 y(n-2) - \cdots - b_M y(n-M) \qquad (5.2)$$

154 Infinite Impulse Response Filters

which represents an infinite impulse response (IIR) filter. The output of the IIR filter depends on the inputs as well as past outputs. The output $y(n)$, at time n, depends not only on the current input $x(n)$, at time n, and on past inputs $x(n-1), x(n-2), \ldots, x(x-N)$, but also on past outputs $y(n-1), y(n-2), \ldots, y(n-M)$.

Assume that zero initial conditions are imposed on (5.2); taking the Z-transform of $y(n)$, we obtain

$$Y(z) = a_0 X(z) + a_1 z^{-1} X(z) + a_2 z^{-2} X(z) + \cdots + a_N z^{-N} X(z)$$
$$- b_1 z^{-1} Y(z) - b_2 z^{-2} Y(z) - \cdots - b_M z^{-M} Y(z) \qquad (5.3)$$

Let $N = M$ in (5.3); the transfer function is

$$H(z) = \frac{Y(z)}{X(z)} = \frac{a_0 + a_1 z^{-1} + a_2 z^{-2} + \cdots + a_N z^{-N}}{1 + b_1 z^{-1} + b_2 z^{-2} + \cdots + b_N z^{-N}} = \frac{N(z)}{D(z)} \qquad (5.4)$$

Multiplying and dividing (5.4) by z^N, $H(z)$ becomes

$$H(z) = \frac{a_0 z^N + a_1 z^{N-1} + a_2 z^{N-2} + \cdots + a_N}{z^N + b_1 z^{N-1} + b_2 z^{N-2} + \cdots + b_N} = C \sum_{i=1}^{N} \frac{z - z_i}{z - p_i} \qquad (5.5)$$

Equation (5.5) describes a transfer function with N zeros and N poles. In Chapter 4, we saw that in order for a system to be stable all the poles must reside inside the unit circle. Hence, for an IIR filter to be stable, the magnitude of each of its poles must be less than 1, or

1. if $|p_i| < 1$, then $h(n) \to 0$, as $n \to \infty$, yielding a stable system
2. if $|p_i| > 1$, then $h(n) \to \infty$, as $n \to \infty$, yielding an unstable system

Note that if $|p_i| = 1$, then the system is referred to as marginally stable and it yields an oscillatory response. Furthermore, multiple-order poles on the unit circle yields an unstable system. If all the coefficients b_j in (5.5) are zero, then there will be poles located only at the origin in the z-plane. With such a system, the output equation $y(n)$ becomes the convolution equation in Chapter 4, representing an FIR filter. The system would then be reduced to a nonrecursive and stable FIR filter.

Sinusoid Generation

From the results in Section 4.1, we can obtain a difference equation that can be programmed in order to generate a sine function. Let the transfer function be

$$H(z) = \frac{Y(z)}{X(z)} = \frac{Cz}{z^2 - Az - B} = \frac{Cz^{-1}}{1 - Az^{-1} - Bz^{-2}}$$

or

$$Y(z)\{1 - Az^{-1} - Bz^{-2}\} = Cz^{-1}X(z)$$

Assume zero initial conditions; using (4.15),

$$y(n) = Ay(n-1) + By(n-2) + Cx(n-1) \tag{5.6}$$

where $A = 2\cos\omega T$, $B = -1$, and $C = \sin\omega T$. Equation (5.6) is a recursive difference equation representing a digital oscillator (IIR filter). The sampling period $T = 1/F_s$, and $\omega = 2\pi f$. For a given sampling frequency, we can calculate A and C to generate a sine function of frequency f. We can apply an impulse at time $n = 1$, such that $x(0) = 1$ and $x(n) = 0$, for $n \neq 0$, and can program this difference equation. Examples 5.6–5.8 implement this digital oscillator using both C and TMS320C30 code.

5.2 IIR FILTER STRUCTURES

We will now discuss several structures that can be used to represent IIR filters.

Direct Form I Structure

Figure 5.1 shows the direct form I structure that can be used to realize the IIR filter in (5.2). For an Nth-order filter, this structure has $2N$ delays

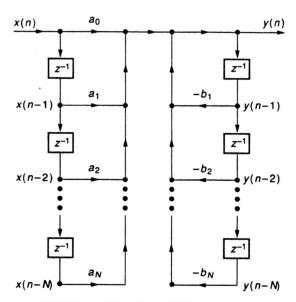

FIGURE 5.1. Direct form I IIR filter structure

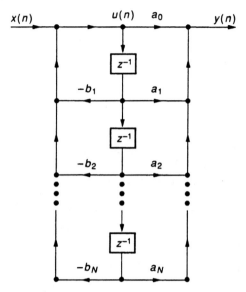

FIGURE 5.2. Direct form II IIR filter structure

represented by z^{-1}. For example, for a second-order filter with $N = 2$, there are four delay elements.

Direct Form II Structure

The direct form II structure (Figure 5.2) is one of the most commonly used structures. It requires half as many delay elements as the direct form I. For example, for a second-order filter, two delay elements z^{-1} are required, as opposed to four with the direct form I. We will now show that (5.2) can be realized with the direct form II. Let the variable $U(z)$ be defined as

$$U(z) = \frac{X(z)}{D(z)} \tag{5.7}$$

where $D(z)$ is the denominator polynomial of the transfer function in (5.4). We can then obtain $Y(z)$ from (5.4) as

$$\begin{aligned} Y(z) &= \frac{N(z)X(z)}{D(z)} = U(z)N(z) \\ &= U(z)\{a_0 + a_1 z^{-1} + a_2 z^{-2} + \cdots + a_N z^{-N}\} \end{aligned} \tag{5.8}$$

5.2 IIR Filter Structures

From (5.7),

$$X(z) = U(z)D(z) = U(z)\{1 + b_1 z^{-1} + b_2 z^{-2} + \cdots + b_N z^{-N}\} \quad (5.9)$$

Taking the inverse Z-transform (IZT) of (5.9),

$$x(n) = u(n) + b_1 u(n-1) + b_2 u(n-2) + \cdots + b_N u(n-N) \quad (5.10)$$

Solving for $u(n)$,

$$u(n) = x(n) - b_1 u(n-1) - b_2 u(n-2) - \cdots - b_N u(n-N) \quad (5.11)$$

The IZT of (5.8) yields

$$y(n) = a_0 u(n) + a_1 u(n-1) + a_2 u(n-2) + \cdots + a_N u(n-N) \quad (5.12)$$

The direct form II IIR structure can be represented by (5.11) and (5.12). Note $u(n)$ at the middle top of Figure 5.2, which satisfies (5.11). The output $y(n)$ in Figure 5.2 satisfies (5.12).

When implementing the filter, we initially set $u(n-1), u(n-2), \ldots$ to zero. After taking a new sample $x(n)$ at time n, we can then solve for $u(n)$. The filter's output at time n (the first output value) becomes

$$y(n) = a_0 u(n) + 0$$

We then update the delay variables in (5.11). At time $n+1$,

$$u(n+1) = x(n+1) - b_1 u(n) - 0$$

where $x(n+1)$ is a newly acquired sample at time $n+1$, and the delay variable $u(n-1)$ is updated to $u(n)$. Equation (5.12) is again used to solve for the output at time $n+1$, or

$$y(n+1) = a_0 u(n+1) + a_1 u(n) + 0$$

and so on. We will see later in this chapter how these two equations are used to program an IIR filter in C and in TMS320C30 code.

Direct Form II Transpose Structure

A modified version of the direct form II structure is its transpose, requiring the same number of delay elements. A second-order direct form II structure has an equivalent transposed structure, as shown in Figure 5.3, obtained from

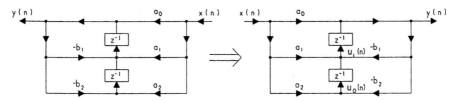

FIGURE 5.3. Direct form II transpose IIR filter structure

the direct form II structure using the following steps:

1. Reverse the directions of all the branches.
2. Reverse the roles of the input and output (input ↔ output).
3. Redraw the structure such that the input node is on the left and the output node is on the right (as is familiar to us).

The transposed structure in Figure 5.3 can be represented by (5.2). To verify this, let $u_0(n)$ and $u_1(n)$ be as shown. The following equations can be obtained using Figure 5.3:

$$u_0(n) = a_2 x(n) - b_2 y(n) \tag{5.13}$$
$$u_1(n) = a_1 x(n) - b_1 y(n) + u_0(n-1) \tag{5.14}$$
$$y(n) = a_0 x(n) + u_1(n-1) \tag{5.15}$$

Using (5.13) to find $u_0(n-1)$, equation (5.14) becomes

$$u_1(n) = a_1 x(n) - b_1 y(n) + [a_2 x(n-1) - b_2 y(n-1)] \tag{5.16}$$

Using (5.16) to solve for $u_1(n-1)$, $y(n)$ in (5.15) becomes

$$y(n) = a_0 x(n)$$
$$+ \{a_1 x(n-1) - b_1 y(n-1) + a_2 x(n-2) - b_2 y(n-2)\} \tag{5.17}$$

which is the same general input–output equation as in (5.2) for a second-order system. This structure is such that it implements the zeros and then the poles, whereas the direct form II structure implements the poles first. The design of an IIR filter using the direct form II (or its transpose) with the Hypersignal-Plus DSP Software, is covered later.

Cascade Structure

The transfer function in (5.5) can be factored as

$$H(z) = CH_1(z) H_2(z) \cdots H_r(z) \tag{5.18}$$

in terms of first- or second-order transfer functions. Figure 5.4 shows the

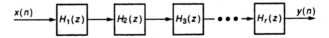

FIGURE 5.4. Cascade form IIR filter structure

5.2 IIR Filter Structures

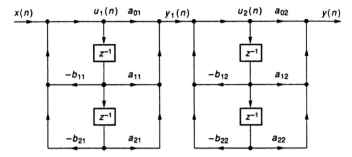

FIGURE 5.5. Fourth-order cascade IIR filter structure with two direct form II sections

cascade form structure. From a practical point of view, the overall transfer function can be represented with cascaded (in series) second-order transfer functions. For each individual section, the direct form II structure or its transpose can be used. In terms of cascaded second-order transfer functions, $H(z)$ can be written as

$$H(z) = \prod_{i=1}^{N/2} \frac{a_{0i} + a_{1i}z^{-1} + a_{2i}z^{-2}}{1 + b_{1i}z^{-1} + b_{2i}z^{-2}} \tag{5.19}$$

with the constant C in (5.18) incorporated into the coefficients. For example, for a fourth-order transfer function, with $N = 4$, (5.19) becomes

$$H(z) = \frac{(a_{01} + a_{11}z^{-1} + a_{21}z^{-2})(a_{02} + a_{12}z^{-1} + a_{22}z^{-2})}{(1 + b_{11}z^{-1} + b_{21}z^{-2})(1 + b_{12}z^{-1} + b_{22}z^{-2})} \tag{5.20}$$

Figure 5.5 shows a fourth-order IIR cascade structure in terms of two direct form II second-order sections. From a mathematical standpoint, the proper ordering of the numerator and denominator factors does not affect the output result. However, from a practical standpoint, proper ordering of each second-order section can minimize quantization noise [1–5]. Note that the output of the first section, $y_1(n)$, becomes the input to the second section. With an intermediate output result stored in one of the TMS320C30 40-bit-wide extended-precision register, the premature truncation of the intermediate output result becomes negligible. We will later implement a sixth-order IIR bandpass filter with three second-order direct form II sections in cascade.

Parallel Form Structure

The transfer function in (5.5) can be represented as

$$H(z) = C + H_1(z) + H_2(z) + \cdots + H_r(z) \tag{5.21}$$

160 Infinite Impulse Response Filters

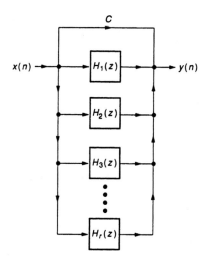

FIGURE 5.6. Parallel form IIR filter structure

which can be obtained using a partial fraction expansion (PFE) on (5.5). Each of the resulting transfer functions shown in Figure 5.6 can be either first- or second-order functions. As with the cascade structure, each section can be a direct form II structure or its transpose. The parallel form structure can be efficiently represented in terms of second-order sections. $H(z)$ can be expressed as

$$H(z) = C + \sum_{i=1}^{N/2} \frac{a_{0i} + a_{1i}z^{-1} + a_{2i}z^{-2}}{1 + b_{1i}z^{-1} + b_{2i}z^{-2}} \qquad (5.22)$$

For example, for a fourth-order transfer function, (5.22) becomes

$$H(z) = C + \frac{a_{01} + a_{11}z^{-1} + a_{21}z^{-2}}{1 + b_{11}z^{-1} + b_{21}z^{-2}} + \frac{a_{02} + a_{12}z^{-1} + a_{22}z^{-2}}{1 + b_{12}z^{-1} + b_{22}z^{-2}} \qquad (5.23)$$

The fourth-order IIR parallel structure, represented in terms of two direct form II sections, is shown in Figure 5.7. Note that, from Figure 5.7, the overall output $y(n)$ can be expressed in terms of the output of each section, or

$$y(n) = Cx(n) + \sum_{i=1}^{N/2} y_i(n) \qquad (5.24)$$

Lattice Structure

Another class of structures is the lattice structure—used in various applications, such as adaptive filtering and speech processing.

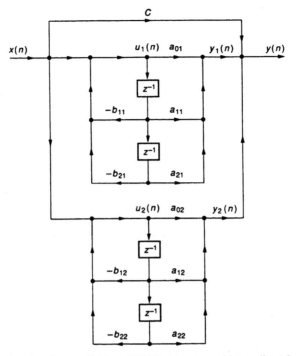

FIGURE 5.7. Fourth-order parallel form IIR filter structure with two direct form II sections

All-Pole Lattice Structure

We discussed the lattice structure in the previous chapter, where we derived the k-parameters for an FIR or "all-zero" filter (except for poles at $z = 0$). Consider now an all-pole lattice structure associated with an IIR filter. This system is the inverse of the all-zero FIR lattice of Figure 4.3, with N poles (except for zeros at $z = 0$). A solution for this system can be developed from the results obtained with the FIR lattice structure. We can solve equations (4.29) and (4.30) backwards, computing $y_{i-1}(n)$ in terms of $y_i(n)$, and so on. For example, (4.29) becomes

$$y_{i-1}(n) = y_i(n) - k_i e_{i-1}(n-1) \tag{5.25}$$

and (4.30) is repeated here as

$$e_i(n) = k_i y_{i-1}(n) + e_{i-1}(n-1) \tag{5.26}$$

Equations (5.25) and (5.26) are represented by the ith section lattice structure in Figure 5.8, which can be extended for a higher-order all-pole IIR lattice structure. For example, given the transfer function of an IIR filter with all poles, the reciprocal would be the transfer function of an FIR filter with

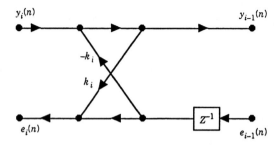

FIGURE 5.8. All-pole IIR lattice filter structure for ith section

all zeros. We want to make sure also that this IIR system is stable, by having all the poles inside the unit circle. It can be shown that this is so if $|k_i| < 1$, $i = 1, 2, \ldots, N$. Therefore, we can test the stability of a system by using the recursive equation (4.46) to find the k-parameters and check that each $|k_i| < 1$.

Example 5.1 *All-Pole Lattice Structure.* The lattice structure for an all-pole system can be found. Let the transfer function be

$$H(z) = \frac{1}{1 + 0.2z^{-1} - 0.5z^{-2}} \tag{5.27}$$

This transfer function is the inverse of the transfer function for the all-zero FIR lattice structure in Example 4.3, where the k-parameters were found to be

$$k_1 = 0.4$$
$$k_2 = -0.5$$

Figure 5.9 shows the IIR lattice structure for this example, extending Figure 5.8 to a two-stage structure.

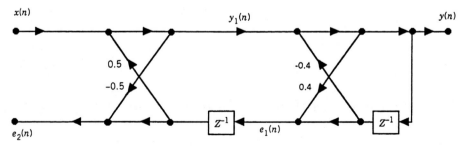

FIGURE 5.9. All-pole IIR lattice filter structure with two sections

5.2 IIR Filter Structures

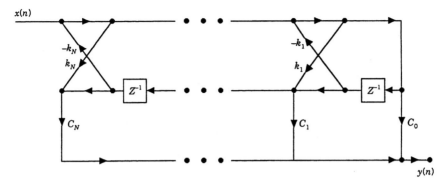

FIGURE 5.10. Nth-order IIR lattice filter structure with both poles and zeros

IIR Lattice Structure with Poles and Zeros

For an IIR lattice structure with poles and zeros, the previous results for all-zero and all-pole structures can be used. The notation used for the coefficients must be changed to reflect both the numerator and denominator polynomials in an IIR system. Figure 5.10 shows the IIR lattice structure with both poles and zeros. It shows a ladder (bottom half) portion added to the all-pole structure. A set of coefficients c_i, expressed in terms of both the numerator (a_i) and denominator (b_i) coefficients, can be computed recursively,

$$c_i = a_i - \sum_{r=i+1}^{N} c_r b_{r(r-i)}, \quad i = 0, 1, \ldots, N \tag{5.28}$$

A more thorough discussion can be found in [6] and [7].

***Example 5.2** Lattice Structure with Poles and Zeros.* This example converts a third-order IIR direct form II structure into a lattice structure. Figure 5.11 shows a third-order IIR filter using the direct form II, and Figure 5.12 shows the equivalent IIR lattice structure. The transfer function from Figure 5.11 is

$$H(z) = \frac{1 + 1.5z^{-1} - 2z^{-2} + z^{-3}}{1 - 0.5z^{-1} + 0.2z^{-2} - 0.1z^{-3}} \tag{5.29}$$

Using the results associated with an all-pole structure, and changing the coefficients a_i into b_i to reflect the denominator polynomial, (4.42) becomes

$$Y_3(z) = 1 + b_{31}z^{-1} + b_{32}z^{-2} + b_{33}z^{-3} = 1 - 0.5z^{-1} + 0.2z^{-2} - 0.1z^{-3}$$

Starting with $r = 3$, we have

$$k_3 = b_{33} = -0.1$$

164   Infinite Impulse Response Filters

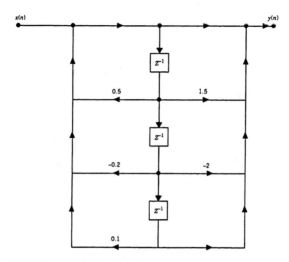

FIGURE 5.11. Third-order IIR direct form II filter structure

Using (4.46), with $r = 3$ and $i = 0$, we have

$$b_{20} = \frac{b_{30} - k_3 b_{33}}{1 - k_3^2} = \frac{1 - (-0.1)(-0.1)}{1 - (-0.1)^2} = 1$$

For $r = 3$ and $i = 1$, we have

$$b_{21} = \frac{b_{31} - k_3 b_{32}}{1 - k_3^2} = \frac{(-0.5) - (-0.1)(0.2)}{1 - (-0.1)^2} = -0.0303$$

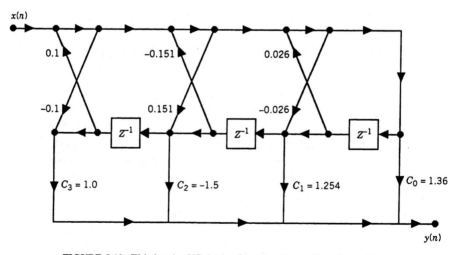

FIGURE 5.12. Third-order IIR lattice filter structure with poles and zeros

and, for $i = 2$,

$$b_{22} = \frac{b_{32} - k_3 b_{31}}{1 - k_3^2} = \frac{(0.2) - (-0.1)(-0.5)}{1 - (-0.1)^2} = 0.1515$$

from which

$$k_2 = b_{22} = 0.1515$$

From (4.42), with $r = 2$ and $i = 1$,

$$Y_2(z) = 1 + b_{21}z^{-1} + b_{22}z^{-2} = 1 + (-0.0303)z^{-1} + (0.1515)z^{-2}$$

From (4.46), with $r = 2$ and $i = 1$,

$$b_{11} = \frac{b_{21} - k_2 b_{21}}{1 - k_2^2} = \frac{(-0.0303) - (0.1515)(-0.0303)}{1 - (0.1515)^2} = -0.0263$$

from which

$$k_1 = b_{11} = -0.0263$$

The k-parameters k_1, k_2, and k_3 provide the solution for the top half of the IIR lattice structure in Figure 5.12. We can now use the recursive relationship in (5.28) to compute the c_i coefficients that will give us the bottom part of the structure in Figure 5.12. We will now use both a's and b's in applying (4.46). Here, from the numerator polynomial (with a_{ri} replaced by a_i) in (5.29),

$$a_0 = 1$$
$$a_1 = 1.5$$
$$a_2 = -2$$
$$a_3 = 1$$

and, from the denominator polynomial in (5.29),

$$b_{31} = -0.5$$
$$b_{32} = 0.2$$
$$b_{33} = -0.1$$

Starting with c_3 and working backwards using (5.28), the coefficients c_i can be found, or

$$c_3 = a_3 = 1$$
$$c_2 = a_2 - \{c_3 b_{31}\} = -2 - 1(-0.5) = -1.5$$
$$\begin{aligned}c_1 &= a_1 - \{c_2 b_{21} + c_3 b_{32}\} \\ &= 1.5 - \{(-1.5)(-0.0303) + (1)(0.2)\} \\ &= 1.2545\end{aligned}$$
$$\begin{aligned}c_0 &= a_0 - \{c_1 b_{11} + c_2 b_{22} + c_3 b_{33}\} \\ &= 1 - \{(1.2545)(-0.0263) + (-1.5)(0.1515) + (1)(-0.1)\} \\ &= 1.3602\end{aligned}$$

The lattice structure can be quite useful for applications in adaptive filtering and speech processing. Although this structure is not as computational efficient as the direct or cascade forms, requiring more multiplication operations, it is less sensitive to quantization effects.

5.3 BILINEAR TRANSFORMATION

The bilinear transformation (BLT) is the most commonly used technique for transforming an analog filter into a discrete filter. It is a one-to-one mapping from the s-plane to the z-plane, using

$$s = \frac{z-1}{z+1} \qquad (5.30)$$

Solving for z,

$$z = \frac{1+s}{1-s} \qquad (5.31)$$

This transformation is such that the following hold:

1. The left region in the s-plane, corresponding to $\sigma < 0$, maps *inside* the unit circle in the z-plane.
2. The right region in the s-plane, corresponding to $\sigma > 0$, maps *outside* the unit circle in the z-plane.
3. The imaginary, or $j\omega$, axis in the s-plane maps *on* the unit circle in the z-plane.

Consider the mapping from the $j\omega$ axis in the s-plane to the unit circle in the z-plane. Let ω_A and ω_D represent the analog and digital frequencies,

5.3 Bilinear Transformation

respectively. With $s = j\omega_A$ and $z = e^{j\omega_D T}$, (5.30) becomes

$$j\omega_A = \frac{e^{j\omega_D T} - 1}{e^{j\omega_D T} + 1} = \frac{e^{j\omega_D T/2}\{e^{j\omega_D T/2} - e^{-j\omega_D T/2}\}}{e^{j\omega_D T/2}\{e^{j\omega_D T/2} + e^{-j\omega_D T/2}\}} \qquad (5.32)$$

Using Euler's expressions for sine and cosine in terms of complex exponential functions, we can solve for ω_A in (5.32), or

$$\omega_A = \tan\frac{\omega_D T}{2} \qquad (5.33)$$

relating the analog frequency ω_A to the digital frequency ω_D. Figure 5.13 is a plot of this relationship for positive values of ω_A and ω_D. The region corresponding to ω_A between 0 and 1 is mapped into the region corresponding to ω_D between 0 and $\omega_s/4$, in a fairly linear fashion, where ω_s is the sampling frequency. However, the entire region of $\omega_A > 1$ is mapped into the region corresponding to ω_D between $\omega_s/4$ and $\omega_s/2$, showing much nonlinearity within this region. This compression within this region is referred to as *frequency warping*. Prewarping is done in order to compensate for this frequency warping. The frequencies ω_A and ω_D are such that

$$H(s)\Big|_{s=j\omega_A} = H(z)\Big|_{z=e^{j\omega_D T}} \qquad (5.34)$$

Note that the shape of the magnitude and phase responses of the transfer function is preserved.

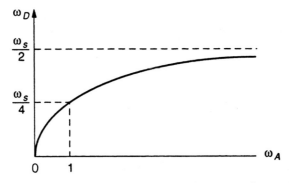

FIGURE 5.13. Relationship between analog and digital frequencies

Design Procedures

The transfer function $H(z)$ can be obtained from a known transfer function $H(s)$ using the following procedure.

1. Obtain a known analog transfer function $H(s)$.
2. Use (5.33) to solve for the analog frequency ω_A.
3. Scale the frequency of the analog transfer function $H(s)$, using

$$H\left(\frac{s}{\omega_A}\right) = H(s)\bigg|_{s=s/\omega_A}$$

4. Use the bilinear transformation equation (5.30) and the scaled transfer function in the previous step to obtain $H(z)$, or

$$H(z) = H\left(\frac{s}{\omega_A}\right)\bigg|_{s=(z-1)/(z+1)}$$

Note that this procedure makes use of a known analog transfer function (step 1) for the design of a discrete-time filter. In the case of bandpass and bandstop filters with lower and upper cutoff frequencies ω_{D1} and ω_{D2}, we will need to solve for two analog frequencies ω_{A1} and ω_{A2}. We will now illustrate the BLT with three examples.

Example 5.3 *First-Order Butterworth Highpass Filter.* Given a first-order highpass analog transfer function $H(s)$, a corresponding discrete-time filter with transfer function $H(z)$ can be obtained. Let the bandwidth or cutoff frequency be 1 radian per second, and let the sampling frequency be $F_s = 5$ Hz.

SOLUTION

1. A first-order highpass filter has the transfer function

$$H(s) = \frac{s}{s+1}$$

where the digital frequency $\omega_D = 1$ radian per second.

2. Use (5.33) to solve for the analog frequency ω_A, prewarping ω_D,

$$\omega_A = \tan\frac{\omega_D T}{2} = \tan\frac{1}{2\times 5} \approx \frac{1}{10}$$

3. Scale $H(s)$ to obtain

$$H\left(\frac{s}{\omega_A}\right) = \frac{(s/\omega_A)}{(s/\omega_A)+1} = \frac{10s}{10s+1}$$

4. The discrete-time transfer function $H(z)$ can now be obtained using the BLT equation in (5.30), or

$$H(z) = H\left(\frac{s}{\omega_A}\right)\bigg|_{s=(z-1)/(z+1)}$$

$$= \frac{10(z-1)/(z+1)}{[10(z-1)/(z+1)]+1} = \frac{10(z-1)}{11z-9}$$

Example 5.4 *Second-Order Butterworth Bandstop Filter.* Given a second-order analog transfer function $H(s)$ for a bandstop filter, a corresponding discrete-time bandstop filter with transfer function $H(z)$ can be found. Let the sampling frequency as well as the lower and upper cutoff frequencies ω_{D1} and ω_{D2} be given.

SOLUTION

1. A second-order bandstop analog filter can be represented by the transfer function

$$H(s) = \frac{s^2 + \omega_r^2}{s^2 + sB + \omega_r^2} \qquad (5.35)$$

where the bandwidth $B = \omega_{A2} - \omega_{A1}$ and the notch, or center, frequency is $\omega_r = \sqrt{\omega_{A1}\omega_{A2}}$. This transfer function can be obtained from the first-order highpass filter $H(s)$ in the previous example, using the transformation $s = (s^2 + \omega_r^2)/sB$.

2. The analog frequencies are

$$\omega_{A1} = \tan\frac{\omega_{D1}T}{2}$$

$$\omega_{A2} = \tan\frac{\omega_{D2}T}{2}$$

3. No scaling on the transfer function is necessary because it is taken care of with the preceding transformation for obtaining the second-order bandstop transfer function $H(s)$ from the first-order highpass $H(s)$. Using ω_{A1} and ω_{A2} in the previous step, the bandwidth B and center frequency ω_r can be found and substituted into $H(s)$ in (5.35).

4. The transfer function $H(z)$ can now be found using

$$H(z) = H(s)\bigg|_{s=(z-1)/(z+1)}$$

170 Infinite Impulse Response Filters

The BLT equation can now be applied to transform $H(s)$ into an equivalent $H(z)$. This transformation can be performed readily using a computer program, as shown in the next section.

5.4 UTILITY PROGRAMS FOR BLT AND MAGNITUDE AND PHASE RESPONSES

The program BLT.BAS (on the accompanying disk) in BASIC allows for the calculation of $H(z)$ from $H(s)$ using the BLT equation $s = (z-1)/(z+1)$. The following example illustrates the use of this program.

Example 5.5 *Second-Order Bandstop Filter Using the Computer Program* BLT.BAS. A second-order bandstop filter is to be designed such that it has lower and upper cutoff frequencies of 950 and 1,050 Hz, respectively. The sampling frequency is 5,000 Hz.

SOLUTION The analog frequencies are

$$\omega_{A1} = \tan\frac{\omega_{D1}T}{2} = \tan\frac{2\pi \times 950}{2 \times 5000} = 0.6796$$

$$\omega_{A2} = \tan\frac{\omega_{D2}T}{2} = \tan\frac{2\pi \times 1050}{2 \times 5000} = 0.7756$$

from which $B = \omega_{A2} - \omega_{A1} = 0.096$, and $\omega_r^2 = (\omega_{A1})(\omega_{A2}) = 0.5271$.
The transfer function is then

$$H(s) = \frac{s^2 + \omega_r^2}{s^2 + sB + \omega_r^2} = \frac{s^2 + 0.5271}{s^2 + 0.096s + 0.5271}$$

The corresponding transfer function $H(z)$ can be obtained by replacing s with $(z-1)/(z+1)$, or

$$H(z) = \frac{\{(z-1)/(z+1)\}^2 + 0.5271}{\{(z-1)/(z+1)\}^2 + 0.096(z-1)/(z+1) + 0.5271}$$

which can be reduced to

$$H(z) = \frac{0.9408 - 0.5827z^{-1} + 0.9408z^{-2}}{1 - 0.5827z^{-1} + 0.8817z^{-2}} \qquad (5.36)$$

These results can be verified using the program BLT.BAS, which calculates $H(z)$ from $H(s)$ using $s = (z-1)/(z+1)$ and which can be quite useful in applying this procedure for higher-order filters.

5.4 Utility Programs for BLT and Magnitude and Phase Responses 171

```
*** Mapping S to Z Plane Using Bilinear Transformation ***


Enter the # of numerator coefficients (30 = Max, 0 = Exit) --> 3
    Enter a(0)s^2  --> 1
    Enter a(1)s^1  --> 0
    Enter a(2)s^0  --> .5271

Enter the # of denominator coefficients --> 3
    Enter b(0)s^2  --> 1
    Enter b(1)s^1  --> .096
    Enter b(2)s^0  --> .5271

Are the above coefficients correct ? (y/n) Y
```

FIGURE 5.14. Coefficients of $H(s)$ for second-order bandstop filter

Conversion of $H(s)$ into $H(z)$ Using Utility Program BLT.BAS

To verify (5.36), run BASIC and load and run the program BLT.BAS. The prompts and the associated data representing $H(s)$ are shown in Figure 5.14. The resulting coefficients of $H(z)$ using the program BLT.BAS are then obtained and are identical to the coefficients in (5.36).

Magnitude and Phase Responses Using Utility Programs AMPLIT.CPP and MAGPHSE.BAS

The magnitude and phase responses of a transfer function $H(z)$ can be obtained using the computer programs AMPLIT.CPP (in C) or MAGPHSE.BAS (in BASIC).

1. AMPLIT.CPP Compile the utility program AMPLIT.CPP (on accompanying disk) using the Turbo C++ compiler and run it. This utility program can be used to find and plot the magnitude and phase responses of a filter with a maximum order of 10. Enter the coefficients of $H(z)$ in (5.36), as shown in Figure 5.15 (use the TAB key to enter the denominator coefficients). Press F10 to obtain the plot of the magnitude of $H(z)$ as shown in Figure 5.16. Press ENTER to plot the phase response as shown in Figure 5.17.

2. MAGPHSE.BAS Another utility program, MAGPHSE.BAS, in BASIC can also be used to find the tabulated magnitude and phase responses. Two resulting output files can be obtained, one containing the frequency and amplitude values, the other containing the frequency and phase values. Run BASIC and load and run the program MAGPHSE.BAS. Figure 5.18 shows the prompts and the coefficients of $H(z)$ to be entered. The resulting file containing the frequency and the associated magnitude is shown in Figure 5.19. Similar tabulated results can be obtained for the phase response versus

Infinite Impulse Response Filters

FILTER COEFFICIENTS	
NUMERATOR	DENOMINATOR
z-0 .9408	z-0 1
z-1 -.5827	z-1 -.5827
z-2 .9408	z-2 .8817
z-3	z-3
z-4	z-4
z-5	z-5
z-6	z-6
z-7	z-7
z-8	z-8
z-9	z-9
z-10	z-10
F1 HELP F5 Quit F10 PLOT	

FIGURE 5.15. Coefficients of $H(z)$ for second-order bandstop filter using AMPLIT.CPP

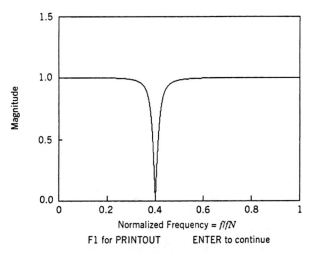

FIGURE 5.16. Magnitude response of $H(z)$ using AMPLIT.CPP

5.4 Utility Programs for BLT and Magnitude and Phase Responses

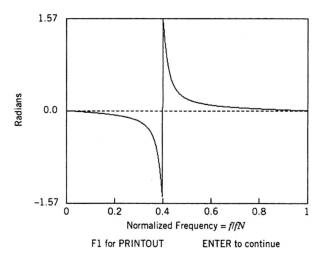

FIGURE 5.17. Phase response of $H(z)$ using AMPLIT.CPP

```
          *** Magnitude and Phase Response of a DT System ***


Enter the # of numerator coefficients (256 = Maximum) --> 3
    Enter a(0)z^-0      --> .9408
    Enter a(1)z^-1      --> -.5827
    Enter a(2)z^-2      --> .9408

Enter the # of denominator coefficients --> 3
    Enter b(0)z^-0      --> 1
    Enter b(1)z^-1      --> -.5827
    Enter b(2)z^-2      --> .8817

Enter the Sampling Frequency in Hz --> 5000

Enter the number of steps desired --> 11

Tabulate (M)agnitude or (P)hase response --> M

Do you want to normalize the magnitude (Y/N) --> Y

Do you want (P)ower or (A)mplitude form   --> A

Are the above entries correct (Y/N) Y
```

FIGURE 5.18. Coefficients to find the magnitude response of second-order bandstop filter using MAGPHSE.BAS

```
              *** AMPLITUDE Response of a DT System ***

                 FREQUENCY                    AMPLITUDE
              ----------------------------------------------
                     0.00                      0.99996
                   250.00                      0.99950
                   500.00                      0.99723
                   750.00                      0.98362
                  1000.00                      0.01114
                  1250.00                      0.98005
                  1500.00                      0.99538
                  1750.00                      0.99840
                  2000.00                      0.99946
                  2250.00                      0.99988
                  2500.00                      1.00000

       Press <Enter> to return to Output Path Menu
```

FIGURE 5.19. Tabulated magnitude response using MAGPHSE.BAS

frequency. These resulting files can be used in conjunction with a plot program (such as PCPLOT) in order to plot the magnitude and phase responses of a filter.

5.5 PROGRAMMING EXAMPLES USING C AND TMS320C30 CODE

We will discuss several examples using C and TMS320C30 code with both simulation and real-time environments. These examples include the implementation of a digital oscillator using a recursive difference equation, and an IIR filter. The Hypersignal package contains a sine generator utility as well as a difference equation utility for generating sine and cosine.

Example 5.6 *Sine Generation Using C Code.* This programming example makes use of Section 4.1, associated with the Z-transform and difference equations, and Section 5.1. It was shown that a sinusoidal function can be represented by a recursive difference equation with poles located *on* the unit circle in the z-plane. The difference equation (5.6) associated with a sine function is repeated here as

$$y(n) = Ay(n-1) + By(n-2) + Cx(n-1) \qquad (5.37)$$

where $A = 2\cos \omega T$, $B = -1$, and $C = \sin \omega T$. We can program this recursive difference equation and generate a sinusoid with a frequency determined by ω. Figure 5.20 shows a listing of the program, SINEC.C, which imple-

```c
/*SINEC.C-SINE GENERATION BY RECURSIVE EQUATION   */
#include <math.h>            /*math library function*/
#define SAMPLE_FREQ 10000    /*sample frequency     */
#define SINE_FREQ 1000       /*desired frequency    */
#define PI 3.14159           /*constant PI          */
volatile int *IO_OUTPUT=(volatile int *)0x804002;

void sinewave(float A,float B,float C)
{
 float Y[3] = {0.0,0.0,0.0};  /*Y[N] array          */
 float X[3] = {0.0,0.0,1.0};  /*X[N] array          */
 int N = 2, i;                /*declare variables   */
 for (i = 0; i < 100; i++)
  {
   Y[N] = A*Y[N-1] + B*Y[N-2] + C*X[N-1]; /*determine Y[N]*/
   *IO_OUTPUT = Y[N]*1000;    /*output Y[N] scaled by 1000 */
   Y[N-2] = Y[N-1];           /*shift Y's back in array    */
   Y[N-1] = Y[N];
   X[N-2] = X[N-1];           /*shift X's back in array    */
   X[N-1] = X[N];
   X[N] = 0.0;                /*set future X's to 0        */
  }
}

main()
{
 float Fs, Fosc, w, T, A, B, C;  /*declare variables       */
 Fs = SAMPLE_FREQ;               /*get sampling frequency  */
 Fosc = SINE_FREQ;               /*get oscillator frequency*/
 T = 1/Fs;                       /*determine sample period */
 w = 2*PI*Fosc;                  /*determine angular freq  */
 A = 2 * cos((w * T));           /*determine coefficient A */
 B = -1.0;                       /*coeff B is constant     */
 C = sin((w * T));               /*determine coefficient B */
    sinewave(A,B,C);             /*call sinewave function  */
}
```

FIGURE 5.20. Digital oscillator program using recursive equation in C code (SINEC.C)

ments (5.37) using C code for the generation of a sinusoid. The sampling frequency (SAMPLE_FREQ) is set at 10 kHz, and the desired frequency (SINE_FREQ) is set at 1 kHz. From the main function in the program, the coefficients A and C are calculated as follows:

$$A = 2\cos \omega T = 2\cos\left(\frac{2\pi \times 1{,}000}{10{,}000}\right)$$

$$C = \sin \omega T = \sin\left(\frac{2\pi \times 1{,}000}{10{,}000}\right)$$

This implementation uses an impulse at time $n = 1$, or $x(n - 1) = 1$ for $n = 1$, and zero otherwise, that is, $x(0) = 1$. Assume that $y(-1) = y(-2) = 0$. From (5.37),

$$\begin{aligned} n = 0; \quad & y(0) = Ay(-1) + By(-2) + Cx(-1) = 0 \\ n = 1; \quad & y(1) = Ay(0) + By(-1) + Cx(0) = C \\ n = 2; \quad & y(2) = Ay(1) + By(0) + 0 = AC \\ \vdots \quad & \vdots \end{aligned} \qquad (5.38)$$

With an impulse at time $n = 1$, the difference equation (5.37) is reduced to

$$y(n) = Ay(n - 1) + By(n - 2), \quad \text{for } n \geq 2 \qquad (5.39)$$

with $y(1) = C$ and $y(0) = 0$. The function sinewave implements this equation. Instead of starting at $n = 0$, and assume an impulse at $n = 1$, the function sinewave starts with time $n = 2$. Note that $x(0) = x(1) = 0$, and $x(2) = 1$ (set in the x array), which produces the same result as in (5.38).

A total of 100 output samples are obtained. Create an appropriate command file and run this program through the simulator to obtain the output sequence (the output file SINEC.DAT) at port address 0x804002h. Previous programming examples in Chapters 1 through 4 provide the necessary background for compiling, linking, simulating, and so on.

Appendix C shows how an output file (in hex) from the simulator can be converted into an appropriate format for plotting, using the File Acquisition option (available from the utilities menu) in the Hypersignal package. Figure 5.21 shows a plot of the sinusoidal waveform. Take the FFT (available with the Hypersignal package) of this output sequence, and verify that a delta function occurs at 1,000 Hz.

An oscillator with a different frequency can be readily obtained. Change the desired frequency to 2,000 (defined by SINE_FREQ) in the program SINEC.C. Recompile, link, and run this program. Verify and generate a sinusoidal waveform of frequency 2,000 Hz.

5.5 Programming Examples Using C and TMS320C30 Code

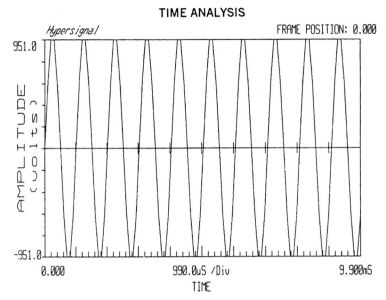

FIGURE 5.21. Time plot of sine waveform generated using a difference equation

Example 5.7 Real-Time Sine Generation Using C Code. This example discusses the real-time implementation of the previous sine-generation example using C code. Figure 5.22 shows the program, SINECR.C, for this example. The AIC data configuration is set for a sampling rate of 10 kHz. Use a command file similar to the one provided in Chapter 1 (Figure 1.5). Run this program and observe a sinusoidal waveform of frequency 1 kHz, using an oscilloscope.

Change the desired frequency SINE_FREQ to 3,000 Hz, and verify that a sinusoidal waveform of frequency 3,000 Hz is generated. Figure 5.23 shows a display (from an HP signal analyzer) of the frequency plot of the output sequence, showing a delta function at 3,000 Hz.

Example 5.8 Sine Generation Using TMS320C30 Code. This program example implements the recursive difference equation (5.37) using TMS320C30 code. This example also makes use of (5.39) for $n \geq 2$. Figure 5.24 shows a listing of the program (SINEASM.ASM) for this example. Note the following from the program:

1. A sampling rate of 10 kHz and a desired frequency of 1 kHz are used for the calculation of the coefficients A and C. Because the values of these coefficients are set in the program, they have to be recalculated for a different desired frequency. In contrast, the C version of this

```c
/*SINECR.C-REAL-TIME SINE GENERATION BY RECURSIVE EQUATION */
#include "aiccom.c"              /*AIC comm routines          */
#include <math.h>                /*math library function      */
#define SAMPLE_FREQ 10000        /*sample frequency           */
#define SINE_FREQ 1000           /*desired frequency          */
#define PI 3.14159               /*constant PI                */
int AICSEC[4] = {0x1428,0x1,0x4A96,0x67}; /*AIC config data*/

void sinewave(float A,float B,float C)
{
  float Y[3] = {0.0,0.0,0.0};    /*Y[N] array                 */
  float X[3] = {0.0,0.0,1.0};    /*X[N] array                 */
  int N = 2, result;             /*declare variables          */
  while(1)
    {
    TWAIT;
    Y[N] = A*Y[N-1] + B*Y[N-2] + C*X[N-1]; /*determine Y[N]*/
    result = (int)(Y[N]*1000);   /*output Y[N] scaled by 1000 */
    PBASE[0x48] = result << 2;   /*output to AIC              */
    Y[N-2] = Y[N-1];             /*shift Y's back in array    */
    Y[N-1] = Y[N];
    X[N-2] = X[N-1];             /*shift X's back in array    */
    X[N-1] = X[N];
    X[N] = 0.0;                  /*set future X's to 0        */
    }
}

main()
{
  float Fs, Fosc, w, T, A, B, C;  /*declare variables          */
  AICSET();                       /*initialize AIC             */
  Fs = SAMPLE_FREQ;               /*get sampling frequency     */
  Fosc = SINE_FREQ;               /*get oscillator frequency*/
  T = 1/Fs;                       /*determine sample period    */
  w = 2*PI*Fosc;                  /*determine angular freq     */
  A = 2 * cos((w * T));           /*determine coefficient A    */
  B = -1.0;                       /*coeff B is constant        */
  C = sin((w * T));               /*determine coefficient B    */
    sinewave(A,B,C);              /*call sinewave function     */
}
```

FIGURE 5.22. Real-time sine-generation program by recursive equation using C code (SINECR.C)

5.5 Programming Examples Using C and TMS320C30 Code

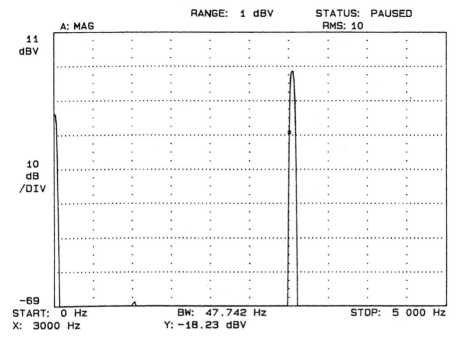

FIGURE 5.23. Real-time frequency plot of 3-kHz sinusoid

example incorporated the calculation of the coefficients within the program.

2. An output routine is set such that the output sequence is scaled by a constant in order to obtain a better plot of the output. The "scaler" constant is set to 1,000.

3. The output subroutine is called three times to output $y(0)$, $y(1)$, and $y(2)$. These output values were first stored in R1 before being scaled within the output routine.

4. For $n \geq 3$, the equation in (5.39) is implemented within a "looped" section of code. For each n, the output subroutine is called to output $y(n)$, $n = 3, 4, \ldots$.

Run the program and create an output sequence SINEASM.DAT. Note the "infinite" loop within the program. Stop the execution of this program after obtaining a reasonable number of data points. A time plot of the output sequence (using the Hypersignal package, for example) shows a sinusoidal waveform with a frequency of 1,000 Hz. Figure 5.25 shows a frequency plot of the output sequence $y(n)$, using the FFT utility in the Hypersignal package.

```
;SINEASM.ASM-SINE GENERATION   Y(N)=A*Y(N-1)+B*Y(N-2)+C*X(N-1)
;                Y(N)=A*Y(N-1)+B*Y(N-2)  ,  FOR N >= 2
          .TITLE    "SINEASM"     ;SINE GENERATOR, Fd = 1 kHz
          .GLOBAL   RESET, BEGIN  ;REF/DEF SYMBOLS
          .SECT     "VECTORS"     ;ASSEMBLE INTO VECTOR SECTION
RESET     .WORD     BEGIN         ;RESET VECTOR
          .DATA                   ;ASSEMBLE INTO DATA SECTION
STACKS    .WORD     809F00H       ;INIT STACK POINTER DATA
IO_ADDR   .WORD     804002H       ;OUTPUT ADDRESS
A         .FLOAT    1.618034      ;A = 2(COS wT), Fs=10 kHz
B         .FLOAT    -1.0          ;B = -1
Y1        .FLOAT    0.587785      ;INITIALLY Y(1)=C=SIN(WT)=.587785
Y0        .FLOAT    0.0           ;INITIALLY Y(0)=0
SCALER    .FLOAT    1000          ;SCALING FACTOR
          .TEXT                   ;ASSEMBLE INTO TEXT SECTION
BEGIN     LDP       STACKS        ;INIT DATA PAGE
          LDI       @STACKS,SP    ;SP-> 0809F00H
          LDI       @IO_ADDR,AR0  ;OUT ADDRESS -> AR0
          LDF       @Y0,R1        ;INITIALLY R1=Y(0)=0
          CALL      OUT           ;OUTPUT Y(0)
          LDF       @Y1,R1        ;INITIALLY R1=Y(1)
          CALL      OUT           ;OUTPUT Y(1)
          LDF       @A,R3         ;R3=A
          MPYF      R3,R1,R1      ;R1=A*Y1
          CALL      OUT           ;OUTPUT Y(2)=A*C
          LDF       @Y1,R0        ;R0=Y2(PREVIOUSLY Y1)DUE TO DELAY
          LDF       @B,R4         ;R4=B
;Y(N) FOR N=>3
LOOP      LDF       R1,R2         ;R2=A*Y1
          MPYF      R3,R1,R1      ;R1=A(A*Y1)
          MPYF      R4,R0,R0      ;R0=B*Y2
          ADDF      R0,R1         ;R1=OUTPUT
          CALL      OUT           ;GO TO SUB FOR OUTPUT
          LDF       R2,R0         ;R0=A*(Y1)   (FOR NEXT N)
          BR        LOOP          ;CONTINUE FOR EACH N
;OUTPUT SUBROUTINE
OUT       LDF       R1,R5         ;SAVE R1
          MPYF      @SCALER,R5    ;SCALE OUTPUT
          FIX       R5,R6         ;R6=INTEGER(R5)
          STI       R6,*AR0       ;OUTPUT @804002H
          RETS                    ;RETURN FROM SUBROUTINE
          .END                    ;END
```

FIGURE 5.24. Sine-generation program using TMS320C30 code (SINEASM.ASM)

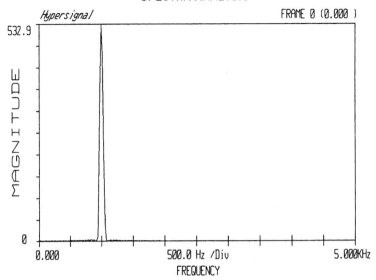

FIGURE 5.25. Frequency plot of 1-kHz sinusoid sequence

Example 5.9 Cosine Generation Using TMS320C30 Code. The difference equation associated with a cosine function is (from Chapter 4)

$$y(n) = Ay(n-1) + By(n-2) + x(n) - (A/2)x(n-1) \quad (5.40)$$

Assume an impulse such that $x(0) = 1$, and

$$y(-1) = y(-2) = 0$$

for $n=0$, $\quad y(0) = Ay(-1) + By(-2) + x(0) - (A/2)x(-1) = 1.0$
for $n=1$, $\quad y(1) = Ay(0) + By(-1) + x(1) - (A/2)x(0) = A - A/2$
for $n=2$, $\quad y(2) = Ay(1) + By(0) + 0 = A(A - A/2) + B$

$\vdots \qquad \vdots \hspace{6cm} (5.41)$

A program similar to the sine generator program `SINEASM.ASM` can be used to implement (5.40). The desired frequency and sampling frequencies are 1 and 10 kHz, respectively. Make the following changes to `SINEASM.ASM` in order to obtain the program `COSINE.ASM` which is on the accompanying disk:

1. `COPY` the program `SINEASM.ASM` as `COSINE.ASM`.
2. Set `Y1` = 0.809017, because $y(1) = A - (A/2)$, and `Y0` = 1.0, because $y(0) = x(0) = 1.0$.

182 Infinite Impulse Response Filters

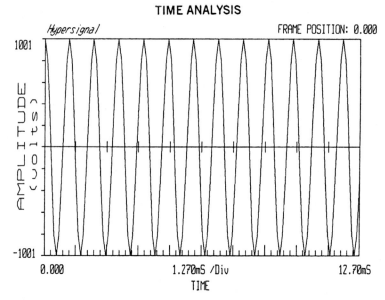

FIGURE 5.26. Time plot of 1-kHz cosine waveform generated using a difference equation

3. Before the third `CALL` instruction to output $y(2)$, add the instruction

```
ADDF    @B,R1
```

Run this `COSINE` program through the simulator to obtain an output sequence `COSINE.DAT`. The time plot of this output sequence is shown in Figure 5.26.

Example 5.10 Sixth-Order IIR Bandpass Filter Using C Code. A sixth-order IIR bandpass filter, with a center frequency of 1250 Hz, was designed with the Hypersignal package in Appendix C. Figure 5.27 shows the program `IIR6BPC.C` to implement this filter. The structure used consists of three second-order direct form II sections in cascade. For each stage, there are three **a** coefficients and two **b** coefficients. The variable `dly` used in the program represents the delay variables **u** that were utilized in Section 5.2 to describe the direct form II structure.

The IIR function implements the following equations:

$$u(n) = x(n) - b_1 u(n-1) - b_2 u(n-2) \qquad (5.42)$$

where $u(n)$ and $x(n)$ are represented in the program by `dly` and `input`, respectively, and

$$y(n) = a_0 u(n) + a_1 u(n-1) + a_2 u(n-2) \qquad (5.43)$$

```c
/*IIR6BPC.C-SIXTH-ORDER IIR BANDPASS FILTER,Fc=1250 Hz  */
#define STAGES 3              /*number of 2nd-order stages*/
float A[STAGES][3]=           {/*numerator coefficients   */
{5.3324E-02, 0.0000E+00, -5.3324E-02},
{5.3324E-02, 0.0000E+00, -5.3324E-02},
{5.3324E-02, 0.0000E+00, -5.3324E-02} };
float B[STAGES][2]=           {/*denominator coefficients */
{-1.4435E+00, 9.4879E-01},
{-1.3427E+00, 8.9514E-01},
{-1.3082E+00, 9.4377E-01} };
float DLY[STAGES][2]= {0};   /*delay samples              */
void IIR(int *IO_in, int *IO_out, int n, int len)
{
 int i, loop = 0;
 float dly, yn, input;
 while (loop < len)
 {
  ++loop;
  input = *IO_in;
  for (i = 0; i < n; i++)
  {
  dly = input-B[i][0]*DLY[i][0]-B[i][1]*DLY[i][1];
  yn  = A[i][2]*DLY[i][1]+A[i][1]*DLY[i][0]+A[i][0]*dly;
  DLY[i][1] = DLY[i][0];
  DLY[i][0] = dly;
  input = yn;
  }
  *IO_out = yn;
 }
}
main()
{
 #define length 345
 volatile int *IO_IN  = (volatile int *) 0x804000;
 volatile int *IO_OUT = (volatile int *) 0x804001;
 IIR((int *)IO_IN, (int *)IO_OUT, STAGES, length);
}
```

FIGURE 5.27. Sixth-order IIR bandpass filter program using C code (IIR6BPC.C)

as discussed earlier in this chapter. Note that the delay variables are updated after the calculation of the two previous equations. The statement

input = yn

specifies that a stage output becomes the input to the next stage. A total of

184 Infinite Impulse Response Filters

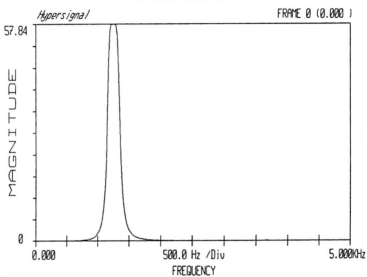

FIGURE 5.28. Frequency response of sixth-order IIR bandpass filter centered at 1,250 Hz

345 samples are taken. Port addresses 804000h and 804001h designate the input and output ports, respectively.

Run this program through the simulator, using an impulse as input (such as the file IMPULSE.DAT on the accompanying disk). The frequency response of the output is plotted in Figure 5.28.

Example 5.11 Real-Time Implementation of Sixth-Order IIR Bandpass Filter Using C Code. Figure 5.29 shows a listing of the program IIR6BPCR.C for the real-time implementation of the sixth-order IIR bandpass filter discussed in the previous example. The coefficients are included in a separate file, IIR6COEF.H, listed in Figure 5.30. A sampling frequency of 10 kHz is set in the AIC configuration data.

Run this program (link also the file VECSIR.OBJ).

Using random noise from an HP analyzer as input, the frequency response of the IIR bandpass filter, with a center frequency of 1,250 Hz, is shown in Figure 5.31.

Example 5.12 Sixth-Order IIR Bandpass Filter Using TMS320C30 Code. The program (IIR6BPA.ASM) listed in Figure 5.32 implements the sixth-order IIR bandpass filter, discussed in the two previous examples, using TMS320C30 code. For each of the three stages, the coefficients are ordered as

5.5 Programming Examples Using C and TMS320C30 Code

```c
/*IIR6BPCR.C-REAL-TIME IIR BANDPASS FILTER IN C      */
#include "aiccom.c"    /*include AIC comm routines*/
#include "iir6coef.h"  /*coefficients file          */
#define VEC_ADDR (volatile int *) 0x00

float DLY[STAGES][2] = {0}; /*delay samples          */
int AICSEC[4] = {0x1428,0x1,0x4A96,0x67}; /*AIC config data*/
int data_in, data_out;

float IIR(int *IO_in, int *IO_out, int n, int len)
{
  int i, loop = 0;
  float dly, yn, input;

  while (loop < len)
  {
    asm("     IDLE    ");
    ++loop;
    input = *IO_in;
    for (i = 0; i < n; i++)
      {
      dly = input - B[i][0] * DLY[i][0] - B[i][1] * DLY[i][1];
      yn = A[i][2] * DLY[i][1] + A[i][1] * DLY[i][0] + A[i][0] * dly;
      DLY[i][1] = DLY[i][0];
      DLY[i][0] = dly;
      input = yn;
      }
    *IO_out = yn;
  }
}

void c_int05()
{
  PBASE[0x48] = data_out << 2;
  data_in = PBASE[0x4C] << 16 >> 18;
}

main()
{
  #define length 345
  volatile int *INTVEC = VEC_ADDR;
  int *IO_OUTPUT, *IO_INPUT;
  IO_INPUT = &data_in;
  IO_OUTPUT = &data_out;
  INTVEC[5] = (volatile int) c_int05;
  AICSET_I();
  for (;;)
    IIR((int *)IO_INPUT, (int *)IO_OUTPUT, STAGES, length);
}
```

FIGURE 5.29. Real-time IIR bandpass filter program using C code (IIR6BPCR.C)

186 Infinite Impulse Response Filters

```
/*IIR6COEF.H-COEFF FOR SIXTH-ORDER IIR BANDPASS FILTER   */
#define STAGES 3              /*number of 2nd-order stages*/
float A[STAGES][3]=           {/*numerator coefficients   */
{5.3324E-02,  0.0000E+00, -5.3324E-02},
{5.3324E-02,  0.0000E+00, -5.3324E-02},
{5.3324E-02,  0.0000E+00, -5.3324E-02} };
float B[STAGES][2]=           {/*denominator coefficients */
{-1.4435E+00,  9.4879E-01},
{-1.3427E+00,  8.9514E-01},
{-1.3082E+00,  9.4377E-01} };
```

FIGURE 5.30. Coefficients file for IIR bandpass filter (IIR6COEF.H)

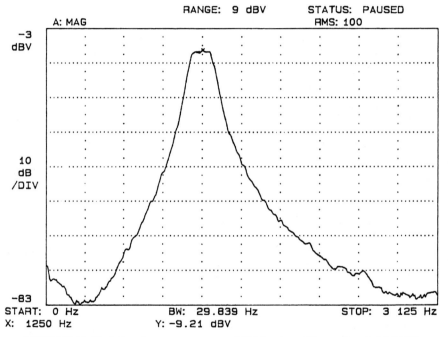

FIGURE 5.31. Real-time frequency response of IIR bandpass filter centered at 1,250 Hz

b_1, b_2, a_1, a_2, a_0. The delay variables $u(n-1)$ and $u(n-2)$, for each stage, are initially set to zero. The input and output port addresses are at 804000h and 804001h, respectively. Note the following from the program:

1. The block of code specified by RPTB up to LOOP, including the parallel instruction ADDF R2,R3, is executed three times (repeated twice).
2. After each output sample, branching back to the function IIR occurs with delay, with the instruction DBNZD AR3,IIR. AR3, which con-

5.5 Programming Examples Using C and TMS320C30 Code

```
*IIR6BPA.ASM-SIXTH-ORDER IIR BANDPASS, Fc=1250 Hz,IMPULSE LENGTH 345
        .GLOBAL  IIR                ;REF/DEF
        .TEXT                       ;ASSEMBLE INTO TEXT
BEGIN   LDI     @IO_IN,AR5          ;POINTER TO IO_INPORT
        LDI     @IO_OUT,AR6         ;POINTER TO IO_OUTPORT
        LDI     LENGTH-1,AR3        ;LENGTH OF RESPONSE
        LDI     @C_ADDR,AR0         ;AR0 POINTS TO COEFF
        LDI     @D_ADDR,AR1         ;AR1 POINTS TO DELAY
        CALL    IIR                 ;CALL SUBROUTINE IIR
END     BR      END                 ;WAIT
IIR     FLOAT   *AR5,R3             ;STAGE INPUT
        MPYF    *AR0++,*AR1++,R0    ;b[i][0]*dly[i][0]
        LDI     STAGES-1, RC        ;INITIALIZE STAGE COUNTER
        RPTB    LOOP                ;REPEAT LOOP RC TIMES
        MPYF    *AR0++,*AR1--,R1    ;b[i][1]*dly[i][1]
||      SUBF    R0,R3               ;input-b[i][0]*dly[i][0]
        MPYF    *AR0++,*AR1++,R0    ;a[i][1]*dly[i][0]
||      SUBF    R1,R3,R2  ;dly=input-b[i][0]*dly[i][0]-b[i][1]*dly[i][1]
        MPYF    *AR0++,*AR1--,R1    ;a[i][2]*dly[i][1]
        ADDF    R0,R1,R3            ;a[i][2]*dly[i][1]+a[i][1]*dly[i][0]
        LDF     *AR1,R4             ;dly[i][2]
||      STF     R2,*AR1++           ;dly[i][0] = dly
        MPYF    R2,*AR0++,R2        ;dly*a[i][0]
||      STF     R4,*AR1++           ;dly[i][1] = dly[i][0]
LOOP    MPYF    *AR0++,*AR1++,R0    ;b[i+1][0]*dly[i+1][0]
||      ADDF    R2,R3               ;STAGE OUTPUT=NEXT STAGE INPUT
        LDI     @C_ADDR,AR0         ;AR0 POINTS TO COEFF
        DBNZD   AR3,IIR             ;DELAYED BRANCH
        FIX     R3,R1               ;FLOAT OUTPUT TO INTEGER
        STI     R1,*AR6             ;OUTPUT TO IO_PORT
        LDI     @D_ADDR,AR1         ;AR1 POINTS TO DELAY
        RETS                        ;RETURN FROM SUBROUTINE
        .DATA         ;coeff:b[i][0],b[i][1],a[i][1],a[i][2],a[i][0]
COEFF   .FLOAT  -1.4435E+00,9.4880E-01,0.0000E+00,-5.3324E-02,5.3324E-02
        .FLOAT  -1.3427E+00,8.9515E-01,0.0000E+00,-5.3324E-02,5.3324E-02
        .FLOAT  -1.3082E+00,9.4378E-01,0.0000E+00,-5.3324E-02,5.3324E-02
DLY     .FLOAT  0.0,0.0,0.0,0.0,0.0,0.0
LENGTH  .SET    345                 ;LENGTH OF IMPULSE
STAGES  .SET    3                   ;NUMBER OF STAGES
C_ADDR  .WORD   COEFF               ;ADDRESS OF COEFF
D_ADDR  .WORD   DLY                 ;ADDRESS OF DELAY
IO_IN   .WORD   804000H             ;POINTER TO IO_INPORT
IO_OUT  .WORD   804001H             ;POINTER TO IO_OUTPORT
        .SECT   "VECTORS"           ;VECTOR SECTION
MAIN    .WORD   BEGIN               ;BEGIN IN VECTOR 0H
        .END                        ;END
```

FIGURE 5.32. Sixth-order IIR bandpass filter program using TMS320C30 code (IIR6BPA.ASM)

tains the number of samples minus 1, is decremented. Branching occurs as long as AR3 is not zero. The condition is first tested for branching, then AR3 is decremented. Note that the three instructions, following the DBNZD instruction, are executed before branching actually occurs.

3. AR0 points to a table containing the coefficients, the AR1 points to a table containing the sample delays as shown:

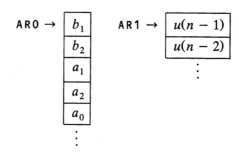

This memory organization shows the coefficients and the delay samples for the first stage only. The coefficient b_1 for the second stage would be after a_0, and so on. Similarly, $u(n-1)$ for the second stage would be after $u(n-2)$, and so on.

4. The newest input sample is stored in R3 using the FLOAT instruction. Initially AR0 points at the starting address of the coefficient table that contains b_1, and AR1 points at $u(n-1)$.

(a) The first multiply operation calculates

$$R0 = b_1 u(n-1)$$

AR0 and AR1 are then incremented to point at b_2 and $u(n-2)$, respectively.

(b) The second multiply operation calculates

$$R1 = b_2 u(n-2)$$

The subtract instruction in parallel calculates

$$R3 = x(n) - b_1 u(n-1)$$

AR0 is then incremented to point at a_1, and AR1 is decremented to point "back" at $u(n-1)$.

(c) The third multiply operation calculates

$$R0 = a_1 u(n-1)$$

5.5 Programming Examples Using C and TMS320C30 Code

The SUBF instruction in parallel yields

$$R2 = x(n) - b_1 u(n-1) - b_2 u(n-2)$$

(d) The fourth multiply operation calculates

$$R1 = a_2 u(n-2)$$

and the ADDF instruction that follows yields

$$R3 = a_1 u(n-1) + a_2 u(n-2)$$

(e) The LDF and STF instructions in parallel update the delay variable $u(n-1)$ to $u(n)$.

(f) The fifth multiply operation calculates

$$R2 = a_0 u(n)$$

The STF instruction in parallel updates $u(n-2)$ to $u(n-1)$

(g) The last multiply operation calculates

$$R0 = b_1 u(n-1)$$

for the SUBSEQUENT stage, and the addition instruction in parallel calculates the stage output

$$R3 = a_0 u(n) + R3 \text{ (obtained from step d)}$$

effectively implementing the equation

$$y(n) = a_0 u(n) + a_1 u(n-1) + a_2 u(n-2)$$

Note that after one output sample is obtained (at the third stage), AR0 and AR1 are reinitialized to point at the beginning addresses of the coefficients and delay samples, respectively. Run this program through the simulator and verify that a frequency response identical to that plotted for the C version in Figure 5.28 is obtained.

It is instructive to determine the execution time of this filter as compared to its C counterpart. Set breakpoints before and after the IIR function, and verify that the TMS320C30-code filter executes slightly faster than its C version [8].

Example 5.13 *Real-Time IIR Bandpass Filter Using TMS320C30 Code.* This example implements in real time the sixth-order IIR bandpass filter discussed previously. Figure 5.33 shows the program (IIR6BPAR.ASM) for this exam-

190 Infinite Impulse Response Filters

```
*IIR6BPAR.ASM-REAL-TIME SIXTH-ORDER IIR BANDPASS, Fc=1250 Hz
        .GLOBAL IIR,AICSET,TWAIT,AICSEC,SPSET,AICIO_P ;REF/DEF
        .TEXT                           ;ASSEMBLE INTO TEXT
BEGIN   LDI     @SPSET,AR7              ;AR7 SPECIFIES SERIAL PORT 0 OR 1
        CALL    AICSET                  ;INIT AIC
REPEAT  LDI     LENGTH-1,AR3            ;LENGTH OF RESPONSE
        LDI     @C_ADDR,AR0             ;AR0 POINTS TO COEFF
        LDI     @D_ADDR,AR1             ;AR1 POINTS TO DELAY
        CALL    IIR                     ;CALL SUBROUTINE IIR
        BR      REPEAT                  ;REPEAT
IIR     CALL    AICIO_P                 ;CALLS AIC FOR POLLING
        FLOAT   R6,R3                   ;STAGE INPUT
        MPYF    *AR0++,*AR1++,R0        ;b[i][0]*dly[i][0]
        LDI     STAGES-1, RC            ;INITIALIZE STAGE COUNTER
        RPTB    LOOP                    ;REPEAT LOOP RC TIMES
        MPYF    *AR0++,*AR1--,R1        ;b[i][1]*dly[i][1]
||      SUBF    R0,R3                   ;input-b[i][0]*dly[i][0]
        MPYF    *AR0++,*AR1++,R0        ;a[i][1]*dly[i][0]
||      SUBF    R1,R3,R2 ;dly=input-b[i][0]*dly[i][0]-b[i][1]*dly[i][1
        MPYF    *AR0++,*AR1--,R1        ;a[i][2]*dly[i][1]
        ADDF    R0,R1,R3                ;a[i][2]*dly[i][1]+a[i][1]*dly[i][0]
        LDF     *AR1,R4                 ;dly[i][2]
||      STF     R2,*AR1++               ;dly[i][0] = dly
        MPYF    R2,*AR0++,R2            ;dly*a[i][0]
||      STF     R4,*AR1++               ;dly[i][1] = dly[i][0]
LOOP    MPYF    *AR0++,*AR1++,R0        ;b[i+1][0]*dly[i+1][0]
||      ADDF    R2,R3                   ;STAGE OUTPUT=NEXT STAGE INPUT
        LDI     @C_ADDR,AR0             ;AR0 POINTS TO COEFF
        DBNZD   AR3,IIR                 ;DELAYED BRANCH
        FIX     R3,R7                   ;FLOAT OUTPUT TO INTEGER
        STI     R7,*AR6                 ;OUTPUT TO IO_PORT
        LDI     @D_ADDR,AR1             ;AR1 POINTS TO DELAY
        RETS                            ;RETURN FROM SUBROUTINE
        .DATA           ;coeff:b[i][0],b[i][1],a[i][1],a[i][2],a[i][0]
COEFF   .FLOAT  -1.4435E+00,9.4880E-01,0.0000E+00,-5.3324E-02,5.3324E-02
        .FLOAT  -1.3427E+00,8.9515E-01,0.0000E+00,-5.3324E-02,5.3324E-02
        .FLOAT  -1.3082E+00,9.4378E-01,0.0000E+00,-5.3324E-02,5.3324E-02
DLY     .FLOAT  0.0,0.0,0.0,0.0,0.0,0.0
LENGTH  .SET    345                     ;LENGTH OF IMPULSE
STAGES  .SET    3                       ;NUMBER OF STAGES
C_ADDR  .WORD   COEFF                   ;ADDRESS OF COEFF
D_ADDR  .WORD   DLY                     ;ADDRESS OF DELAY
AICSEC  .WORD   1428h,1h,4A96h,67h      ;AIC CONGIG DATA
        .SECT   "VECTORS"               ;VECTOR SECTION
MAIN    .WORD   BEGIN                   ;BEGIN IN VECTOR 0H
        .END                            ;END
```

FIGURE 5.33. Real-time IIR bandpass filter program using TMS320C30 code (IIR6BPAR.ASM)

ple. The AIC configuration data is set for a 10-kHz sampling rate. The main program calls the AIC communication routines in `AICCOMA.ASM`, discussed in Chapter 3. Serial port 0 or 1 is selected with `SPSET`. In the AIC communication program, `SPSET` is set or defaulted to zero. `SPSET` can be set to 10 in the main program in order to select serial port 1.

Run this program, linking the main program `IIR6BPAR.OBJ` with `AICCOMA.OBJ`. Verify that the same frequency response can be obtained as in Figure 5.31 with the previous C version.

Exercise 5.1 Real-Time Fourth-Order Bandpass Filter. A fourth-order Butterworth bandpass filter can be derived from a second-order lowpass filter using the transformation $s = (s^2 + \omega_r^2)/sB$, or

$$H(s) = H_{LP}(s)\Big|_{s=(s^2+\omega_r^2)/sB}$$

$$= \frac{s^2 B^2}{s^4 + (\sqrt{2}B)s^3 + (2\omega_r^2 + B^2)s^2 + (\sqrt{2}B\omega_r^2)s + \omega_r^4}$$

where $\omega_r = \omega_{A1}\omega_{A2}$ is the center frequency and $B = \omega_{A2} - \omega_{A1}$ is the bandwidth.

Let the lower and upper cutoff frequencies be 1,000 and 1,500 Hz, respectively, and let the sampling frequency be 10 kHz.

1. Show that

$$H(s) = \frac{0.034 s^2}{s^4 + 0.261 s^3 + 0.365 s^2 + 0.043 s + 0.027}$$

2. Use the program `BLT.BAS` to apply the bilinear transformation on $H(s)$. Show that

$$H(z) = \frac{0.02 - 0.04 z^{-2} + 0.02 z^{-4}}{1 - 2.55 z^{-1} + 3.202 z^{-2} - 2.036 z^{-3} + 0.641 z^{-4}}$$

3. Use the program `MAGPHSE.BAS` in order to find the magnitude of $H(z)$ versus frequency. Plot this magnitude (using a plot program) and verify that the center frequency is at 1250 Hz. Note that a magnitude plot can also be obtained using `AMPLIT.CPP` (see Figure 5.15).

4. Implement this filter using both simulation and real time, and verify the results in the previous step.

Exercise 5.2 Pass / Fail Alarm Generator. An alarm generator can be achieved using sinusoidal generation (Example 5.8). The pseudorandom

192 Infinite Impulse Response Filters

```
;ALARMR.ASM - PASS/FAIL ALARM GENERATOR USING EVM/AIC
        .TITLE    "ALARMR"             ;PASS/FAIL ALARM GEN.
        .GLOBAL   RESET,BEGIN,AICSET_I,AICSEC,AICIO_I,SPSET
        .SECT     "VECTORS"            ;ASSEMBLE INTO VECTOR SECT
RESET   .WORD     BEGIN                ;RESET VECTOR
        .SPACE    4                    ;SKIP 4 WORDS
TIMER0  .WORD     LOOP_S               ;TINTO VECTOR LOCATION @ 5H
        .SPACE    58                   ;REMAINDER OF VECTOR SECTION
        .DATA                          ;ASSEMBLE INTO DATA SECTION
AICSEC  .WORD     1428H,1H,4A96H,67H
LENGTH  .SET      8                    ;# OF REPETITIONS OF PROGRAM
TIME_P  .SET      3000                 ;LENGTH OF PASS SIGNAL
TIME_R  .SET      4                    ;# OF REPS OF FAIL SIGNAL
TIME_F  .SET      1500                 ;LENGTH OF FAIL SIGNAL
SEED    .WORD     7E521603H            ;INITIAL SEED VALUE
A1      .FLOAT    +1.618034            ;A COEFFICIENT FOR 1-kHz
A2      .FLOAT    +0.618034            ;A COEFFICIENT FOR 2-kHz
A4      .FLOAT    -1.618034            ;A COEFFICIENT FOR 4-kHz
Y1      .FLOAT    +0.5877853           ;C COEFFICIENT FOR 1-kHz
Y2      .FLOAT    +0.9510565           ;C COEFFICIENT FOR 2-kHz
Y4      .FLOAT    +0.5877853           ;C COEFFICIENT FOR 4-kHz
B       .FLOAT    -1.0                 ;B COEFFICIENT
Y0      .FLOAT    0.0                  ;INITIAL CONDITION
SCALER  .FLOAT    1000                 ;SCALING FACTOR
        .TEXT                          ;ASSEMBLE INTO TEXT SECTION
BEGIN   LDP       SPSET                ;INIT DATA PAGE
        CALL      AICSET_I             ;INIT AIC
        LDI       @SEED,R0             ;R0 = INITIAL SEED VALUE
        PUSH      R0                   ;PUSH R0 ONTO STACK
        LDI       LENGTH+1,R5          ;R5 = # OF REPS + 1
LOOP_N  SUBI      1,R5                 ;DECREMENT R5
        BZ        END                  ;END AFTER LENGTH SAMPLES
        POP       R0                   ;POP R0 FROM STACK
        LDI       R0,R4                ;PUT SEED IN R4
        LSH       -31,R4               ;MOVE BIT 31 TO LSB    =>R4
        LDI       R0,R2                ;R2=R0=SEED
        LSH       -30,R2               ;MOVE BIT 30 TO LSB    =>R2
        ADDI      R2,R4                ;ADD BITS (31+30)      =>R4
        LDI       R0,R2                ;R2=R0=SEED
        LSH       -28,R2               ;MOVE BIT 28 TO LSB    =>R2
        ADDI      R2,R4                ;ADD BITS (31+30+28)   =>R4
        LDI       R0,R2                ;R2=R0=SEED
        LSH       -17,R2               ;MOVE BIT 17 TO LSB    =>R2
        ADDI      R2,R4                ;ADD BITS(31+30+28+17)=>R4
        AND       1,R4                 ;MASK LSB OF R4
```

FIGURE 5.34. Real-time pass/fail alarm generator program (ALARMR.ASM)

5.5 Programming Examples Using C and TMS320C30 Code

```
            LSH      1,R0            ;SHIFT SEED LEFT BY 1
            OR       R4,R0           ;PUT R4 INTO LSB OF R0
            PUSH     R0              ;PUSH R0 ONTO STACK
            LDI      R4,R4           ;STORE INTEGER R4
            BNZ      LOOP_P          ;TO PASS LOOP IF # 0
            BZ       LOOP_F          ;TO FAIL LOOP IF 0
;SEQUENCE FOR PASS SIGNAL => 4 kHz
LOOP_P      PUSH     R4              ;PUSH R4 ONTO STACK
            LDI      TIME_P,R6       ;LENGTH OF PASS SIGNAL
            PUSH     R6              ;PUSH R6 ONTO STACK
            LDF      @Y0,R1          ;INITIALLY R1=Y(0)=0
            LDF      @Y4,R1          ;INITIALLY R1=Y(1)
            LDF      @A4,R3          ;R3=A
            MPYF     R3,R1,R1        ;R1=A x Y(1)
            LDF      @Y4,R0          ;R0=Y2 (PREV Y1) DUE TO DELAY
            LDF      @B,R4           ;R4=B
            BR       WAIT            ;GO TO OUTPUT ROUTINE
;SEQUENCE FOR 2-kHz FAIL SIGNAL
LOOP_F      PUSH     R4              ;PUSH R4 ONTO STACK
            LDI      TIME_R,R6       ;# OF REPS OF FAIL SIGNAL
            PUSH     R6              ;PUSH R6 ONTO STACK
LOOP_F2     LDI      TIME_F,R6       ;LENGTH OF FAIL SIGNAL
            LDF      @Y0,R1          ;INITIALLY R1=Y(0)=0
            LDF      @Y2,R1          ;INITIALLY R1=Y(1)
            LDF      @A2,R3          ;R3=A
            MPYF     R3,R1,R1        ;R1=A x Y(1)
            LDF      @Y2,R0          ;R0=Y2 (PREVY1) DUE TO DELAY
            LDF      @B,R4           ;R4=B
            BR       WAIT            ;GO TO OUTPUT ROUTINE
;SEQUENCE FOR 1-kHz FAIL SIGNAL
LOOP_F1     LDI      TIME_F,R6       ;LENGTH OF FAIL SIGNAL
            LDF      @Y0,R1          ;INITIALLY R1=Y(0)=0
            LDF      @Y1,R1          ;INITIALLY R1=Y(1)
            LDF      @A1,R3          ;R3=A
            MPYF     R3,R1,R1        ;R1=A x Y(1)
            LDF      @Y1,R0          ;R0=Y2 (PREV Y1) DUE TO DELAY
            LDF      @B,R4           ;R4=B
;Y(N) FOR N >= 3
WAIT        IDLE                     ;WAIT FOR INTERRUPT
            BR       WAIT            ;BRANCH TO WAIT
LOOP_S      LDF      R1,R2           ;R2=A x Y1
            MPYF     R3,R1,R1        ;R1=A(A x Y1)
            MPYF     R4,R0,R0        ;R0=B x Y2
            ADDF     R0,R1           ;R1=OUTPUT
```

FIGURE 5.34. (*Continued*)

```
;OUTPUT ROUTINE
        PUSH    R6              ;SAVE R6
        LDF     R1,R7           ;STORE R1 INTO R7
        MPYF    @SCALER,R7      ;SCALE OUTPUT
        FIX     R7              ;CONVERT R7 INTO INTEGER
        CALL    AICIO_I         ;AIC I/O ROUTINE.OUTPUT R7
        LDF     R2,R0           ;R0=A x Y1    (FOR NEXT N)
        POP     R6              ;RESTORE R6
        SUBI    1,R6            ;DECREMENT TIME COUNTER
        BZ      CONT            ;CONTINUE IF TIME_( ) = 0
        RETI                    ;RETURN TO INTERRUPT
CONT    POP     R6              ;POP INTER VALUE FROM STACK
        POP     R6              ;POP NEXT STACK VALUE TO R6
        POP     R4              ;POP NEXT STACK VALUE TO R4
        BNZ     LOOP_N          ;BRANCH FOR NEXT SAMPLE
        PUSH    R4              ;PUSH R4 ONTO STACK
        SUBI    1,R6            ;DECREMENT REPS COUNTER
        LDI     R6,R1           ;LOAD R6 INTO R1
        PUSH    R6              ;PUSH R6 ONTO STACK
        AND     1,R1            ;LOGICAL AND OF 1 & R1
        BNZ     LOOP_F1         ;GO TO 1kHz FAIL LOOP
        POP     R6              ;POP R6 FROM STACK
        BNZD    LOOP_F2         ;BRANCH TO 2 kHz FAIL SIG.
        PUSH    R6              ;PUSH R6 ONTO STACK
        NOP                     ;NO OPERATION
        NOP                     ;NO OPERATION
        POP     R6              ;POP R6 FROM STACK
        POP     R4              ;POP R4 FROM STACK
        BZ      LOOP_N          ;GET NEXT SAMPLE
END     .END                    ;END
```

FIGURE 5.34. *(Continued)*

noise generator (Example 3.2) produces a 1 or a 0 and is used to determine the frequency of the sinusoid to be generated. The scheme is to associate a 1 with an acceptable device and a 0 with a defective. When the noise generator output is a 1, a 4-kHz sinusoid is generated, and when the noise generator output is a 0, a 2-kHz sinusoid followed by a 1-kHz sinusoid are generated. Figure 5.34 shows a listing of the real-time program ALARMR.ASM for this exercise. The coefficients A and C ($B = -1$) in the recursive difference equation (5.37) are calculated for a sampling frequency of 10 kHz (interrupt-driven) and are set in the program. The variable LENGTH determines the number of times the outcome is repeated. For example, with LENGTH = 8, the sequence $\{1, 1, 1, 1, 0, 1, 0, 1\}$ is obtained from the noise generator (see Figure 3.7).

1. Connect a small audio speaker at the output of the EVM and run the program (ALARMR). Verify the following sequence of tones with frequencies: 4 kHz, 4 kHz, 4 kHz, 4 kHz (due to the first four values of 1s), followed by 2 kHz, 1 kHz, 2 kHz, 1 kHz (due to the fifth value of 0). TIME_R determines the number of times the 2-kHz–1-kHz sequence is repeated. Complete the sequence of tones generated for the remaining three values {1, 0, 1} from the noise generator.
2. Obtain a simulated version of the ALARMR.ASM program so that the output sequence can be captured into a file. Set TIME_P to 10 and TIME_F to 8. Verify that the 4-kHz "pass" signal consists of 10 output points and the "fail" signals of 2 and 1 kHz each consists of 8 output points.

REFERENCES

[1] L. B. Jackson, *Digital Filters and Signal Processing*, Kluwer Academic, Norwell, Mass., 1986.

[2] L. B. Jackson, Roundoff noise analysis for fixed-point digital filters realized in cascade or parallel form, *IEEE Transactions on Audio and Electroacoustics*, **Au-18**, June 1970, pp. 107–122.

[3] L. B. Jackson, An analysis of limit cycles due to multiplicative rounding in recursive digital filters, in *Proceedings of the Seventh Allerton Conference on Circuit System Theory*, 1969, pp. 69–78.

[4] L. B. Lawrence and K. V. Mina, A new and interesting class of limit cycles in recursive digital filters, in *Proceedings of the IEEE International Symposium on Circuit Systems*, April 1977, pp. 191–194.

[5] R. Chassaing and D. W. Horning, *Digital Signal Processing with the TMS320C25*, Wiley, New York, 1990.

[6] A. H. Gray and J. D. Markel, Digital lattice and ladder filter synthesis, *IEEE Transactions on Acoustics, Speech, and Signal Processing*, **ASSP-21**, December 1973, p. 491–500.

[7] A. H. Gray and J. D. Markel, A normalized digital filter structure, *IEEE Transactions on Acoustics, Speech, and Signal Processing*, **ASSP-23**, June 1975, pp. 268–277.

[8] R. Chassaing, B. Bitler, and D. W. Horning, Real-time digital filters in C, in *Proceedings of the 1991 ASEE Annual Conference*, June, 1991.

[9] A. V. Oppenheim and R. Schafer, *Discrete-Time Signal Processing*, Prentice-Hall, Englewood Cliffs, N.J., 1989.

[10] N. Ahmed and T. Natarajan, *Discrete-Time Signals and Systems*, Reston, Reston, Va., 1983.

[11] R. D. Strum and D. E. Kirk, *First Principles of Discrete Systems and Digital Signal Processing*, Addison Wesley, Reading, Mass., 1988.

[12] T. W. Parks and C. S. Burrus, *Digital Filter Design*, Wiley, New York, 1987.

[13] W. D. Stanley, *Digital Signal Processing*, Reston, Reston, Va., 1975.

[14] D. J. DeFatta, J. G. Lucas, and W. S. Hodgkiss, *Digital Signal Processing: A System Approach*, Wiley, New York, 1988.

[15] J. G. Proakis and D. G. Manolakis, *Introduction to Signal Processing*, Macmillan, New York, 1988.

[16] L. R. Rabiner and B. Gold, *Theory and Application of Digital Signal Processing*, Prentice-Hall, Englewood Cliffs, N.J., 1975.

[17] *Digital Filter Design Package (DFDP)*, Atlanta Signal Processors, Inc., Atlanta, Ga., 1991.

[18] S. Waser and M. Flynn, *Introduction to Arithmetic for Digital Systems Designers*, Holt, Rinehart and Winston, New York, 1982.

[19] V. C. Hamacher, Z. G. Vranesic, and S. G. Zaky, *Computer Organization*, McGraw-Hill, New York, 1978.

[20] J. F. Kaiser, Some practical considerations in the realization of linear digital filters, in *Proceedings of the Third Allerton Conference on Circuit System Theory*, October 1965, pp. 621–633.

[21] P. Papamichalis (editor), *Digital Signal Processing Applications with the TMS320 Family—Theory, Algorithms, and Implementations*, Vol. 3, Texas Instruments, Inc., Dallas, Tex., 1990.

[22] D. W. Horning and R. Chassaing, IIR filter scaling for real-time digital signal processing, *IEEE Transactions on Education*, March 1991, pp. 108–112.

[23] *Hypersignal-Plus DSP Software*, Hyperception, Inc., Dallas, Tex.

[24] *Digital Signal Processing Applications with the TMS320C30 Evaluation Module—Selected Applications Notes*, Texas Instruments, Inc., Dallas, Tex., 1991.

[25] *Proceedings of the TMS320 1991 Educators Conference*, Texas Instruments, Inc., July 1991.

6
Fast Fourier Transform

CONCEPTS AND PROCEDURES
- *Development of the fast Fourier transform (FFT) using radix-2 and radix-4*
- *Development of the fast Hartley transform (FHT)*
- *Programming examples of the FFT using C and TMS320C30 code*

The discrete Fourier transform (DFT) is useful for the analysis of discrete-frequency representation of a discrete-time sequence. The fast Fourier transform is an efficient algorithm that can be used to obtain this discrete-frequency representation with fewer computations than the DFT [1–4]. Two different procedures for computing an FFT are presented: the decimation-in-frequency and the decimation-in-time. Examples are provided to illustrate the FFT. Programming examples using both C and TMS320C30 code are included.

The development of the fast Hartley transform (FHT) as a variant of the FFT is also covered.

6.1 INTRODUCTION

The discrete Fourier transform (DFT) is used to convert a time-domain sequence into an equivalent sequence in the frequency domain. The inverse discrete Fourier transform can be used to convert a sequence in the frequency domain back into its time-domain representation. The fast Fourier transform (FFT) is a very efficient algorithm based on the discrete Fourier transform. The FFT is one of the operations most commonly used in digital signal processing to provide a frequency spectrum analysis.

Several variants of the FFT are available, such as the Winograd transform [5, 6], the discrete cosine transform (DCT) [7–9], and the discrete Hartley transform (DHT) [10–16]. The development of the DHT is presented later.

6.2 DEVELOPMENT OF THE FFT ALGORITHM: RADIX-2

The fast Fourier transform (FFT) reduces considerably the computational requirements of the discrete Fourier transform (DFT). The Fourier transform of an analog signal $x(t)$ is

$$X(\omega) = \int_{-\infty}^{\infty} x(t) e^{-j\omega t} \, dt \qquad (6.1)$$

The discrete Fourier transform of a discrete-time signal $x(nT)$ is

$$X(k) = \sum_{n=0}^{N-1} x(n) W^{nk}, \qquad k = 0, 1, \ldots, N-1 \qquad (6.2)$$

where the sampling period T is implied in $x(n)$ and the frame length is N. The constants W are referred to as the twiddle factors, which represent the phase and are defined as

$$W = e^{-j2\pi/N} \qquad (6.3)$$

Note that W is a function of the length N. To obtain $X(k)$ for a specific k, approximately N complex additions and N complex multiplications are required, because

$$X(k) = x(0) + x(1)W^k + x(2)W^{2k} + x(3)W^{3k} + \cdots + x(N-1)W^{(N-1)k} \qquad (6.4)$$

for $k = 0, 1, \ldots, N-1$. To compute $X(0), X(1), \ldots, X(N-1)$, a total of approximately N^2 complex additions and N^2 complex multiplications are required. Hence, the computational requirements of the DFT can be quite intensive, especially for large N.

The FFT takes advantage of the periodicity and symmetry of the twiddle factor W for the calculation of (6.2). Figure 6.1 illustrates the periodicity and symmetry of W, using a length of $N = 8$. From the periodicity of W,

$$W^k = W^{k+N} \qquad (6.5)$$

and, from the symmetry of W,

$$W^k = -W^{k+N/2} \qquad (6.6)$$

6.3 Decimation-in-Frequency FFT Algorithm

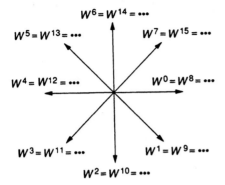

FIGURE 6.1. Periodicity and symmetry of twiddle factor W

The FFT decomposes an N-point DFT into smaller transforms. An N-point DFT is decomposed into two $(N/2)$-point DFTs. Each $(N/2)$-point DFT is further decomposed into two $(N/4)$-point DFTs, and so on. The last decomposition consists of $(N/2)$ two-point DFTs. The smallest transform is determined by the radix of the FFT. For a radix-2 FFT, N must be a power of 2, and the smallest transform is the two-point DFT. To compute an FFT, two procedures are available: the decimation-in-frequency (DIF) and the decimation-in-time (DIT).

6.3 DECIMATION-IN-FREQUENCY FFT ALGORITHM

Let a sequence $x(n)$ be separated into two halves:

$$x(0), x(1), \ldots, x\left(\frac{N}{2} - 1\right)$$

and

$$x\left(\frac{N}{2}\right), x\left(\frac{N}{2} + 1\right), \ldots, x(N - 1)$$

The DFT equation (6.2) can be separated into two summations:

$$X(k) = \sum_{n=0}^{(N/2)-1} x(n)W^{nk} + \sum_{n=N/2}^{N-1} x(n)W^{nk} \quad (6.7)$$

Let $n = n + N/2$ in the second summation, then (6.7) becomes

$$X(k) = \sum_{n=0}^{(N/2)-1} x(n)W^{nk} + W^{kN/2} \sum_{n=0}^{(N/2)-1} x\left(n + \frac{N}{2}\right)W^{nk} \quad (6.8)$$

where $W^{kN/2}$ is taken out of the second summation because it is not a function of n. Furthermore,

$$W^{kN/2} = e^{-jk\pi} = (e^{-j\pi})^k = (\cos \pi - j \sin \pi)^k = (-1)^k$$

Equation (6.8) then becomes

$$X(k) = \sum_{n=0}^{(N/2)-1} \left[x(n) + (-1)^k x\left(n + \frac{N}{2}\right) \right] W^{nk} \qquad (6.9)$$

Because $(-1)^k = 1$ for even k and -1 for odd k, (6.9) can be separated for even and odd k, or

$$X(k) = \sum_{n=0}^{(N/2)-1} \left[x(n) + x\left(n + \frac{N}{2}\right) \right] W^{nk} \quad \text{for } k \text{ even} \qquad (6.10)$$

and

$$X(k) = \sum_{n=0}^{(N/2)-1} \left[x(n) - x\left(n + \frac{N}{2}\right) \right] W^{nk} \quad \text{for } k \text{ odd} \qquad (6.11)$$

Substituting $k = 2k$ for even k, and $k = 2k + 1$ for odd k, we can write (6.10) and (6.11) as

$$X(2k) = \sum_{n=0}^{(N/2)-1} \left[x(n) + x\left(n + \frac{N}{2}\right) \right] W^{2nk},$$

$$k = 0, 1, \ldots, \left(\frac{N}{2}\right) - 1 \qquad (6.12)$$

and

$$X(2k+1) = \sum_{n=0}^{(N/2)-1} \left[x(n) - x\left(n + \frac{N}{2}\right) \right] W^n W^{2nk},$$

$$k = 0, 1, \ldots, \left(\frac{N}{2}\right) - 1 \qquad (6.13)$$

Because W is a function of the length N, it can be written as W_N. Then W_N^2 can be written as $W_{N/2}$, so that (6.12) and (6.13) can be more clearly written as two $(N/2)$-point DFTs. Furthermore, let

$$a(n) = x(n) + x\left(n + \frac{N}{2}\right) \qquad (6.14)$$

$$b(n) = x(n) - x\left(n + \frac{N}{2}\right) \qquad (6.15)$$

6.3 Decimation-in-Frequency FFT Algorithm

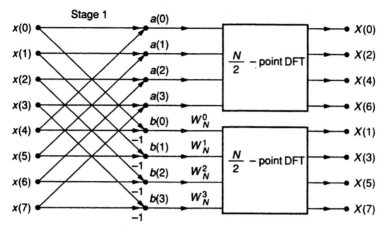

FIGURE 6.2. Decomposition of N-point DFT into two $(N/2)$-point DFTs, for $N = 8$

in (6.12) and (6.13):

$$X(2k) = \sum_{n=0}^{(N/2)-1} a(n) W_{N/2}^{nk} \tag{6.16}$$

$$X(2k+1) = \sum_{n=0}^{(N/2)-1} b(n) W_N^n W_{N/2}^{nk} \tag{6.17}$$

Figure 6.2 shows the decomposition of an N-point DFT into two $(N/2)$-point DFTs, for $N = 8$. Due to the decomposition process, shown in Figure 6.2, the X's are even in the upper half and odd in the lower half. The total number of complex multiplications and additions is now reduced to approximately $N + 2(N/2)^2$. The periodicity and symmetry properties of the twiddle factor W in (6.5) and (6.6) can provide additional reduction in the number of computations.

The decomposition or decimation process can now be repeated. From Figure 6.2, each of the $(N/2)$-point DFTs can be computed using two $(N/4)$-point DFTs, as shown in Figure 6.3. The resulting output sequence $X(0), X(4), X(2), \ldots$ is ordered as in Figure 6.3 due to the decomposition process. From the upper part of the output sequence in Figure 6.2, $X(0)$ and $X(4)$ are ordered as the even values, and $X(2)$ and $X(6)$ as the odd values. Similarly, from the lower part of Figure 6.2, $X(1)$ and $X(5)$ are ordered as the even values, and $X(3)$ and $X(7)$ as the odd values. The final order of the output sequence is shown to be scrambled as follows: $X(0), X(4), X(2), X(6), X(1), X(5), X(3), X(7)$. We will need to unscramble, or reorder, this output sequence. This can be done using a special instruction available with the TMS320C30, as is shown later in this chapter. The output sequence $X(k)$ represents the DFT of the time sequence $x(n)$.

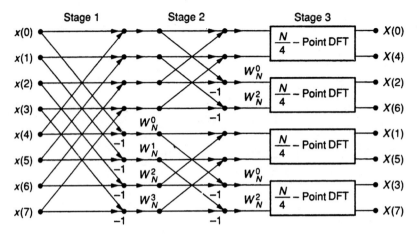

FIGURE 6.3. Decomposition of N-point DFT into four $(N/4)$-point DFTs, for $N = 8$

This last decomposition is as far as we can go because we have now a set of $(N/2)$ two-point DFTs. For the two-point DFT, (6.2) can be written as

$$X(k) = \sum_{n=0}^{1} x(n) W^{nk}, \qquad k = 0, 1 \qquad (6.18)$$

or

$$X(0) = x(0)W^0 + x(1)W^0 = x(0) + x(1) \qquad (6.19)$$

and

$$X(1) = x(0)W^0 + x(1)W^1 = x(0) - x(1) \qquad (6.20)$$

because, for $N = 2$, $W^1 = e^{-j2\pi/2} = -1$.

Equations (6.19) and (6.20) can be represented by the flow graph in Figure 6.4, usually referred to as a butterfly. Figure 6.5 shows the final flow graph for an eight-point FFT. This algorithm is referred to as decimation-in-frequency (DIF) because the output sequence $X(k)$ is decomposed, or decimated, into successfully smaller subsequences. This decomposition process goes through M stages, where $N = 2^M$.

For a complex FFT, $2N$ memory locations are required in order to store both the real and imaginary values. The same memory locations allocated for

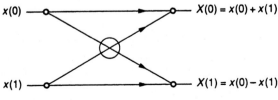

FIGURE 6.4. Two-point FFT butterfly

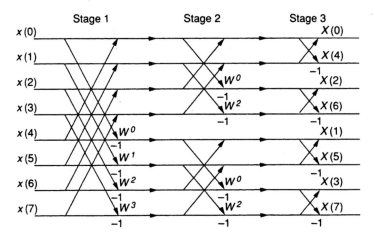

FIGURE 6.5. Eight-point FFT flow graph using decimation-in-frequency

the input samples $x(n)$ also can be used for the output sequence $X(k)$. Intermediate results (after each stage) can be stored in these memory locations.

Example 6.1 *Eight-Point FFT Using Decimation-in-Frequency.* Let the input sequence $x(n)$ represent a rectangular waveform $x(0) = x(1) = x(2) = x(3) = 1$ and $x(4) = x(5) = x(6) = x(7) = 0$. The output sequence $X(k)$ can be found. Figure 6.5 can be used to compute the eight-point FFT. For $N = 8$, the twiddle factors are

$$W^0 = 1$$
$$W^1 = e^{-j2\pi/8} = \cos(\pi/4) - j\sin(\pi/4) = 0.707 - j0.707$$
$$W^2 = e^{-j4\pi/8} = -j$$
$$W^3 = e^{-j6\pi/8} = -0.707 - j0.707$$

Consider the output values (from Figure 6.5) after each stage.

1. After stage 1:
$$x(0) + x(4) = 1 \rightarrow x(0)$$
$$x(1) + x(5) = 1 \rightarrow x(1)$$
$$x(2) + x(6) = 1 \rightarrow x(2)$$
$$x(3) + x(7) = 1 \rightarrow x(3)$$
$$[x(0) - x(4)]W^0 = 1 \rightarrow x(4)$$
$$[x(1) - x(5)]W^1 = 0.707 - j0.707 \rightarrow x(5)$$
$$[x(2) - x(6)]W^2 = -j \rightarrow x(6)$$
$$[x(3) - x(7)]W^3 = -0.707 - j0.707 \rightarrow x(7)$$

The intermediate values (first-stage output) $x(0), x(1), \ldots, x(7)$ become the input to the second stage.

2. After stage 2 (using the intermediate output results from stage 1 as the inputs into stage 2):

$$x(0) + x(2) = 2 \rightarrow x(0)$$
$$x(1) + x(3) = 2 \rightarrow x(1)$$
$$[x(0) - x(2)]W^0 = 0 \rightarrow x(2)$$
$$[x(1) - x(3)]W^2 = 0 \rightarrow x(3)$$
$$x(4) + x(6) = 1 - j \rightarrow x(4)$$
$$x(5) + x(7) = (0.707 - j0.707) + (-0.707 - j0.707)$$
$$= -j1.41 \rightarrow x(5)$$
$$[x(4) - x(6)]W^0 = 1 + j \rightarrow x(6)$$
$$[x(5) - x(7)]W^2 = -j1.41 \rightarrow x(7)$$

The intermediate values (second-stage output) $x(0), x(1), \ldots, x(7)$ become the input to the third stage.

3. After stage 3 (for the final output sequence):

$$X(0) = x(0) + x(1) = 4$$
$$X(4) = x(0) - x(1) = 0$$
$$X(2) = x(2) + x(3) = 0$$
$$X(6) = x(2) - x(3) = 0$$
$$X(1) = x(4) + x(5) = (1 - j) + (-j1.41) = 1 - j2.41$$
$$X(5) = x(4) - x(5) = 1 + j0.41$$
$$X(3) = x(6) + x(7) = (1 + j) + (-j1.41) = 1 - j0.41$$
$$X(7) = x(6) - x(7) = 1 + j2.41$$

The values $X(0), X(4), X(2), \ldots, X(7)$ form the scrambled output sequence. We will show shortly how to reorder this output sequence using a bit-reversal technique. The magnitude of the output sequence can then be taken and plotted.

Figure 6.6 shows an alternative flow graph for the decimation-in-frequency, which can be obtained from Figure 6.5 by interchanging appropriate nodes.

Example 6.2 *16-Point FFT Using Decimation-in-Frequency.* Given $x(0) = x(1) = \cdots = x(7) = 1$, and $x(8) = x(9) = \cdots = x(15) = 0$ as the input sequence, the output sequence can be found with a 16-point flow graph FFT using the DIF algorithm. Verify that the resulting output sequence

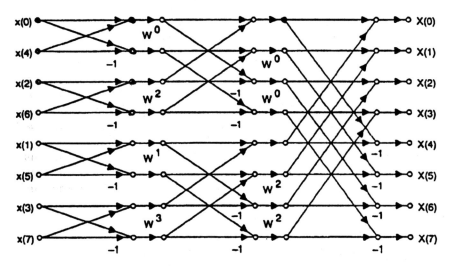

FIGURE 6.6. Alternative eight-point FFT flow graph using decimation-in-frequency

$X(0), X(1), \ldots, X(15)$ is as shown in Figure 6.7. Reorder the output sequence, and verify that the output magnitude, which represents a sync function, is as plotted in Figure 6.8. Note that $X(8)$ represents the magnitude at the Nyquist frequency.

6.4 DECIMATION-IN-TIME FFT ALGORITHM

The decimation-in-frequency (DIF) process decomposes the output sequence into smaller subsequences. Another procedure is the decimation-in-time (DIT), which decomposes the input sequence into smaller subsequences. Let the input sequence be decomposed into an even sequence

$$x(0), x(2), \ldots, x(2n)$$

and an odd sequence

$$x(1), x(3), \ldots, x(2n+1)$$

We can apply (6.2) to obtain

$$X(k) = \sum_{n=0}^{(N/2)-1} x(2n) W^{2nk} + \sum_{n=0}^{(N/2)-1} x(2n+1) W^{(2n+1)k} \quad (6.21)$$

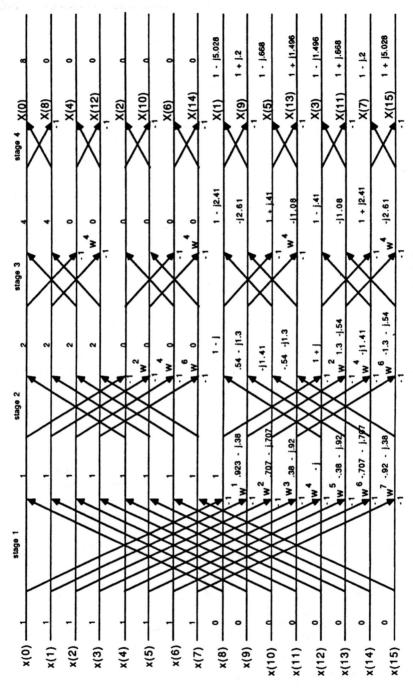

FIGURE 6.7. 16-Point FFT using decimation-in-frequency

6.4 Decimation-in-Time FFT Algorithm

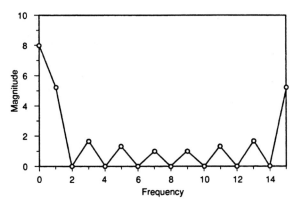

FIGURE 6.8. Output magnitude for 16-point FFT

Using $W_N^2 = W_{N/2}$ in (6.21),

$$X(k) = \sum_{n=0}^{(N/2)-1} x(2n)W_{N/2}^{nk} + W_N^k \sum_{n=0}^{(N/2)-1} x(2n+1)W_{N/2}^{nk} \quad (6.22)$$

Equation (6.22) represents the summation of two $(N/2)$-point DFTs. Let

$$C(k) = \sum_{n=0}^{(N/2)-1} x(2n)W_{N/2}^{nk} \quad (6.23)$$

$$D(k) = \sum_{n=0}^{(N/2)-1} x(2n+1)W_{N/2}^{nk} \quad (6.24)$$

Then (6.22) can be written in terms of $C(k)$ and $D(k)$ as

$$X(k) = C(k) + W_N^k D(k) \quad (6.25)$$

Equation (6.25) needs to be interpreted for $k > (N/2) - 1$. Using the symmetry property (6.6) of the twiddle factor, $W^{k+N/2} = -W^k$,

$$X\left(k + \frac{N}{2}\right) = C(k) - W^k D(k), \quad k = 0, 1, \ldots, \frac{N}{2} - 1 \quad (6.26)$$

For $N = 8$, (6.25) and (6.26) become

$$X(k) = C(k) + W^k D(k), \quad k = 0, 1, 2, 3 \quad (6.27)$$

$$X(k+4) = C(k) - W^k D(k), \quad k = 0, 1, 2, 3 \quad (6.28)$$

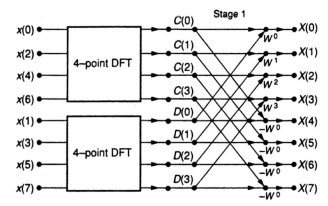

FIGURE 6.9. Decomposition of eight-point DFT into two four-point DFTs using decimation-in-time

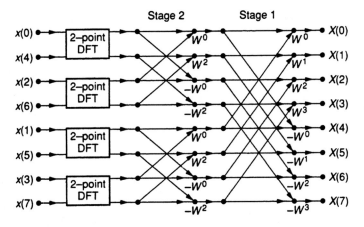

FIGURE 6.10. Decomposition of two four-point DFTs into four two-point DFTs using decimation-in-time

Figure 6.9 shows the decomposition of an eight-point DFT into two four-point DFTs using a decimation-in-time. The decimation process is repeated so that each four-point DFT is further decomposed into two two-point DFTs, as shown in Figure 6.10. This is as far as the decomposition process goes. Figure 6.11 shows the final flow graph for an eight-point FFT using a decimation-in-time. The input sequence $x(n)$ is shown to be scrambled in Figure 6.11, in the same manner as the output sequence $X(k)$ was scrambled during the decimation-in-frequency process. Note that now, using a scrambled input sequence $x(n)$, the output sequence $X(k)$ becomes properly ordered.

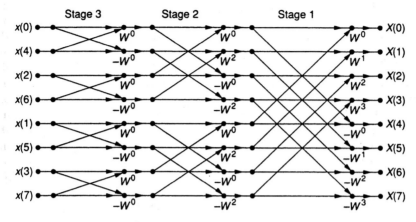

FIGURE 6.11. Eight-point FFT flow graph using decimation-in-time

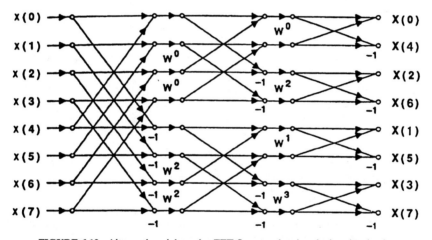

FIGURE 6.12. Alternative eight-point FFT flow graph using decimation-in-time

Figure 6.12 shows an alternative flow graph for the decimation-in-time, which can be obtained from Figure 6.11 by interchanging appropriate nodes.

6.5 BIT REVERSAL FOR UNSCRAMBLING

A bit-reversal procedure is used to reorder a scrambled sequence. This is accomplished by swapping the bits in the following fashion. For $N = 8$, represented by three bits, the first and third bits are swapped. For example, 100 is replaced by 001, and 110 is replaced by 011. In this fashion, the output sequence (Figure 6.5) using the decimation-in-frequency, or the input sequence (Figure 6.11) using the decimation-in-time can be reordered. This bit-reversal procedure also can be applied for larger values of N. For example, for $N = 64$, represented by six bits, the first and sixth bits, the second and fifth bits, and the third and fourth bits, would be swapped.

Bit Reversal with Indirect Addressing

Swapping memory locations is no longer necessary if the bit-reversed addressing mode available with the TMS320C30 is used. We will now illustrate this indirect addressing mode with reversed carry, for $N = 8$. Given a set of data $x0, x1, x2, \ldots, x7$, we wish to resequence it as $x0, x4, x2, x6, x1, x5, x3, x7$.

Set the index register IR0 to one-half the length of the FFT, or IR0 = 4. Let auxiliary register AR1 contain the base address. Choose a base address of zero or $(0000)_b$, for illustration purpose. The instruction

```
NOP    *AR1++(IR0)B
```

is an indirect mode of addressing instruction for bit reversal. On execution, AR1 is incremented to point at memory address 4, which is the base address offset by IR0. The second time the instruction is executed AR1 will point at memory address 2. We arrive at this address by adding $(0100)_b + (0100)_b = (0010)_b$ with reversed carry, or

$$\begin{array}{r} 0100 \\ + \ 0100 \\ \hline 0010 \end{array}$$

This is obtained because the carry bit of 1 is in the reversed (i.e., to the right) direction, caused by the B in the instruction NOP   *AR1++(IR0)B. The third time the instruction is executed, AR1 will contain the memory address 6, obtained by adding the offset value of IR0 = 4 to the previously obtained address of 2. The fourth time, AR1 will contain the memory address 1, because $(0110)_b + (0100)_b = (0001)_b$ with reversed carry, and so on.

We have used this indirect mode of addressing with reverse carry on the input sequence. We can use a similar procedure on the output sequence by loading the auxiliary register AR1 with the last or highest address, then post decrementing, or

```
NOP    *AR1--(IR0)B
```

This procedure is also applicable for a higher-length FFT. For a complex FFT, the real components of the input sequence can be arranged in even-numbered addresses and the imaginary components in odd-numbered addresses. In such cases, the index (offset) register IR0 is set to N (instead of $N/2$).

We will discuss later a programming FFT example in C using bit reversal for swapping addresses. A real-time 512-point real-valued FFT is also implemented incorporating the bit-reversal indirect addressing mode using TMS320C30 code.

6.6 DEVELOPMENT OF THE FFT ALGORITHM: RADIX-4 DIF

A radix-4 (base 4) algorithm is used to increase the execution speed of the FFT. Discussions on higher radices can be found in references [17]–[21]. The butterfly of a radix-4 algorithm consists of four inputs and four outputs. The FFT length is 4^M, where M is the number of stages. For a 16-point FFT, there are only two stages (or iterations) as compared to four stages with the radix-2 algorithm. The development of the radix-4 algorithm using decimation-in-frequency follows. The DFT equation (6.2) is decomposed into four summations (instead of two summations), as follows:

$$X(k) = \sum_{n=0}^{(N/4)-1} x(n)W^{nk} + \sum_{n=N/4}^{(N/2)-1} x(n)W^{nk} + \sum_{n=N/2}^{(3N/4)-1} x(n)W^{nk}$$

$$+ \sum_{n=3N/4}^{N-1} x(n)W^{nk} \tag{6.29}$$

Let $n = n + N/4$, $n = n + N/2$, $n = n + 3N/4$ in the second, third, and fourth summations, respectively. Then (6.29) can be written as

$$X(k) = \sum_{n=0}^{(N/4)-1} x(n)W^{nk} + W^{kN/4} \sum_{n=0}^{(N/4)-1} x\left(n + \frac{N}{4}\right)W^{nk}$$

$$+ W^{kN/2} \sum_{n=0}^{(N/4)-1} x\left(n + \frac{N}{2}\right)W^{nk} + W^{3kN/4} \sum_{n=0}^{(N/4)-1} x\left(n + \frac{3N}{4}\right)W^{nk}$$

$$\tag{6.30}$$

Using

$$W^{kN/4} = \left(e^{-j2\pi/N}\right)^{kN/4} = e^{-jk\pi/2} = (-j)^k$$

$$W^{kN/2} = e^{-jk\pi} = (-1)^k$$

$$W^{3kN/4} = (j)^k$$

equation (6.30) becomes

$$X(k) = \sum_{n=0}^{(N/4)-1} \left[x(n) + (-j)^k x\left(n + \frac{N}{4}\right) + (-1)^k x\left(n + \frac{N}{2}\right) \right.$$
$$\left. + (j)^k x\left(n + \frac{3N}{4}\right) \right] W^{nk} \quad (6.31)$$

To arrive at a four-point DFT decomposition, let $W_N^4 = W_{N/4}$. Equation (6.31) can then be written as four four-point DFTs, or

$$X(4k) = \sum_{n=0}^{(N/4)-1} \left[x(n) + x\left(n + \frac{N}{4}\right) + x\left(n + \frac{N}{2}\right) + x\left(n + \frac{3N}{4}\right) \right] W_{N/4}^{nk}$$
$$(6.32)$$

$$X(4k+1) = \sum_{n=0}^{(N/4)-1} \left[x(n) - jx\left(n + \frac{N}{4}\right) - x\left(n + \frac{N}{2}\right) + jx\left(n + \frac{3N}{4}\right) \right] W_N^n W_{N/4}^{nk}$$
$$(6.33)$$

$$X(4k+2) = \sum_{n=0}^{(N/4)-1} \left[x(n) - x\left(n + \frac{N}{4}\right) + x\left(n + \frac{N}{2}\right) - x\left(n + \frac{3N}{4}\right) \right] W_N^{2n} W_{N/4}^{nk}$$
$$(6.34)$$

$$X(4k+3) = \sum_{n=0}^{(N/4)-1} \left[x(n) + jx\left(n + \frac{N}{4}\right) - x\left(n + \frac{N}{2}\right) - jx\left(n + \frac{3N}{4}\right) \right] W_N^{3n} W_{N/4}^{nk}$$
$$(6.35)$$

for $k = 0, 1, \ldots, (N/4) - 1$. Equations (6.32) through (6.35) represent a decomposition process of four $(N/4)$-point DFTs. The flow graph for a 16-point, radix-4 decimation in frequency FFT is shown in Figure 6.13. Note the four-point butterfly in the flow graph (as far as the decomposition process goes). The $\pm j$ and -1 are not shown in Figure 6.13. The values shown in Figure 6.13 are for the following example.

Example 6.3 *16-Point Radix-4 FFT Using Decimation-in-Frequency.* Let $x(n)$ represent the rectangular sequence $x(0) = x(1) = \cdots = x(7) = 1$, and $x(8) = x(9) = \cdots = x(15) = 0$. The values of the twiddle factors are shown in Table 6.1.

6.6 Development of the FFT Algorithm: Radix-4 DIF

From Figure 6.13, the intermediate stage results can be found. For example, after stage 1:

$x(0)$:
$$[x(0) + x(4) + x(8) + x(12)]W^0 = 1 + 1 + 0 + 0 = 2 \rightarrow x(0)$$
$x(1)$:
$$[x(1) + x(5) + x(9) + x(13)]W^0 = 1 + 1 + 0 + 0 = 2 \rightarrow x(1)$$
$x(2)$:
$$[x(2) + x(6) + x(10) + x(14)]W^0 = 1 + 1 + 0 + 0 = 2 \rightarrow x(2)$$
$x(3)$:
$$[x(3) + x(7) + x(11) + x(15)]W^0 = 1 + 1 + 0 + 0 = 2 \rightarrow x(3)$$
$x(4)$:
$$[x(0) - jx(4) - x(8) + jx(12)]W^0 = 1 - j - 0 + 0 = 1 - j \rightarrow x(4)$$
$\vdots$

$x(11)$:
$$[x(3) - x(7) + x(11) - x(5)]W^6 = 0 \rightarrow x(11)$$
$x(12)$:
$$[x(0) + jx(4) - x(8) - jx(12)]W^0 = 1 + j - 0 - 0 = 1 + j \rightarrow x(12)$$
$\vdots$

$x(15)$:
$$[x(3) + jx(7) - x(11) - jx(15)]W^9 = [1 + j - 0 - 0](-W^1)$$
$$= -1.307 - j0.541 \rightarrow x(15)$$

After stage 2 (for example):

$$X(3) = (1 + j) + (1.307 - j0.541) + (-j1.414)$$
$$+ (-1.307 - j0.541) = 1 - j1.496$$

and

$$X(15) = (1 + j)(1) + (1.307 - j0.541)(-j) + (-j1.414)(1)$$
$$+ (-1.307 - j0.541)(-j) = 1 + j5.028$$

Note that the output sequence is scrambled. This output can be properly resequenced using digit reversal, in a similar fashion as a bit reversal in a

214 Fast Fourier Transform

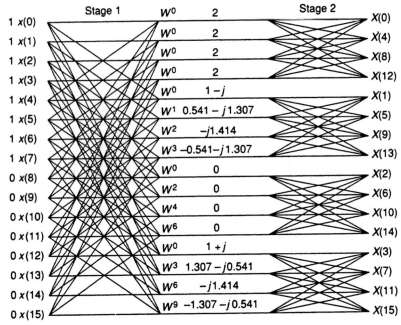

FIGURE 6.13. 16-Point radix-4 FFT flow graph using decimation-in-frequency

radix-2 algorithm. The radix-4 (base 4) uses the digits 0, 1, 2, 3. For example, $X(8)$ would be swapped with $X(2)$ because $(8)_{10}$ in decimal is $(20)_4$. Digits 0 and 1 would be reversed to yield $(02)_4$ which is also $(02)_{10}$ in decimal. Similarly, $X(4)$ would be swapped with $X(1)$, $X(12)$ with $X(3)$, and so on.

Using the flow graph in Figure 6.13, verify that the output sequence $X(0), X(1), \ldots, X(15)$ is identical to that obtained for the 16-point radix-2 shown in Figure 6.7.

Although higher radices can provide further reduction in computation, programming considerations become more complex. As a result, the radix-2 based FFT is still the most widely used, followed by the radix-4.

TABLE 6.1 Twiddle factors for 16-point FFT, radix-4

m	W_N^m	$W_{N/4}^m$
0	1	1
1	$0.9238 - j0.3826$	$-j$
2	$0.707 - j0.707$	-1
3	$0.3826 - j0.9238$	$+j$
4	$0 - j$	1
5	$-0.3826 - j0.9238$	$-j$
6	$-0.707 - j0.707$	-1
7	$-0.9238 - j0.3826$	$+j$

6.7 INVERSE FAST FOURIER TRANSFORM

The inverse discrete Fourier transform (IDFT) is defined as

$$x(n) = \frac{1}{N} \sum_{k=0}^{N-1} X(k) W^{-nk}, \quad n = 0, 1 \ldots, N-1$$

Comparing this equation with the DFT equation (6.2), we see that we can use the same FFT (forward transform) algorithm to find the IDFT (reverse transform), with the following changes:

1. adding a scaling factor of $1/N$
2. replacing W^{nk} by its complex conjugate W^{-nk}

With these changes, the same FFT flow graphs also can be used for the inverse fast Fourier transform (IFFT).

6.8 FAST HARTLEY TRANSFORM

Whereas complex additions and multiplications are required for an FFT, the Hartley transform [10–16] requires only real multiplications and additions. The FFT maps a real function of time into a complex function of frequency, whereas the fast Hartley transform (FHT) maps the same real-time function into a real function of frequency. The FHT can be particularly useful in cases where the phase is not a concern.

The discrete Hartley transform (DHT) of a time sequence $x(n)$ is defined as

$$H(k) = \sum_{n=0}^{N-1} x(n) \operatorname{cas}\left(\frac{2\pi nk}{N}\right), \quad k = 0, 1, \ldots, N-1 \quad (6.36)$$

where

$$\operatorname{cas} u = \cos u + \sin u \quad (6.37)$$

In a similar development to the FFT, (6.36) can be decomposed as

$$H(k) = \sum_{n=0}^{(N/2)-1} x(n) \operatorname{cas}\left(\frac{2\pi nk}{N}\right) + \sum_{n=N/2}^{N-1} x(n) \operatorname{cas}\left(\frac{2\pi nk}{N}\right) \quad (6.38)$$

216 Fast Fourier Transform

Let $n = n + N/2$ in the second summation of (6.38),

$$H(k) = \sum_{n=0}^{(N/2)-1} \left\{ x(n)\operatorname{cas}\left(\frac{2\pi nk}{N}\right) + x\left(n + \frac{N}{2}\right)\operatorname{cas}\left(\frac{2\pi k[n + N/2]}{N}\right)\right\} \tag{6.39}$$

Using (6.37) and the identities

$$\sin(A + B) = \sin A \cos B + \cos A \sin B$$
$$\cos(A + B) = \cos A \cos B - \sin A \sin B, \tag{6.40}$$

For odd k,

$$\begin{aligned}\operatorname{cas}\left(\frac{2\pi k[n + N/2]}{N}\right) &= \cos\left(\frac{2\pi nk}{N}\right)\cos(\pi k) - \sin\left(\frac{2\pi nk}{N}\right)\sin(\pi k) \\ &\quad + \sin\left(\frac{2\pi nk}{N}\right)\cos(\pi k) + \cos\left(\frac{2\pi nk}{N}\right)\sin(\pi k) \\ &= -\cos\left(\frac{2\pi nk}{N}\right) - \sin\left(\frac{2\pi nk}{N}\right) \\ &= -\operatorname{cas}\left(\frac{2\pi nk}{N}\right)\end{aligned} \tag{6.41}$$

and, for even k,

$$\operatorname{cas}\left(\frac{2\pi k[n + N/2]}{N}\right) = \cos\left(\frac{2\pi nk}{N}\right) + \sin\left(\frac{2\pi nk}{N}\right) = \operatorname{cas}\left(\frac{2\pi nk}{N}\right) \tag{6.42}$$

Using (6.41) and (6.42), (6.39) becomes

$$H(k) = \sum_{n=0}^{(N/2)-1} \left[x(n) + x\left(n + \frac{N}{2}\right)\right]\operatorname{cas}\left(\frac{2\pi nk}{N}\right), \quad \text{for even } k \tag{6.43}$$

and

$$H(k) = \sum_{n=0}^{(N/2)-1} \left[x(n) - x\left(n + \frac{N}{2}\right)\right]\operatorname{cas}\left(\frac{2\pi nk}{N}\right), \quad \text{for odd } k \tag{6.44}$$

6.8 Fast Hartley Transform

Let $k = 2k$ for even k, and let $k = 2k + 1$ for odd k. Equations (6.43) and (6.44) become

$$H(2k) = \sum_{n=0}^{(N/2)-1} \left[x(n) + x\left(n + \frac{N}{2}\right) \right] \text{cas}\left(\frac{2\pi n 2k}{N}\right) \quad (6.45)$$

$$H(2k+1) = \sum_{n=0}^{(N/2)-1} \left[x(n) - x\left(n + \frac{N}{2}\right) \right] \text{cas}\left(\frac{2\pi n[2k+1]}{N}\right) \quad (6.46)$$

Furthermore, using (6.40),

$$\text{cas}\left(\frac{2\pi n[2k+1]}{N}\right) = \cos\left(\frac{2\pi n}{N}\right)\left\{\cos\left(\frac{2\pi n 2k}{N}\right) + \sin\left(\frac{2\pi n 2k}{N}\right)\right\}$$
$$+ \sin\left(\frac{2\pi n}{N}\right)\left\{\cos\left(\frac{2\pi n 2k}{N}\right) - \sin\left(\frac{2\pi n 2k}{N}\right)\right\}$$

and

$$\sin\left(\frac{2\pi kn}{N}\right) = -\sin\left(\frac{2\pi k[N-n]}{N}\right)$$

$$\cos\left(\frac{2\pi kn}{N}\right) = \cos\left(\frac{2\pi k[N-n]}{N}\right)$$

Equation (6.46) becomes

$$H(2k+1) = \sum_{n=0}^{(N/2)-1} \left\{ \left[x(n) - x\left(n + \frac{N}{2}\right) \right] \cos\left(\frac{2\pi n}{N}\right) \text{cas}\left(\frac{2\pi n 2k}{N}\right) \right.$$
$$\left. + \sin\left(\frac{2\pi n}{N}\right) \text{cas}\left(\frac{2\pi 2k[N-n]}{N}\right) \right\} \quad (6.47)$$

Substituting $N/2 - n$ for n in the second summation, (6.47) becomes

$$H(2k+1) = \sum_{n=0}^{(N/2)-1} \left\{ \left[x(n) - x\left(n + \frac{N}{2}\right) \right] \cos\left(\frac{2\pi n}{N}\right) \right.$$
$$\left. + \left[x\left(\frac{N}{2} - n\right) - x(N-n) \right] \sin\left(\frac{2\pi n}{N}\right) \right\} \text{cas}\left(\frac{2\pi n 2k}{N}\right) \quad (6.48)$$

Let

$$a(n) = x(n) + x\left(n + \frac{N}{2}\right)$$

$$b(n) = \left[x(n) - x\left(n + \frac{N}{2}\right) \right] \cos\left(\frac{2\pi n}{N}\right)$$
$$+ \left[x\left(\frac{N}{2} - n\right) - x(N-n) \right] \sin\left(\frac{2\pi n}{N}\right)$$

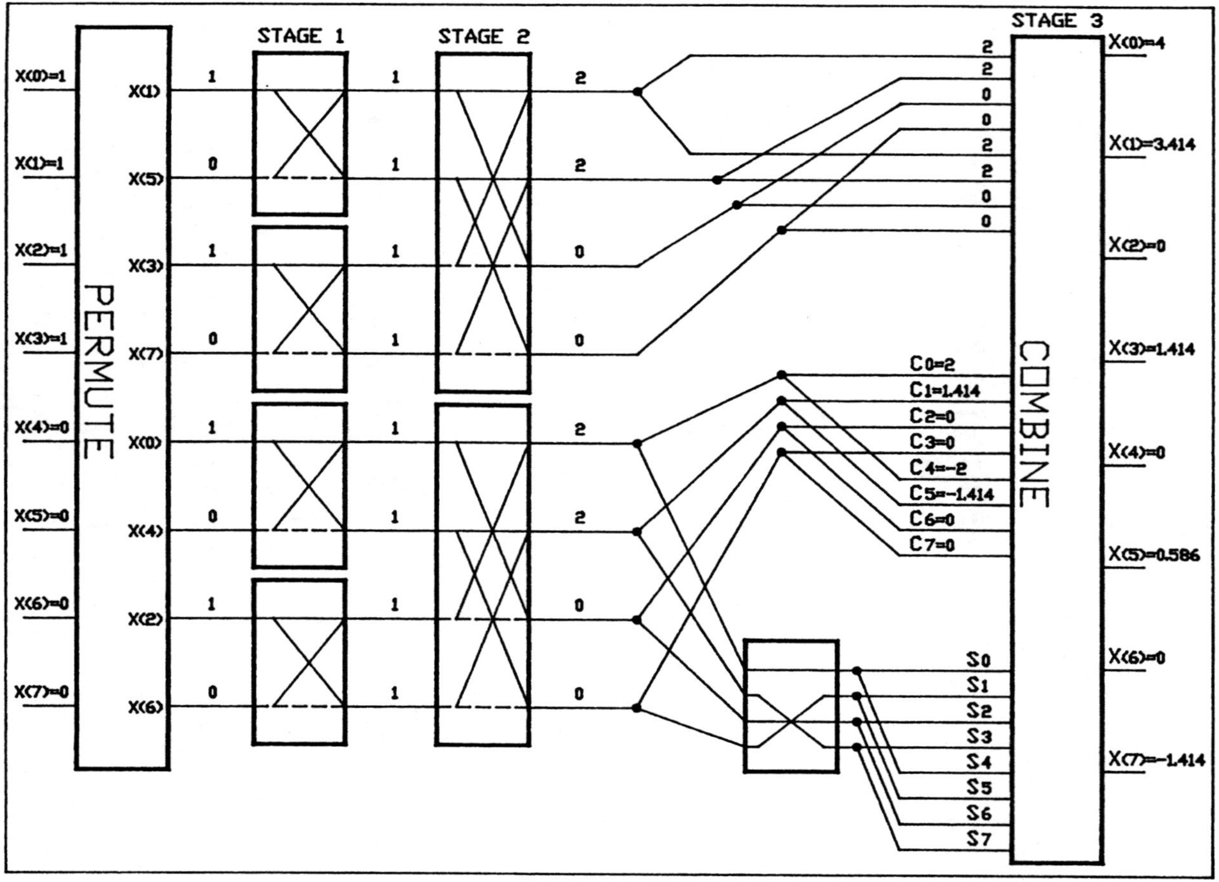

FIGURE 6.14. Eight-point FHT flow graph

Equations (6.45) and (6.48) become

$$H(2k) = \sum_{n=0}^{(N/2)-1} a(n)\text{cas}\left(\frac{2\pi n 2k}{N}\right) \qquad (6.49)$$

$$H(2k+1) = \sum_{n=0}^{(N/2)-1} b(n)\text{cas}\left(\frac{2\pi n 2K}{N}\right) \qquad (6.50)$$

A more complete development of the FHT can be found in [12]. We now illustrate the FHT with two examples: an eight-point FHT and a 16-point FHT. We will then readily verify these results from the previous FFT examples.

Example 6.4 *Eight-Point Fast Hartley Transform.* Let the rectangular sequence $x(n)$ be represented by $x(0) = x(1) = x(2) = x(3) = 1$, and $x(4) = x(5) = x(6) = x(7) = 0$. The flow graph in Figure 6.14 is used to find $X(k)$. We will now use $X(k)$ instead of $H(k)$. The sequence is first permuted and the intermediate results after the first two stages are as shown in Figure 6.14. The coefficients Cn and Sn are (with $N = 8$)

$$Cn = \cos(2\pi n/N)$$
$$Sn = \sin(2\pi n/N)$$

The output sequence $X(k)$ after the final stage 3 is also shown in Figure 6.14. For example,

$$X(0) = 2 + 2C0 + 2S0 = 2 + 2(1) + 2(0) = 4$$
$$X(1) = 2 + 2C1 + 2S1 = 2 + 1.414 + 0 = 3.41$$
$$\vdots$$
$$X(7) = 0 + 0(C7) + 2S7 = -1.414 \qquad (6.51)$$

This resulting output sequence can be verified from the $X(k)$ obtained with the FFT, using

$$\text{DHT}\{x(n)\} = \text{Re}\{\text{DFT}[x(n)]\} - \text{Im}\{\text{DFT}[x(n)]\} \qquad (6.52)$$

For example, from the eight-point FFT in Example 6.1, $X(1) = 1 - j2.41$, and

$$\text{Re}\{X(1)\} = 1$$
$$\text{Im}\{X(1)\} = -2.41$$

220 Fast Fourier Transform

Using (6.52),

$$\text{DHT}\{x(1)\} = X(1) = 1 - (-2.41) = 3.41$$

as in (6.51). Conversely, the FFT can be obtained from the FHT using

$$\text{Re}\{\text{DFT}[x(n)]\} = \tfrac{1}{2}\{\text{DHT}[x(N-n)] + \text{DHT}[x(n)]\}$$
$$\text{Im}\{\text{DFT}[x(n)]\} = \tfrac{1}{2}\{\text{DHT}[x(N-n)] - \text{DHT}[x(n)]\} \quad (6.53)$$

For example, using (6.53) to obtain $X(1) = 1 - j2.41$ from the FHT,

$$\text{Re}\{X(1)\} = \tfrac{1}{2}\{X(7) + X(1)\} = \tfrac{1}{2}\{-1.41 + 3.41\} = 1$$
$$\text{Im}\{X(1)\} = \tfrac{1}{2}\{X(7) - X(1)\} = \tfrac{1}{2}\{-1.41 - 3.41\} = -2.41 \quad (6.54)$$

where the left-hand side of (6.53) is associated with the FFT and the right-hand side with the FHT.

Example 6.5 *16-Point Fast Hartley Transform.* Let the rectangular sequence $x(n)$ be represented by $x(0) = x(1) = \cdots = x(7) = 1$, and $x(8) = x(9) = \cdots = x(15) = 0$. A 16-point FHT flow graph can be arrived at, building on the 8-point FHT. The permutation of the input sequence before the first stage is as follows for the first (upper) eight-point FHT: $x(0), x(8), x(4), x(12), x(2), x(10), x(6), x(14)$ and for the second (lower) eight-point FHT: $x(1), x(9), x(5), x(13), x(3), x(11), x(7), x(15)$. After the third stage, the intermediate output results for the upper and the lower eight-point FHTs are as obtained in the previous eight-point FHT example. Figure 6.15 shows the flow graph of the fourth stage for the 16-point FHT. The intermediate output results from the third stage become the input to the fourth stage in Figure 6.15. The output sequence $X(0), X(1), \ldots, X(15)$ from Figure 6.15 can be verified using the results obtained with the 16-point FFT in Example 6.2. For example, using

$$C_n = \cos\frac{2\pi n}{N} = \cos\frac{\pi n}{8}$$

$$S_n = \sin\frac{2\pi n}{N} = \sin\frac{\pi n}{8}$$

with $N = 16$, $X(1)$ can be obtained from Figure 6.15:

$$X(1) = 3.414 + 3.414C1 - 1.414S1 = 3.414 + 3.154 - 0.541 = 6.027$$

as in Figure 6.15. Equation (6.53) can be used to verify $X(1) = 1 - j5.028$, as

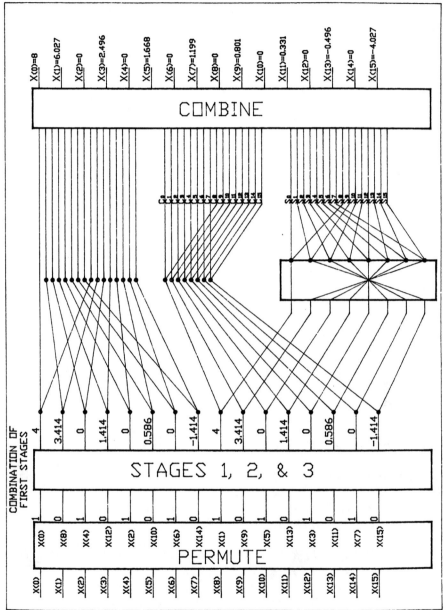

FIGURE 6.15. 16-Point FHT flow graph

222 Fast Fourier Transform

```
/*FFT8.C-MAIN FFT.RECTANGULAR INPUT DATA.CALLS FFT.C*/
#include <complex.h>    /*complex structure definition*/
extern void FFT();       /*FFT function                */
volatile int *out_addr=(volatile int *)0x804001;

main()
  {                  /*complex input samples */
  COMPLEX y[8]={10.0,0.0,10.0,0.0,10.0,0.0,10.0,0.0,
                0.0,0.0,0.0,0.0,0.0,0.0,0.0,0.0};
  int i, n=8;

  FFT(y,n);               /*calls generic FFT function*/
  for (i=0;i<n;i++)
    {
    *out_addr=(y[i]).real;
    *out_addr=(y[i]).imag;
    }
  }
```

FIGURE 6.16. Main program for eight-point FFT using C code (FFT8.C)

obtained using the FFT in Example 6.2. Note that, for example,

$$X(15) = -1.414 + (-1.414C15) + (3.414S15)$$
$$= -1.414 - 1.306 - 1.306$$
$$= -4.0269$$

as shown in Figure 6.15.

6.9 FFT PROGRAMMING EXAMPLES USING C AND TMS320C30 CODE

We will further illustrate the fast Fourier transform (FFT) using C code and mixed C and TMS320C30 code. A number of programs associated with the discrete cosine transform (DCT), the fast Hartley transform (FHT), and the fast Fourier transform (FFT) can be found in [21].

Example 6.6 Complex Eight-Point FFT Using C Code. This example implements with the simulator a decimation-in-frequency, radix-2 complex FFT, using C code. Figure 6.16 shows the main program FFT8.C for an eight-point complex FFT. The input samples represent a rectangular sequence, $x(0) = \cdots = x(3) = 10 + j0$ and $x(4) = \cdots = x(7) = 0 + j0$. The main program calls a generic FFT function FFT.C, passing the addresses of the input data (from y) and the FFT length (from n). The output port address is at 0x804001. A header file (COMPLEX.H), which contains the complex structure definition is called. Both the real and the imaginary components of the output sequence are obtained. Create the following header file, COMPLEX.H, which defines the complex structure:

```
struct cmpx
  {
  double real;
  double imag;
  };
  typedef struct cmpx COMPLEX
```

```
/*FFT.C-FFT RADIX-2 USING DIF. FOR UP TO 512 POINTS     */
#include <complex.h>  /*complex structure definition    */
#include <twiddle.h>  /*header file with twiddle constants*/
void FFT(COMPLEX *Y, int N) /*input sample array, # of points   */
  {
  COMPLEX temp1,temp2;    /*temporary storage variables       */
  int i,j,k;              /*loop counter variables            */
  int upper_leg, lower_leg; /*index of upper/lower butterfly leg */
  int leg_diff;           /*difference between upper/lower leg */
  int num_stages=0;       /*number of FFT stages, or iterations */
  int index, step;        /*index and step between twiddle factor*/
/* log(base 2) of # of points = # of stages  */
  i=1;
  do
    {
    num_stages+=1;
    i=i*2;
    } while (i!=N);
/* starting difference between upper and lower butterfly legs*/
  leg_diff=N/2;
/* step between values in twiddle factor array twiddle.h    */
  step=512/N;
/* For N-point FFT                                          */

  for (i=0;i<num_stages;i++)
    {
    index=0;
    for (j=0;j<leg_diff;j++)
      {
      for (upper_leg=j;upper_leg<N;upper_leg+=(2*leg_diff))
        {
        lower_leg=upper_leg+leg_diff;
        temp1.real=(Y[upper_leg]).real + (Y[lower_leg]).real;
        temp1.imag=(Y[upper_leg]).imag + (Y[lower_leg]).imag;
        temp2.real=(Y[upper_leg]).real - (Y[lower_leg]).real;
        temp2.imag=(Y[upper_leg]).imag - (Y[lower_leg]).imag;
        (Y[lower_leg]).real=temp2.real*(w[index]).real-temp2.imag*(w[index]).imag;
        (Y[lower_leg]).imag=temp2.real*(w[index]).imag+temp2.imag*(w[index]).real;
        (Y[upper_leg]).real=temp1.real;
        (Y[upper_leg]).imag=temp1.imag;
        }
      index+=step;
```

FIGURE 6.17. FFT function using C code (FFT.C)

```
        }
      leg_diff=leg_diff/2;
      step*=2;
      }
/* bit reversal for resequencing data */
   j=0;
   for (i=1;i<(N-1);i++)
      {
      k=N/2;
      while (k<=j)
         {
         j=j-k;
         k=k/2;
         }
      j=j+k;
      if (i<j)
         {
         temp1.real=(Y[j]).real;
         temp1.imag=(Y[j]).imag;
         (Y[j]).real=(Y[i]).real;
         (Y[j]).imag=(Y[i]).imag;
         (Y[i]).real=temp1.real;
         (Y[i]).imag=temp1.imag;
         }
      }
   return;
   }
```

FIGURE 6.17. *(Continued)*

The generic FFT function FFT.C is listed in Figure 6.17. A header file, TWIDDLE.H (on the accompanying disk), contains the twiddle constants and is called from FFT.C. Consider the following from the generic FFT function FFT.C, using $N = 8$ to illustrate (see also Figure 6.5):

1. The loop counter variable $i = 0$ represents the first stage (iteration). The value of leg_diff $= 4$ specifies the difference between the upper and the lower butterfly legs. For example, at stage 1, we need to calculate $y(0) + y(4)$ and $y(0) - y(4)$, where $y(0)$ is designated with upper_leg and $y(4)$ with lower_leg. This is an in-place FFT, in which case the memory locations used to store the input can again be used to store the intermediate and final output data. For example, temp1 $= y(0) + y(4) \rightarrow y(0)$. Similarly, temp2 $= y(0) - y(4)$. The calculation of $y(4)$ after the first stage involves complex operations involving the complex twiddle constant W. Note that $(A + jB)(C + jD) = (AC - BD) + j(BC + AD)$, where $j = \sqrt{-1}$. The constant W can be represented by $C + jD$, with a real and an imaginary component. These calculations are performed with the counter variable $j = 0$. When $j = 1$, the upper_leg and lower_leg specify $y(1)$ and $y(5)$, respectively. Then temp1 $= y(1) + y(5) \rightarrow y(1)$ and temp2 $= y(1) - y(5)$. With $j = 2$, $y(2) + y(6)$ becomes $y(2)$ after the first stage, because the upper_leg and lower_leg specify now $y(2)$ and $y(6)$, respectively.

6.9 FFT Programming Examples Using C and TMS320C30 Code

With $j = 3$, `temp1` $= y(3) + y(7) \to y(3)$. Note that the calculations of $y(5)$, $y(6)$, and $y(7)$ after the first stage contain complex operations involving the constant W. The different values for W are obtained from the file `TWIDDLE.H`, selected with the variable `step`. The variable `index` is used in `W(index)` to represent the W's.

2. The loop counter `i = 1` represents the second stage, and `leg_diff = 2`. With $j = 0$, `upper_leg` and `lower_leg` specify $y(0)$ and $y(2)$. The intermediate output results $y(0)$ and $y(2)$ are calculated in a procedure similar to that in step 1. Then `upper_leg` and `lower_leg` will specify $y(4)$ and $y(6)$, respectively. With $j = 1$, they will specify $y(1)$ and $y(3)$, then $y(5)$ and $y(7)$. The intermediate results after stage 2, $y(0), \ldots, y(7)$, can now be obtained.

3. The loop counter variable `i = 2` represents the third and final stage, and `leg_diff = 1`. The variables `upper_leg` and `lower_leg` specify $y(0)$ and $y(1)$, then they will specify $y(2)$ and $y(3)$, then $y(4)$ and $y(5)$, and finally $y(6)$ and $y(7)$. For each set of `upper_leg` and `lower_leg` values, similar calculations are performed to obtain the output from stage 3, or the final output sequence.

The last section of the program produces the proper sequencing or reorder-

```
/*FFT8.CMD           COMMAND FILE FOR LINKING    */
-c                   /*USING C CONVENTION        */
vecsir               /*INTERR/RESET DEF          */
fft8.obj             /*MAIN PROGRAM              */
fft.obj              /*FFT function              */
-O fft8.out          /*LINKED COFF OUTPUT FILE   */
-l rts30.lib         /*RUN-TIME LIBRARY SUPPORT*/
MEMORY
{
 VECS:   org = 0         len = 0x40    /*INTERRUPT VECTORS*/
 SRAM:   org = 0x40      len = 0x3FC0  /*USER STATIC RAM   */
 RAM:    org = 0x809800  len = 0x800   /*INTERNAL RAM      */
 IO:     org = 0x804000  len = 0x200   /*I/O               */
}
SECTIONS
{
 .text:   {} > SRAM       /*CODE                         */
 .cinit:  {} > RAM        /*INITIALIZATION TABLES        */
 .stack:  {} > RAM        /*SYSTEM STACK                 */
 vecs:    {} > VECS       /*VECTOR SECTION               */
 OUT_ADDR 0x804001: {} > IO    /*OUTPUT PORT ADDRESS    */
}
```

FIGURE 6.18. Command file for FFT example (`FFT8.CMD`)

```
/*FFTGEN.C-GENERATES TWIDDLE CONSTANTS */
#include <math.h>
#include <stdio.h>
#define N 256    /*to generate 256 complex points*/
main()
  {
  FILE *fptr;
  double sinval[N];
  double cosval[N];
  double arg;
  int i;
  fptr=fopen("twid256.h","w");
  arg=2*3.141592654/512;
  for(i=0;i<N;i++)
     {
     cosval[i]=(float)cos((i*arg));
     sinval[i]=-(float)sin((i*arg));
     }
  fprintf(fptr,"struct\n");
  fprintf(fptr,"    {\n");
  fprintf(fptr,"    double real;\n");
  fprintf(fptr,"    double imag;\n");
  fprintf(fptr,"    }");
  fprintf(fptr," w[]={");
  for(i=0;i<N;i++)
     {
     fprintf(fptr,"%8.5f,%8.5f,\n",cosval[i],sinval[i]);
     fprintf(fptr,"        ");
     }
  fclose(fptr);
  }
```

FIGURE 6.19. Program to generate twiddle constants (FFTGEN.C)

ing of the data. The command file FFT8.CMD for this example is listed in Figure 6.18. Compile and link using (see Chapter 1)

```
CL30 -sq -o2 -iinclude FFT8.C -z FFT8.CMD
```

The -i option is necessary when using a subdirectory include for the header files. Create the output file FFT8.DAT at port address 0x804001. Run this program through the simulator and verify that the same results (with a gain of 10) are obtained as in Example 6.1.

The program FFTGEN.C, listed in Figure 6.19, can be used to generate the twiddle constants. With $N = 512$, a total of 256 sets of complex values

6.9 FFT Programming Examples Using C and TMS320C30 Code

are generated. Compile (with Turbo C++, for example) and run this program. The output file contains the twiddle constants, stored in the file TWID256.H. Edit the file TWID256.H. At the end of this file, replace the last comma (,) by a closing brace (}) and a semicolon (;) to end this file. The appropriate twiddle values within the file TWID256 are selected depending on the value of n (the FFT length) set in the main program. The address of the length of the FFT is passed as N to the generic function FFT.C. The variable step in FFT.C selects (steps through) the appropriate twiddle values.

Example 6.7 *Eight-Point FFT with Real-Valued Input, Using Mixed C and TMS320C30 Code with the Simulator.* This example illustrates the use of real-valued input FFT as opposed to the more general complex FFT. Even

```c
/*FFT8MC.C-8-POINT FFT.CALLS FUNCTION FFT_RL.ASM IN TMS320C30 CODE*/
#include <math.h>
#define N 8            /*FFT length   */
#define M 3            /*# of stages  */
float data[N] = {1,1,1,1,0,0,0,0};    /*real-valued input samples*/
float real1, img1;
extern void fft_rl(int, int, float *);   /*generic FFT function  */
volatile int *IO_OUT = (volatile *) 0x804001; /*output port addr*/

main()
{
  int loop;
  fft_rl(N, M, (float *)data);
  *IO_OUT = (int)(data[0]*1000);
  for (loop = 1; loop < N/2; loop++)
  {
    real1 = data[loop];
    img1 = data[N-loop];
    *IO_OUT = (int)(real1*1000);
    *IO_OUT = (int)(img1*1000);
  }
  *IO_OUT = (int)(data[N/2]*1000);
  for (loop = N/2+1; loop < N; loop++)
  {
    real1 = data[N-loop];
    img1 = data[loop];
    *IO_OUT = (int)(real1*1000);
    *IO_OUT = (int)(img1*(-1000));
  }
}
```

FIGURE 6.20. Main program for eight-point real-valued input FFT using mixed code (FFT8MC.C)

though the input signal $x(n)$ is a real sequence, the output sequence $X(k)$ will still be complex. In this case, a further reduction in computational requirements can be achieved. The real (with real input values only) FFT can be executed in about half the time as the complex FFT. The main program in C code FFT8MC.C, listed in Figure 6.20, calls a real-valued FFT function FFT_RL.ASM. The input sequence is a real-valued rectangular sequence. The resulting output sequence is scaled by 1,000. The output sequence $X(k) = X_R(k) + jX_I(k)$ is such that the following conditions will be met [with $x(n)$ real]:

$$X_R(k) = X_R(N - k), \qquad k = 1, 2, \ldots, \frac{N}{2} - 1$$

$$X_I(k) = -X_I(N - k), \qquad k = 1, 2, \ldots, \frac{N}{2} - 1$$

$$X_I(0) = X_I\left(\frac{N}{2}\right) = 0 \qquad (6.55)$$

Note that these conditions are met in Example 6.1, because the imaginary components of the input sequence are zero. A discussion of real-valued input FFT can be found in [22]. The function FFT_RL.ASM (on accompanying disk) is listed in [21]. Due to the input sequence being real, the memory arrangement of the output sequence follows [21]:

$$X_R(0)$$
$$X_R(1)$$
$$\vdots$$
$$X_R\left(\frac{N}{2}\right) = X_R(4)$$
$$X_I\left(\frac{N}{2} - 1\right) = X_R(3)$$
$$X_I\left(\frac{N}{2} - 2\right) = X_R(2)$$
$$\vdots$$
$$X_I(1)$$

The function FFT_RL.ASM is in TMS320C30 code and is based on the Fortran version in [22]. The bit reversal is done on the input data. To ensure that the data is properly aligned, a few instructions have been added (appropriately commented in FFT_RL.ASM, on accompanying disk) within the bit-reversal routine, based on a design tip by D. G. Chandler and N. L. Chang [23]. An alternative approach is to specify the alignment process directly within the C program, as suggested by L. Brenman in [24].

```
;TWID8.ASM-TWIDDLE CONSTANTS FOR REAL-VALUED FFT.
        .global    _sine
        .data
_sine   .float     0.000000
        .float     0.707107
        .float     1.000000
        .float     0.707107
```

FIGURE 6.21. Twiddle constants for eight-point real-valued input FFT (TWID8.ASM)

Due to the memory arrangement of the output sequence, taking advantage of (6.55) in the function FFT_RL.ASM, this output sequence is printed in the following order:

$$X_R(0), X_R(1), X_I(1), X_R(2), X_I(2), X_R(3), X_I(3), X_R(4),$$
$$X_R(5), X_I(5), X_R(6), X_I(6), X_R(7), X_I(7)$$

Note that $X_I(0) = X_I(4) = 0$ from (6.55). The twiddle constants for the eight-point FFT is listed as TWID8.ASM in Figure 6.21. Use an appropriate command file to link the main program FFT8MC.OBJ, the generic FFT function FFT_RL.OBJ, TWID8.OBJ, and the run-time library support file rts30.lib. Download the linked output file into the simulator, and create the simulator output file at port address 0x804001h. This output file (FFT8MC.DAT) is shown in Figure 6.22. The resulting output sequence can be verified from the results obtained in Example 6.1.

Example 6.8 *Real-Time 512-Point Real-Valued Input FFT, Using Mixed C and TMS320C30 Code.* This example is a real-time version of the previous example. The main program in the previous example (Figure 6.20) is modified to incorporate a real-time implementation with the AIC. The resulting

```
0x00000fa0
0x000003e8
0xffffff692
0x00000000
0x00000000
0x000003e8
0xfffffe62
0x00000000
0x000003e8
0x0000019e
0x00000000
0x00000000
0x000003e8
0x0000096e
```

FIGURE 6.22. Output file for eight-point real-valued input FFT (FFT8MC.DAT)

```c
/*FFT512R.C-REAL-TIME FFT WITH 512 POINTS. CALLS FFT_RL*/
#include <math.h>                    /*std library func    */
#include "aiccom.c"                  /*AIC com routines    */
#define N 512                        /*size of FFT         */
#define M 9                          /*number of stages    */
volatile int index = 0;              /*input_output index*/
float *IO_buffer, *data, *temp;      /*pointers to array buffers*/
int AICSEC[4] = {0x1428,0x1,0x4A96,0x67};  /*AIC config data    */
extern void fft_rl(int, int, float *);     /*fft function protype*/
volatile int *INTVEC = (volatile int *) 0x00; /*addr of inter vecs */
void c_int05()                       /*interrupt handler function */
{
  PBASE[0x48] = ((int)(IO_buffer[index])) << 2;      /*output data */
  IO_buffer[index] = (float)(PBASE[0x4C] << 16 >> 18); /*input data */
  if (++index >= N) index = 0;       /*increment index, reset = N */
}
main()
{
  int loop;                          /* declare variable   */
  float real, img;                   /* declare variables  */
  INTVEC[5] = (volatile int) c_int05;  /*install inter function */
  AICSET_I();                        /*config AIC for interrupt*/
  IO_buffer = (float *) calloc(N, sizeof(float));  /*input_output buffer */
  data = (float *) calloc(N, sizeof(float));       /* fft data buffer   */
  while (1)                          /* create endless loop */
  {
    fft_rl(N, M, (float *)data);     /*call FFT function    */
    data[0] = sqrt(data[0]*data[0])/N;  /*magnitude of X(0)  */
    for (loop = 1; loop < N/2; loop++)  /*calculate X(1)..X(N/2-1)*/
    {
      real = data[loop];             /*real part            */
      img = data[N-loop];            /*imaginary part       */
      data[loop] = sqrt(real*real+img*img)/N; /*find magnitude */
    }
    data[N/2] = sqrt(data[N/2]*data[N/2])/N;  /*magnitude of X(N/2) */
    for (loop = N/2+1; loop < N; loop++)       /*X(N/2+1).. X(N-1) */
      data[loop] = data[N-loop];     /*use symetry          */
    while (index);                   /*wait till IO_buffer is empty*/
    temp = data;                     /*temp pointer => data buffer */
    data = IO_buffer;                /*IO_buffer becomes new data buffer */
    IO_buffer = temp;                /*data buffer becomes new IO_buffer */
    IO_buffer[0] = -2048;            /*frame sync pulse (negative spike) */
  }
}
```

FIGURE 6.23. Real-time main program for 512-point real-valued input FFT (FFT512R.C)

program FFT512R.C for the real-time implementation of a 512-point real-valued input FFT is shown in Figure 6.23. This main program calls the same FFT function FFT_RL.ASM discussed in the previous example. The interrupt rate is set for 10 kHz in the AIC data configuration. This scheme uses two buffers and allows a pointer to be switched from one buffer to another. Such scheme is more efficient than switching the data from one buffer to another buffer.

6.9 FFT Programming Examples Using C and TMS320C30 Code

Figure 6.24 shows a C program SINEGEN.C that can be used to generate the twiddle constants for a 512-point real-valued input FFT. A total of $N/2$ (one-half of a sine sequence), or 256, points are required [for a complex FFT, $(5/4)N$ points are required for W]. These twiddle constants are in such a format as to optimize the execution speed, even at a slight cost of memory size. Compile (with Turbo C++ for example) and run SINEGEN.C. The 256 twiddle constants, for the 512-point real-valued FFT, are stored in the file twid512.asm on accompanying disk. Change N from 512 to 8 and twid512.asm to twid8.asm, in the program SINEGEN.C, in order to generate and store in a file (twid8.asm) the required four twiddle constants for an eight-point real-valued FFT. Note that this source file needs to be assembled before linking.

Link the appropriate files (including VECSIR.OBJ for the reset vector). Run the linked output file FFT512R.OUT using the EVM. Input to the EVM a 2-kHz sinusoidal signal. The magnitude of the output sequence, obtained from the EVM output, is plotted in Figure 6.25 using a Hewlett Packard signal analyzer, HP3561A. A similar output can be observed with an oscilloscope. Figure 6.25 shows a "delta" function at the frequency of the input sinusoid (2 kHz), and a second delta function at its folded frequency. The distance between the two negative spikes represents a range of 10 kHz, which is the sampling rate set in the AIC configuration data in the main program

```
/*SINEGEN.C-GENERATES SINE VALUES FOR REAL FFT*/
#include <math.h>
#include <stdio.h>
#define N 512
#define pi 3.141592654

main( )
{
  FILE *stream;
  int n;
  float result;
  stream = fopen("twid512.asm", "w+");
  fprintf(stream, "\n%s", "            .global    _sine");
  fprintf(stream, "\n%s", "            .data");
  fprintf(stream, "\n%s%7f", "_sine       .float    ", 0.0000000);
  for (n = 1; n < N/2; n++)
  {
    result = sin(n*2*pi/N);
    fprintf(stream, "\n%s%7f", "            .float    ", result);
  }
  fclose(stream);
}
```

FIGURE 6.24. Program to generate twiddle constants for real-valued input FFT (SINEGEN.C)

232 Fast Fourier Transform

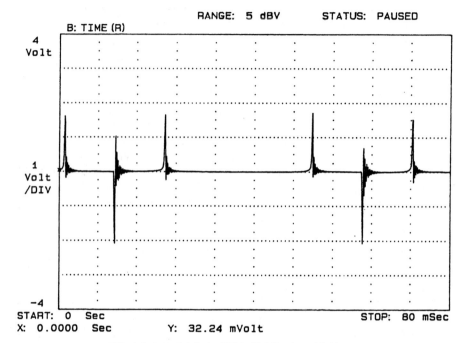

FIGURE 6.25. 512-Point FFT of 2-kHz sinusoidal input

FFT512R.C, listed in Figure 6.23. Note that aliasing would occur when the two delta functions meet at 5 kHz (at the middle of the two negative spikes). The negative spike is produced by the main C program with the last statement:

IO_buffer [0] = -2048

This negative spike, used for reference or range, is repeated after every frame. Note that the sampling period is $T = 1/F_s = 0.1$ ms. Because there are 512 points, the distance between the start of each frame (between negative spikes) is 512 × 0.1 ms = 51.2 ms. This distance can be verified from Figure 6.25.

Exercise 6.1 Eight-Point IFFT. Let the output sequence $X(0) = 4$, $X(1) = 1 - j2.41, \ldots,$ $X(7) = 1 + j2.41$ obtained in Example 6.1 become the input to the flow graph in Figure 6.5. Verify that with those two changes the resulting output becomes the rectangular sequence $x(0) = 1$, $x(1) = 1, \ldots,$ $x(7) = 0$.

Exercise 6.2 Implementation of a 16-Point Complex FFT. Change the main program FFT8.C to use 16 sets of input samples (with COMPLEX Y(16) and $n = 16$). No changes are required in the generic FFT function FFT.C or in the header file that contains the twiddle constants. Change the command file accordingly in order to obtain FFT16.OUT. With a rectangular sequence as input, verify that the output sequence is the same (with a gain of 10) as in Example 6.2.

REFERENCES

[1] J. W. Cooley and J. W. Tukey, An algorithm for machine computation of complex Fourier series, *Mathematics of Computation*, **19**, 1965, pp. 297–301.

[2] C. S. Burrus and T. W. Parks, *DFT/FFT and Convolution Algorithms: Theory and Implementation*, Wiley, New York, 1988.

[3] G. D. Bergland, A guided tour of the fast Fourier transform, *IEEE Spectrum*, **6**, July 1969, pp. 41–52.

[4] E. O. Brigham, *The Fast Fourier Transform*, Prentice-Hall, Englewood Cliffs, N.J., 1974.

[5] S. Winograd, On computing the discrete Fourier transform, *Mathematics of Computation*, **32**, No. 141, January 1978, pp. 175–199.

[6] H. F. Silverman, An introduction to programming the Winograd Fourier transform algorithm (WFTA), *IEEE Transactions on Acoustics, Speech, and Signal Processing*, **ASSP-25**, April 1977, pp. 152–165.

[7] N. Ahmed, T. Natarajan, and K. R. Rao, Discrete cosine transform, *IEEE Transactions on Computers*, **C-23**, January 1974, pp. 90–93.

[8] H. S. Hou, A fast recursive algorithm for computing the discrete cosine transform, *IEEE Transactions on Acoustics, Speech, and Signal Processing*, **ASSP-35**, October 1987, pp. 1455–1461.

[9] B. G. Lee, A fast cosine transform, *Proceedings of the 1984 Conference on ASSP*, March 1984.

[10] R. N. Bracewell, The fast Hartley transform, *Proceedings of the IEEE*, **72**, August 1984, pp. 1010–1018.

[11] R. N. Bracewell, Assessing the Hartley transform, *IEEE Transactions on Acoustics, Speech, and Signal Processing*, **ASSP-38**, December 1990, pp. 2174–2176.

[12] R. N. Bracewell, *The Hartley transform*, Oxford University Press, New York, 1986.

[13] H. S. Hou, The fast Hartley transform algorithm, *IEEE Transactions on Computers*, **C-36**, February 1987, pp. 147–156.

[14] H. S. Hou, Correction to "The fast Hartley transform algorithms," *IEEE Transactions on Computers*, **C-36**, September 1987, pp. 1135–1136.

[15] A. Zakhor and A. V. Oppenheim, Quantization errors in the computation of the discrete Hartley transform, *IEEE Transactions on Acoustics, Speech, and Signal Processing*, **ASSP-35**, October 1987, pp. 1592–1601.

[16] H. V. Sorensen, D. L. Jones, C. S. Burrus, and M. T. Heideman, On computing the discrete Hartley transform, *IEEE Transactions on Acoustics, Speech, and Signal Processing*, **ASSP-33**, October 1985, pp. 1231–1238.

[17] P. E. Papamichalis and C. S. Burrus, Conversion of digit-reversed to bit-reversed order in FFT algorithms, *Proceedings of ICASSP 89, USA*, May 1989, pp. 984–987.

[18] M. Vetterli and P. Duhamel, Split-radix algorithms for length-P^m DFT's, *IEEE Transactions on Acoustics, Speech, and Signal Processing*, **ASSP-37**, January 1989, pp. 57–64.

[19] H. V. Sorensen, M. T. Heideman, and C. S. Burrus, On computing the split-radix FFT, *IEEE Transactions on Acoustics, Speech, and Signal Processing*, **ASSP-34**, February 1986, pp. 152–156.

[20] C. S. Burrus, Unscrambling for fast DFT algorithms, *IEEE Transactions on Acoustics, Speech, and Signal Processing*, **ASSP-36**, July 1988, pp. 1086–1087.

[21] P. E. Papamichalis (editor), *Digital Signal Processing Applications with the TMS320 Family—Theory, Algorithms, and Implementations*, Vol. 3, Texas Instruments, Inc., Dallas, Tex., 1990.

[22] H. Sorensen, D. L. Jones, M. T. Heideman, and C. S. Burrus, Real-valued fast Fourier transform algorithms, *IEEE Transactions on Acoustics, Speech, and Signal Processing*, **ASSP-35**, June 1987, pp. 849–863.

[23] *Details on Signal Processing*, Texas Instruments, Inc., Dallas Tex., Fall 1990.

[24] *Details on Signal Processing*, Texas Instruments, Inc., Dallas, Tex., Winter 1992.

[25] DSP Committee, IEEE ASSP (editors), *Programs for Digital Signal Processing*, IEEE Press, New York, 1979.

[26] G. Goertzel, An algorithm for the evaluation of finite trigonometric series, *American Mathematics Monthly*, **65**, January 1958, pp. 34–35.

[27] S. Kay and R. Sudhaker, A zero crossing spectrum analyzer, *IEEE Transactions on Acoustics, Speech, and Signal Processing*, **ASSP-34**, February 1986, pp. 96–104.

[28] L. R. Rabiner and B. Gold, *Theory and Application of Digital Signal Processing*, Prentice-Hall, Englewood Cliffs, N.J., 1975.

[29] F. Harris, On the use of windows for harmonic analysis with the discrete Fourier transform, *Proceedings of the IEEE*, **66**, January 1978, pp. 51–83.

[30] N. Ahmed and T. Natarajan, *Discrete-Time Signals and Systems*, Reston, Reston, Va., 1983.

[31] K. S. Lin (editor), *Digital Signal Processing Applications with the TMS320 Family—Theory, Algorithms, and Implementations*, Vol. 1, Texas Instruments, Inc., Dallas, Tex., 1987.

[32] P. E. Papamichalis (editor), *Digital Signal Processing Applications with the TMS320 Family—Theory, Algorithms, and Implementations*, Vol. 2, Texas Instruments, Inc., Dallas, Tex., 1989.

[33] DSP Committee, IEEE ASSP (editors), *Selected Papers in Digital Signal Processing II*, IEEE Press, New York, 1976.

[34] *TMS3203X User's Guide*, Texas Instruments, Inc., Dallas, Tex., 1991.

[35] P. M. Embree and B. Kimble, *C Language Algorithms for Digital Signal Processing*, Prentice-Hall, Englewood Cliffs, N.J., 1990.

[36] A. V. Oppenheim and R. Schafer, *Discrete-Time Signal Processing*, Prentice-Hall, Englewood Cliffs, N.J., 1989.

[37] W. M. Gentleman and G. Sande, Fast Fourier transforms for fun and profit, *1966 Fall Joint Computer Conference, AFIPS Proceedings*, Vol. 29, Spartan, Washington, D.C., 1966, pp. 563–578.

7
Adaptive Filters

CONCEPTS AND PROCEDURES
* *Adaptive structures*
* *The least mean square* (LMS) *algorithm*
* *Programming examples using C and TMS320C30*

    Adaptive filters are best used in cases where signal conditions or system parameters are slowly changing and the filter is to be adjusted to compensate for this change. The least mean square (LMS) *criterion is the search algorithm used to provide the strategy for adjusting the filter coefficients. Programming examples using both C and TMS320C30 code are included to give a basic intuitive understanding of adaptive filters.*

7.1 INTRODUCTION

In conventional digital filters (FIR and IIR) it is assumed that all the process parameters to determine the filter characteristics are known. They may vary with time, but the nature of the variation is assumed to be known. In many practical problems, there may be a large uncertainty in some parameters because of inadequate prior test data about the process. Some parameters might be expected to change with time, but the exact nature of the change is not predictable. In such cases, it is highly desirable to design the filter to be self-learning, so that it can adapt itself to the situation at hand.

    The coefficients of an adaptive filter are adjusted to compensate for changes in input signal, output signal, or system parameters. Instead of being rigid, an adaptive system can learn the signal characteristics and track slow

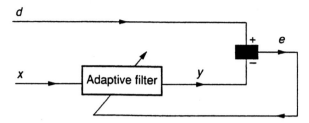

FIGURE 7.1. General adaptive structure

changes. An adaptive filter can be very useful when there is uncertainty about the characteristics of a signal or when these characteristics change. Intuitive ideas and the programming examples included in this chapter can provide a good understanding of adaptive systems. Figure 7.1 shows a basic adaptive structure where the adaptive filter's output y is compared with a desired signal d to yield an error signal e, which is fed back to the adaptive filter. The coefficients of the adaptive filter are adjusted, or optimized, using a least mean square (LMS) algorithm based on the error signal e. Although there are several strategies for performing adaptive filtering, we will discuss here only the LMS searching algorithm with a linear combiner (FIR filter) structure. The LMS is based on the following equation for the adaptation of the filter's coefficients:

$$w_k(n+1) = w_k(n) + \beta e(n) x_k(n) \tag{7.1}$$

Equation (7.1) can also be written as

$$w(n+1, k) = w(n, k) + \beta e(n) x(n, k) \tag{7.2}$$

where $k = 0, 1, \ldots$ and w represents the weights or the coefficients of the adaptive filter. The input to the adaptive filter is $x(n)$, the rate of convergence and accuracy of the adaptation process (adaptive step size) is β, and the error signal

$$e(n) = d(n) - y(n)$$

is the difference between the desired signal $d(n)$ and the filter's output $y(n)$. The output of the adaptive filter is

$$y(n) = \sum_{k=0}^{N} w(n, k) x(n-k) \tag{7.3}$$

For a specific time n each coefficient, or weight, $w_k(n)$ is updated or replaced by a new coefficient, as described in (7.1), unless the error signal $e(n)$ is zero.

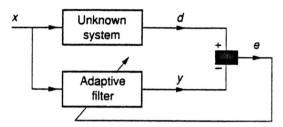

FIGURE 7.2. Adaptive structure for system identification

After the filter's output $y(n)$, the error signal $e(n)$ and each of the coefficients $w_k(n)$ are updated for a specific time n, a new sample is acquired (from an analog-to-digital converter) and the adaptation process is repeated for a different time. Note that from (7.1), the weights are not updated when $e(n)$ becomes zero.

7.2 ADAPTIVE STRUCTURES

A number of adaptive structures have been used for different applications in adaptive filtering [1, 2], including the following structures for system identification and noise cancellation:

1. *For system identification.* Figure 7.2 shows an adaptive structure that can be used for system identification. The error signal e is the difference between the response of the unknown system d and the response of the adaptive filter y. This error signal is fed back to the adaptive filter and is used to update the adaptive filter's coefficients, until $y = d$. When this happens, the adaptation process is finished, and e approaches zero. This scheme is such that the adaptive filter models the unknown system.
2. *For noise cancellation.* Figure 7.3 shows an adaptive structure that can be used for noise cancellation. The desired signal d is corrupted by uncorrelated additive noise n. The input to the adaptive filter is a noise

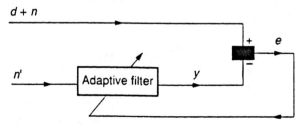

FIGURE 7.3. Adaptive structure for noise cancellation

n' that is correlated with the noise n. The noise n' could come from the same source as n but modified by the environment. The adaptive filter's output y is adapted to the noise n. When this happens, the error signal approaches the desired signal d. If n is uncorrelated with n', then $E[e^2]$ is minimized, where $E[\ ]$ is the expected value. Because the expected value is generally unknown, it is usually approximated using, for example, a running average.

Applications with additional structures are discussed in the programming examples, Section 7.4, as well as in Chapter 8.

7.3 LINEAR ADAPTIVE COMBINER AND THE LMS ALGORITHM

The linear adaptive combiner is one of the most useful adaptive filter structures and is an adjustable FIR filter. Whereas the coefficients of the frequency-selective FIR filter discussed in Chapter 4 are fixed, the coefficients, or weights, of the adaptive FIR filter can be adjusted based on a changing environment such as an input signal. The linear adaptive combiner is just an FIR filter with adjustable weights, or coefficients. Adaptive IIR filters (which are not discussed in this text) also can be used. A major problem with an adaptive IIR filter is that its poles may be updated (during the adaptation process) to values outside the unit circle, making the filter unstable.

The delayed input of the linear adaptive combiner is weighted and summed to yield the output $y(n)$ in (7.3). Equation (7.3) can also be written as vectors,

$$y(n) = \mathbf{W}(n)\mathbf{X}^T(n) = \mathbf{X}(n)\mathbf{W}^T(n) \tag{7.4}$$

where

$$\mathbf{W}(n) = \begin{bmatrix} w(0,n) & w(1,n) & \cdots & w(N,n) \end{bmatrix} \tag{7.5}$$

$$\mathbf{X}(n) = \begin{bmatrix} x(n) & x(n-1) & \cdots & x(n-N) \end{bmatrix} \tag{7.6}$$

and the superscript T represents the transpose.

Performance Measure

Consider again the basic adaptive filter structure in Figure 7.1. A performance measure is needed to determine how good the filter is. This measure is based on the error signal,

$$e(n) = d(n) - y(n) \tag{7.7}$$

The weights are adjusted such that a mean squared error function is minimized. This mean squared error function is $E[e^2(n)]$, where E represents the expected value, or

$$E[e^2(n)] = E[d^2(n)] - 2E[d(n)y(n)] + E[y^2(n)] \tag{7.8}$$

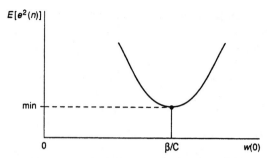

FIGURE 7.4. Performance curve for one weight

Consider the case of one weight only. Then (7.8) becomes

$$E[e^2(n)] = E[d^2(n)] - 2E[d(n)x(n)]w(0) + E[x^2(n)]w^2(0) \quad (7.9)$$

If $d(n)$ and $x(n)$ are statistically independent, then $E[d(n)x(n)] = E[d(n)]E[x(n)]$. However, this is not generally the case. If the signals d and x are statistically time-invariant, then the expected values of their products are constants. In this case, (7.9) can be written as

$$E[e^2(n)] = A - 2\beta w(0) + Cw^2(0) \quad (7.10)$$

The performance measure for one weight is then

$$w(0) = \beta/C \quad (7.11)$$

which represents the value at which $E[e^2(n)]$ is a minimum. A performance surface of this result is shown in Figure 7.4. For two weights, the performance function of Figure 7.4 becomes a bowl.

The weight is updated in the following fashion:

$$w(0, n+1) = w(0, n) - \beta \frac{dE[e^2(n)]}{dw(0)} \quad (7.12)$$

where β is a constant that determines the rate of convergence. The minus sign in (7.12) is associated with the direction of the step. From Figure 7.4, it can be seen that if the current value of the weight is on the left-hand side of the minimum value β/C, the step must be positive (incremented). This would specify the negative slope region of the performance function. On the other hand, if the current value of the weight is on the right-hand side of the minimum value β/C, the step must be negative (decremented). This condition would then specify the positive slope region of the performance function. Hence the step direction is proportional to the negative of the slope.

To extend this one-dimensional search from one weight to $N + 1$ weights requires taking the gradient of (7.8), which is a vector of the first derivatives

7.3 Linear Adaptive Combiner and the LMS Algorithm

with respect to the weights $w(0), w(1), \ldots, w(N)$. However, instead of using the gradient of $E[e^2(n)]$, an estimate can be found, using instead the gradient of $e^2(n)$, or

$$\begin{aligned}\operatorname{grad}\{e^2(n)\} &= 2e(n)\operatorname{grad}\{e(n)\} \\ &= 2e(n)\operatorname{grad}\{d(n) - \mathbf{X}^T(n)\mathbf{W}(n)\} \\ &= -2e(n)\mathbf{X}(n)\end{aligned} \quad (7.13)$$

Similar to (7.12), for multiple weights,

$$\mathbf{W}(n+1) = \mathbf{W}(n) + 2\beta e(n)\mathbf{X}(n) \quad (7.14)$$

which represents the LMS algorithm [1–3]. Equation (7.14) provides a simple but powerful and efficient means of updating the weights, or coefficients, without the need for averaging or differentiating and will be used for implementing adaptive FIR filters. This equation will be used in the next section with the programming examples and rewritten with $2\beta = \beta$, or,

$$w_k(n+1) = w_k(n) + \beta e(n)x(n-k) \quad (7.15)$$

The LMS is well suited for a number of applications, including adaptive echo and noise cancellation, prediction, and equalization.

Other variants of the LMS algorithm have been employed, such as the sign–error LMS, the sign–data LMS, and the sign–sign LMS.

1. For the sign–error LMS algorithm, (7.15) becomes

$$w_k(n+1) = w_k(n) + \beta \operatorname{sgn}[e(n)]x(n-k) \quad (7.16)$$

where sgn is the signum function,

$$\operatorname{sgn}[e(n)] = \begin{cases} 1, & \text{if } e(n) \geq 0 \\ -1, & \text{if } e(n) < 0 \end{cases} \quad (7.17)$$

2. For the sign–data LMS algorithm, (7.15) becomes

$$w_k(n+1) = w_k(n) + \beta e(n)\operatorname{sgn}[x(n-k)] \quad (7.18)$$

3. For the sign–sign LMS algorithm, (7.15) becomes

$$w_k(n+1) = w_k(n) + \beta \operatorname{sgn}[e(n)]\operatorname{sgn}[x(n-k)] \quad (7.19)$$

which can be further written as

$$w_k(n+1) = \begin{cases} w_k(n) + \beta, & \text{if sgn}[e(n)] = \text{sgn}[x(n-k)] \\ w_k(n) - \beta & \text{otherwise} \end{cases} \quad (7.20)$$

The sign–sign variant of the basic LMS algorithm is more concise from a mathematical viewpoint, because no multiplication operation is required for this algorithm.

The implementation of these variants does not exploit the pipeline features of the TMS320C30 processor. The execution speed on the TMS320C30 for these variants can be expected to be slower than for the basic LMS algorithm due to additional decision-type instructions required for testing conditions involving the sign of the error signal or the data sample. Hence, the programming examples covered in this book only use the basic LMS algorithm.

The LMS algorithm has been quite useful in adaptive equalizers, telephone cancellers, and so forth. Other methods, such as the recursive least squares (RLS) algorithm [4], can offer faster convergence than the LMS but at the expense of more computations. The RLS algorithm is based on starting with the optimal solution and then using each input sample to update the impulse response in order to maintain that optimality. The right step size and direction are defined over each time sample.

A discussion of adaptive algorithms for restoring signal properties can also be found in [4]. Such algorithms become useful when an appropriate reference signal is not available. The filter is adapted in such a way as to restore some property of the signal lost before reaching the adaptive filter. Instead of the desired waveform as a template, as in the LMS or RLS algorithms, this property is used for the adaptation of the filter. When the desired signal is available, the conventional approach such as the LMS can be used; otherwise a priori knowledge about the signal is used.

7.4 PROGRAMMING EXAMPLES USING C AND TMS320C30 CODE

Several examples will illustrate various applications of adaptive signal processing using the least mean square algorithm. Two projects on adaptive filters are discussed in Chapter 8.

Example 7.1 *Adaptation Using C Code Without the TI Simulator.* This example illustrates the LMS algorithm using C code. It simulates the adaptive structure shown in Figure 7.1. The program for this example ADAPTC.C is shown in Figure 7.5. The desired signal $d = 2\cos[2n\pi f/F_s)]$ is a cosine waveform with an amplitude of 2 and a frequency of 1,000 Hz. The noise

7.4 Programming Examples using C and TMS320C30 Code

```c
/*ADAPTC.C-ADAPTATION USING LMS IN C WITHOUT THE TI SIMULATOR*/
#include <stdio.h>
#include <math.h>
#define beta 0.05                       //step factor for coeff update
#define N    20                         //order of filter (N+1)
#define NS   40                         //number of samples
#define FS   8000                       //sampling frequency
#define Pi   3.1415926
#define DESIRED 2*cos(2*Pi*T*1000/FS)   //desired signal
#define NOISE sin(2*Pi*T*1000/FS)       //noise signal

main()
{
  long I, T;
  double D, Y, E;
  double W[N+1] = {0.0};
  double X[N+1] = {0.0};
  FILE *desired, *Y_out, *error, *noise;
  desired = fopen ("DESIRED", "w++"); //open file for desired samples
  Y_out = fopen ("Y_OUT", "w++");     //open file for Y output samples
  error = fopen ("ERROR", "w++");     //open file for error samples
  noise = fopen ("NOISE", "w++");     //open file for noise samples
  for (T = 0; T < NS; T++)            //start adaptive alogrithm
  {
    X[0] = NOISE;                     //new noise sample
    D = DESIRED;                      //desired signal
    Y = 0;                            //set output of filter to zero
    for (I = 0; I <= N; I++)
      Y += (W[I] * X[I]);             //calculate filter output
    E = D - Y;                        //calculate error signal
    for (I = N; I >= 0; I--)          //convolve the coeff with noise samples
    {
      W[I] = W[I] + (2*beta*E*X[I]);  //update filter coefficients
      if (I != 0)
        X[I] = X[I-1];                //move data sample
    }
    fprintf (desired, "\n%10g  %10f", (float) T/FS, D);
    fprintf (Y_out, "\n%10g  %10f", (float) T/FS, Y);
    fprintf (error, "\n%10g  %10f", (float) T/FS, E);
    fprintf (noise, "\n%10g  %10f", (float) T/FS, X[0]);
  }
  fclose (desired);
  fclose (Y_out);
  fclose (error);
  fclose (noise);
}
```

FIGURE 7.5. Program in C for general adaptive structure (ADAPTC.C)

reference signal, which is the input to the adaptive filter, is $x = \sin[2n\pi f/F_s]$. The sampling frequency F_s is set at 8,000 Hz. The order of the adaptive filter is set for a total of 21 weights or coefficients (with $N = 20$). The constant β, set to 0.05, determines the rate of convergence or the step size. A large value of β can affect the stability of the adaptation process. The number of samples is chosen as 40.

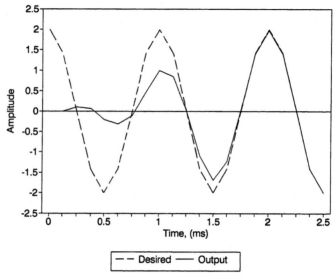

FIGURE 7.6. Plot of output signal $y(n)$ from C program converging to desired signal $d(n)$

The coefficients or weights are updated using the LMS equation (7.15) in the program. Figure 7.6 shows a plot of the adaptive filter's output $y(n)$ converging to the desired cosine signal $d(n)$.

A version of the program in Figure 7.5, with graphics and interactive capabilities to plot the adaptation process for different values of β is on the accompanying disk as ADAPTIVE.C. This program version uses a desired signal $d(n) = \cos[2n\pi f/F_s]$, a noise reference $x(n) = \sin[2n\pi f/F_s]$, and 31 coefficients for the adaptive filter. Compile and run the program ADAPTIVE.C. When prompted to enter a value for the rate of adaptation β, enter 0.01. The plot shown in Figure 7.7(a) will then be displayed on the PC. Press F2 to enter a new value of $\beta = 0.05$ for a faster rate of convergence and obtain the plot in Figure 7.7(b).

Example 7.2 Adaptive Filter for Noise Cancellation, Using C Code. The adaptive filter structure in Figure 7.3 is used for this example to illustrate the cancellation of an additive noise. Figure 7.8 shows the program listing ADAPTDNC.C in C code, which implements a 50-coefficient adaptive filter. The output of the 50-coefficient adaptive filter Y is calculated from

$$Y\mathrel{+}= W[I]*Delay[I], \quad I = 0, 1, \ldots, N-1$$

in the program, which is the convolution equation representing an FIR filter. The coefficients are updated from

$$W[I] = W[I] + \beta*E*Delay[I], \quad I = 0, 1, \ldots, N-1$$

7.4 Programming Examples using C and TMS320C30 Code

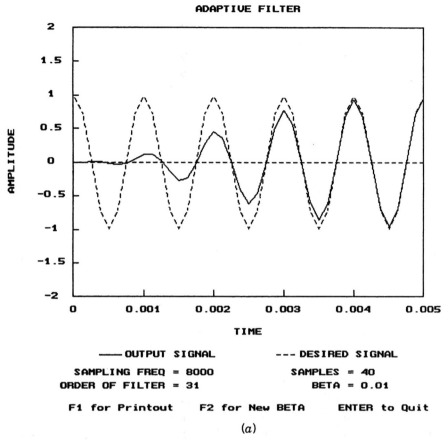

FIGURE 7.7. Plot of output signal $y(n)$ converging to desired signal $d(n)$ using interactive capability with program ADAPTIVE.C: (a) $\beta = 0.01$; (b) $\beta = 0.05$.

where Delay[I] represents the input noise samples to the 50-coefficient adaptive filter and is updated from

Delay[I] = Delay[I-1]

The overall output of the adaptive structure $E = D - Y$ is the difference between the primary input D (desired signal plus noise) and the adaptive filter output Y.

To test the program ADAPTDNC.C, proceed with the following:

1. Choose a 1-kHz sine function as the desired signal d. It is on the accompanying disk as SIN1000. A utility is available with the Hypersignal package (Appendix C) that can generate a sine or cosine function

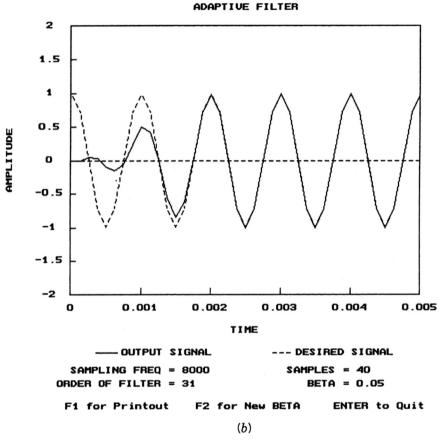

FIGURE 7.7. *(Continued)*

of a specified frequency. For example, from the DIFFERENCE EQUA-TIONS option,

SIN(2*PI*N*1000 / 10000)

generates a 1,000-Hz sine function with a sampling frequency F_s of 10,000 Hz. The number of points per cycle is NP = F_s/f = 10. The resulting file needs to be converted to decimal format using the BINARY to ASCII Hypersignal utility (after deleting the first 10 values). The Hypersignal package also contains a sine generator utility.

2. An additive sinusoidal "noise" n is chosen to be a 312-Hz sine function (on the accompanying disk as SIN312). The desired signal d is added to n to form the primary input D to the adaptive structure. The resulting file is on the accompanying disk as DPLUSN.

```c
/*ADAPTDNC.C-50-COEFFICIENT ADAPTIVE FILTER IN C         */
#define beta 2.5E-10      /*rate of convergence         */
#define N 50              /*# of coefficients           */
#define NS 128            /*# of output sample points */

main()
{
  int I,T;
  float Y, E, D;
  float W[N+1];
  float Delay[N+1];
  volatile int *IO_INPUT = (volatile int*) 0x804000;
  volatile int *IO_INPUT1= (volatile int*) 0x804001;
  volatile int *IO_OUTPUT= (volatile int*) 0x804002;
  for (T=0; T < N; T++)
  {
    W[T] = 0.0;
    Delay[T] = 0.0;
  }
  for (T=0; T < NS; T++)   /*NS is # of output samples */
  {
    Delay[0] = *IO_INPUT1; /*noise signal n             */
    D = *IO_INPUT;         /*desired signal+noise d+n  */
    Y = 0;
    for (I = 0; I < N; I++)
      Y += (W[I] * Delay[I]);           /*filter output */
    E = D - Y;                          /*error signal  */
    for (I = N; I > 0; I--)
    {
      W[I] = W[I] + (beta*E*Delay[I]); /*update coeffs */
      if (I != 0)
      Delay[I] = Delay[I-1];            /*update samples*/
    }
    *IO_OUTPUT = E;
  }
}
```

FIGURE 7.8. Adaptive filter program for noise cancellation using C code (ADAPTDNC.C)

3. The reference input to the 50-coefficient adaptive filter, is chosen as a 312-Hz cosine function (on the accompanying disk as COS312).

4. The primary input DPLUSN ($d + n$) is specified at port 0x804000, the reference input COS312 at port 0x804001, and the resulting output E is captured from port 0x804002.

248 Adaptive Filters

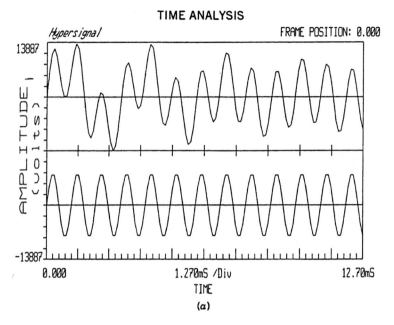

(a)

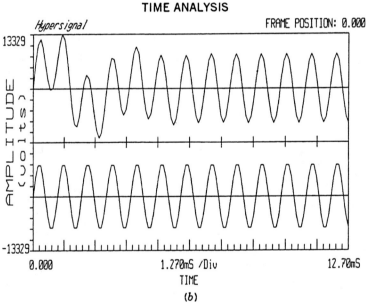

(b)

FIGURE 7.9. Plot of adaptive filter output (upper graph) converging to desired input (lower graph): (a) $\beta = 0.25 \times 10^{-10}$; (b) $\beta = 1.0 \times 10^{-10}$.

7.4 Programming Examples using C and TMS320C30 Code

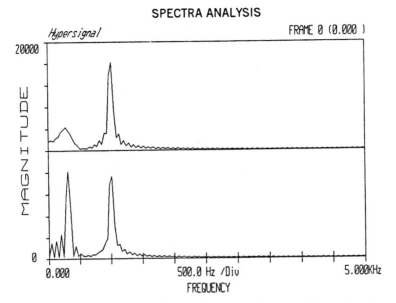

FIGURE 7.10. Frequency response of adaptive filter output (upper graph) showing reduction of 312-Hz additive sinusoidal noise for $\beta = 1.0 \times 10^{-10}$

Compile and run the program ADAPTDNC. Figure 7.9(a) (upper graph) shows the output converging to the desired 1,000-Hz sinusoid (lower graph) using a convergence rate of $\beta = 0.25 \times 10^{-10}$. Figure 7.9(b) illustrates a faster rate of convergence with $\beta = 1.0 \times 10^{-10}$. The frequency response of the output for $\beta = 1.0 \times 10^{-10}$ is displayed in Figure 7.10. The lower graph contains two "delta" functions representing the 312-Hz and the 1,000-Hz sinusoids at the primary input and the upper graph illustrates the reduction of the 312-Hz noise.

EXECUTION TIME
The execution time of the adaptive filter can be found. To find the main processing time for *one* sample, set a breakpoint (using the simulator) at

```
FLOAT *AR7,R5
```

which is the equivalent assembly code (after compiling) of the instruction

```
Delay[0] = *IO_INPUT1
```

for acquiring an input noise sample (input to the 50-coefficient adaptive filter) from port 0x804001. Execute and run the program until the set breakpoint, and obtain a clock value of CLK = 97(hex). Execute and run the

250 Adaptive Filters

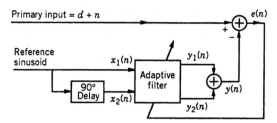

FIGURE 7.11. Two-weight adaptive notch structure

program again. Execution should again halt at the set breakpoint, producing a new clock value of CLK = 37A(hex). The difference (37A − 97) times the instruction cycle time of 60 ns yields 44.34 µs, which is the execution time of the adaptive filter for one sample. This 50-coefficient adaptive filter allows for a maximum sampling frequency of $F_s = 1/44.34$ µs $\simeq 22.6$ kHz, which is quite satisfactory for many applications. A (lower) sampling frequency would permit a higher-order adaptive filter.

An equivalent adaptive filter in assembly code produces an execution time of 13.62 µs for one sample, as discussed in Example 7.5.

Example 7.3 *Adaptive Notch Filter with Two Weights, Using TMS320C30 Code.* Figure 7.11 shows a two-weight adaptive notch structure that can be used to illustrate the cancellation of a sinusoidal interference. This structure as a notch filter has been discussed in [1, 2]. The primary input is a desired signal d with additive interference noise n. The reference input to the adaptive filter consists of $x_1(n)$ and $x_2(n)$ (which is $x_1(n)$ delayed by 90°). The adaptive filter consists only of two weights or coefficients. The output of the adaptive filter $y(n) = y_1(n) + y_2(n) = w_1(n)x_1(n) + w_2(n)x_2(n)$. The output signal $e(n)$ is the difference between the signal at the primary input $(d + n)$ and the output of the adaptive filter $y(n)$.

To simulate this notch filter, the desired signal d is chosen to be a sine function with a frequency of 1 kHz and the noise n a sine function with a frequency of 312 Hz. The reference input noise to the adaptive filter $x_1(n)$ is also a sine function with a frequency of 312 Hz. The second input to the adaptive filter $x_2(n)$ is a cosine function of the same frequency as $x_1(n)$. The files are all on the accompanying disk: SIN312 for n and $x_1(n)$, COS312 for $x_2(n)$, and DPLUSN for $d + n$. These files can be generated using the second-generation fixed-point simulator (a small program can be written to add d and n). They can also be generated with the Hypersignal package using the DIFFERENCE EQUATIONS utility option. For example, $x_2(n)$ can be generated using the DIFFERENCE EQUATIONS option (in Hypersignal) with COS(2*PI*N*312 / 10000), where 312 represents the frequency of the cosine function and 10000 is the sampling frequency F_s. The resulting data file (the first 10 values should be deleted) can be converted to decimal format using the BINARY to ASCII utility (in Hypersignal). Note that the number of points per cycle is NP = F_s/f = 32 points per cycle.

The program NOTCH2W.ASM, listed in Figure 7.12 implements the two-weight adaptive notch filter. Create the following port configuration file to contain

File Name	Port Address
SIN312	804000h
COS312	804001h
DPLUSN	804002h
NOTCH2W.DAT	804003h

Observe the following from the program NOTCH2W.ASM:

1. AR2 and AR5 contain the port addresses for $x_1(n)$ and $x_2(n)$, respectively. AR1 contains the first memory address in internal RAM. The instruction FLOAT *AR2,R3 inputs a sample value of $x_1(n)$ (from SIN312). The sample value is then stored in internal RAM at 809800h (pointed by AR1). An input sample of $x_2(n)$ (from COS312) is obtained using the instruction FLOAT *AR5,R3. The cosine sample is then stored in RAM at 809801h (pointed by AR1). AR1 is then incremented in a circular fashion to point "back" at 809800h. The desired signal plus noise sample, from the file DPLUSN at 804002h, is obtained and loaded in R4. This sample will be used for the calculation of the error signal. The starting address (809802h) of the two coefficients is loaded into AR0 before the FIR filter subroutine is called.

2. The filter subroutine finds $y(n)$ = H1(n)$x_1(n)$ + H2(n)$x_2(n)$, where H1(n) and H2(n) represent the coefficients at time n. Chapter 4 contains many examples for implementing an FIR filter using TMS320C30 code. In this case, there are only two coefficients and two samples. The first sample is the sine (SINE312) sample $x_1(n)$, which is one of the two inputs to the adaptive filter, and the second sample is the cosine (COS312) sample $x_2(n)$. For faster execution, the code for the filter subroutine can be placed directly where it was called, hence eliminating the instructions to call the filter subroutine (CALL FILT) and to return from this subroutine (RETSU). This also can be done with the adaptation subroutine ADAPT. The separate subroutines were used here to make the program easier to follow.

3. R0 contains a sample of the adaptive filter's output $y(n)$ for a specific time n. The error signal (for this same specific time n) is calculated using the instruction SUBF R0,R4,R0. R4 contains the sample $d + n$ obtained previously in step 1 with the instruction FLOAT *AR3,R4. This error signal $e(n)$ is converted to integer, then output to port address 804003h pointed by AR4. This first error signal sample value is not meaningful because the output of the adaptive filter is zero. Note that the two coefficients, initialized to zero (before the LOOP section of code) are not yet updated.

252 Adaptive Filters

```
;NOTCH2W.ASM-TWO-WEIGHT ADAPTIVE NOTCH FILTER
            .TITLE    "NOTCH2W"           ;2-WEIGHT ADAPTIVE NOTCH FILTER
            .GLOBAL   RESET,BEGIN,FILT,ADAPT        ;REF/DEF SYMBOLS
            .SECT     "VECTORS"           ;ASSEMBLE INTO VECT SECTION
RESET       .WORD     BEGIN               ;RESET VECTOR
            .DATA                         ;ASSEMBLE INTO DATA SECTION
SIN_ADDR    .WORD     804000H             ;SINE PORT ADDRESS
COS_ADDR    .WORD     804001H             ;COSINE PORT ADDRESS
DN_ADDR     .WORD     804002H             ;D+N PORT ADDRESS
OUT_ADDR    .WORD     804003H             ;OUTPUT PORT ADDRESS
SC_ADDR     .WORD     SC                  ;RAM ADDR OF SINE+COSINE SAMPLES
COEFF_ADDR  .WORD     COEFF               ;ADDR OF COEFF H(N-1)
ERF_ADDR    .WORD     ERR_FUNC            ;ADDR OF ERROR FUNCTION
BETA        .FLOAT    5.0E-10             ;RATE OF ADAPTATION
LENGTH      .SET      2                   ;FILTER LENGTH N = 2
            .BSS      SC,LENGTH           ;N SPACE FOR SINE+COSINE SAMPLES
            .BSS      COEFF,LENGTH        ;N SPACE FOR COEFFICIENTS
            .BSS      ERR_FUNC,1          ;1 SPACE FOR ERROR FUNTION
            .TEXT                         ;ASSEMBLE INTO TEXT SECTION
BEGIN       LDI       @SIN_ADDR,AR2       ;SINE ADDR        -> AR2
            LDI       @COS_ADDR,AR5       ;COSINE ADDR      -> AR5
            LDI       @DN_ADDR,AR3        ;D+N ADDR         -> AR3
            LDI       @OUT_ADDR,AR4       ;OUTPUT ADDR      -> AR4
            LDI       @ERF_ADDR,AR6       ;ERROR FUNC ADDR  -> AR6
            LDI       LENGTH,BK           ;FILTER LENGTH N  -> BK
            LDI       @COEFF_ADDR,AR0     ;COEFF H(N-1) ADDRESS    -> AR0
            LDI       @SC_ADDR,AR1        ;SAMPLE SINE+COSINE ADDR -> AR1
            LDF       0,R0                ;INIT R0=0
            RPTS      LENGTH-1            ;PERFORM NEXT 2 INST. N TIMES
            STF       R0,*AR0++           ;INIT COEFF TO 0, IN // WITH
||          STF       R0,*AR1++%          ;INIT SINE+COSINE SAMPLES TO 0
LOOP        FLOAT     *AR2,R3             ;INPUT SINE SAMPLE INTO R3
            STF       R3,*AR1++%          ;STORE SINE SAMPLE IN RAM
            FLOAT     *AR5,R3             ;INPUT COSINE SAMPLE INTO R3
            STF       R3,*AR1++%          ;STORE COSINE SAMPLE IN RAM
            FLOAT     *AR3,R4             ;INPUT SIGNAL+NOISE(D+N) INTO R4
            LDI       @COEFF_ADDR,AR0     ;H(N-1) ADDR -> AR0
            CALL      FILT                ;CALL FIR SUBROUTINE FILT
            SUBF      R0,R4,R0            ;ERROR = DN - Y --> R0
            FIX       R0,R1               ;CONVERT R0 TO INTEGER -> R1
            STI       R1,*AR4             ;STORE R1 (ERROR) INTO OUTPUT PORT
            MPYF      @BETA,R0            ;R0 = ERROR FUNCTION = BETA*ERROR
            STF       R0,*AR6             ;STORE ERROR FUNCTION
            LDI       @COEFF_ADDR,AR0     ;H(N-1) ADDR -> AR0
            CALL      ADAPT               ;CALL ADAPTING SUBROUTINE
```

FIGURE 7.12. Program to implement two-weight adaptive notch filter using TMS320C30 code (NOTCH2W.ASM)

```
                BR      LOOP                  ;REPEAT WITH NEXT SAMPLE
;FIR FILTER SUBROUTINE
FILT            MPYF    *AR0++,*AR1++%,R0     ;H1(n)*x1(n)=y1(n) -> R0
                LDF     0,R2                  ;R2 = 0
                MPYF    *AR0++,*AR1++%,R0     ;H2(n)*x2(n)=y2(n) -> R0
||              ADDF    R0,R2,R2              ;R2 = y1(n)
                ADDF    R0,R2,R0              ;y1(n)+y2(n)=y(n) -> R0
                RETSU                         ;RETURN FROM SUBROUTINE
;ADAPTATION SUBROUTINE
ADAPT           MPYF    *AR6,*AR1++%,R0       ;ERROR FUNCTION*x1(n) -> R0
                LDF     *AR0,R3               ;H1(n) -> R3
                MPYF    *AR6,*AR1++%,R0       ;ERROR FUNCTION*x2(n) -> R0
||              ADDF    R3,R0,R2              ;H1(n)+ERROR FUNCTION*x1(n)->R2
                LDF     *+AR0,R3              ;H2(n) -> R3
||              STF     R2,*AR0++             ;H1(n+1)=H1(n)+ERROR FUNCTION*x1(n)
                ADDF    R3,R0,R2              ;H2(n)+ERROR FUNCTION*x2(n)->R2
                STF     R2,*AR0               ;H2(n+1)=H2(n)+ERROR FUNCTION*x2(n)
                RETSU                         ;RETURN FROM SUBROUTINE
                .END                          ;END
```

FIGURE 7.12. *(Continued)*

The resulting value in R0, which now contains $e(n)$, is multiplied by the constant BETA to yield the error function stored in memory (809804h) pointed by AR6. AR0 is reinitialized to the starting address (809802h) of the coefficients before the adaptation subroutine is called.

4. Within the adaptive subroutine ADAPT, the first multiply instruction yields R0, which contains the value $\beta e(n)x_1(n)$ for a specific time n. Note that AR6 contains the memory address where the error function $\beta e(n)$ is stored. The second multiply instruction yields R0, which contains the value $\beta e(n)x_2(n)$. This multiply instruction is in parallel with an ADDF instruction in order to update the first coefficient H1(n). This first coefficient was loaded into R3 before the second multiply operation. The parallel addition instruction

```
||    ADDF  R3,R0,R2
```

adds R3 (which contains the first coefficient H1(n)) and R0 [which contains $\beta e(n)x_1(n)$ from the first multiply operation]. R2 contains now the updated coefficient H1(n+1). The instruction

```
LDF  *+AR0(1),R3
```

loads the second coefficient H2(n) into R3. Note that AR0 is preincremented *without* modification to the address (809803h) of the second coefficient. The parallel instruction

```
||    STF    R2,*AR0++
```

stores the updated coefficient R2 = H1(n+1) in memory pointed by AR0 = 809802h. AR0 is then postincremented to point at the address (809803h) of the second coefficient. The second ADDF instruction is similar to the first one and updates the second coefficient H2(n+1).

For each time n the preceding steps are repeated. For example, for the next unit of time, a new sample is acquired for the sine (SIN312), cosine (COS312), and $d + n$ (DPLUSN). The output of the adaptive filter $y(n)$ is calculated using the newly acquired sine and cosine samples and the previously updated coefficients. The second error signal sample is then calculated and output. The coefficients are then updated again.

Run this program. The time plot of the primary input $d + n$ (DPLUSN) is plotted in Figure 7.13(a), which is the addition of the desired signal d of frequency 1,000 Hz and the noise n of frequency 312 Hz. Figure 7.13(b) shows a frequency plot of d and n, displaying two delta functions at 1,000 and 312 Hz, respectively. Figure 7.14(a) shows the output "error" signal $e(n)$ converging to the desired signal $d(n)$. Figure 7.14(b) shows a delta function at the desired signal of frequency 1,000 Hz. It also shows a reduction of the 312-Hz noise.

A real-time version of this example is included in Chapter 8 to illustrate the reduction of a 60-Hz artifact in an electrocardiogram (ECG) signal.

Example 7.4   Adaptive Predictor Using TMS320C30 Code. This example describes an adaptive predictor structure, shown in Figure 7.15. This structure can be used for coding a speech signal. The primary input is the desired signal $d(n)$. This signal is delayed and becomes the input to the adaptive filter. The output of the adaptive filter $y(n)$ is adapted to the desired signal $d(n)$. When this is accomplished, the error signal $e(n)$ becomes zero.

To test this structure, the desired signal is chosen to be a sine function with a frequency of 312 Hz. The data file of this sine function is on the accompanying disk as SIN312 (used in the previous examples). The reference input to the adaptive filter is a cosine function with the same frequency as the sine function (delayed) and one-half its amplitude. This file is on the accompanying disk as HCOS312.

The program used for this example ADAPTP.ASM in TMS320C30 code is listed in Figure 7.16. The length of the filter is set to 41 coefficients or weights. This program has many similar features as the program NOTCH2W.ASM discussed in the previous example with the two-weight adaptive notch structure (Figure 7.12). See also the FIR filter examples in Chapter 4. Note that the filter and adaptation subroutines can be placed where they

7.4 Programming Examples using C and TMS320C30 Code

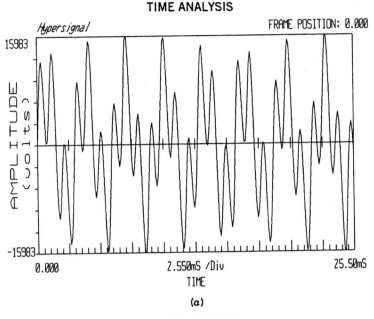

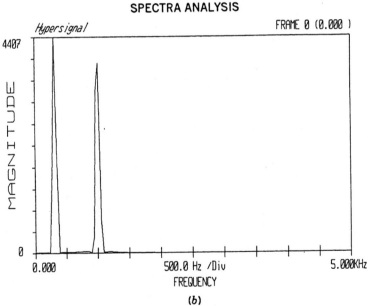

FIGURE 7.13. Plot of desired signal plus noise ($d + n$) in the adaptive notch filter structure: (a) time domain; (b) frequency domain

256 Adaptive Filters

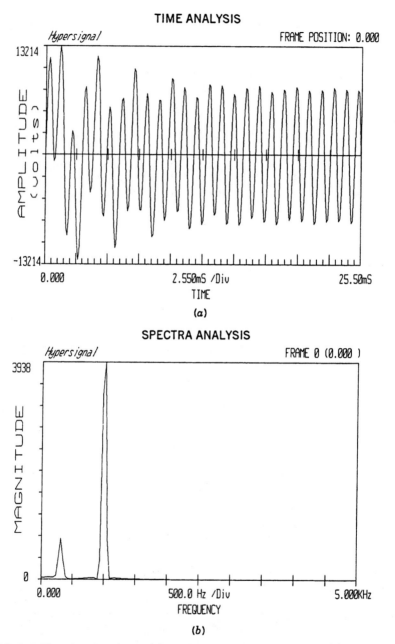

FIGURE 7.14. Plot of output signal $e(n)$ of adaptive notch filter structure: (a) time domain; (b) frequency domain

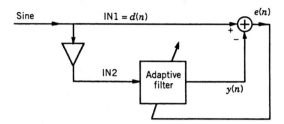

FIGURE 7.15. Adaptive predictor structure

are called. This would eliminate the two `CALL` and the two `RETSU` instructions, resulting in a faster execution (as shown in the next example). This was not done, in order to make it easier to follow the program flow.

Consider the program `ADAPTP.ASM` in Figure 7.16:

1. `DN_ADDR` specifies the port address (804001h) for the input (`HCOS312`) to the 41-coefficient adaptive filter. `DNS_ADDR` specifies the memory address of the "last" (higher-memory) sample `HCOS312`. Note that 41 memory locations are reserved in the `BSS` section for the `HCOS312` samples that are used as the input to the adaptive filter.
2. The FIR filter subroutine performs a total of 41 multiplications to yield the adaptive filter's output $y(n)$ for a specific time n.
3. Before the adaptation subroutine is called, the repeat counter register `RC` is loaded with length-2, or `RC` = 39. This is in order to use the repeat block instruction `RPTB`. The block of code between the instruction `RPTB LOOP_END` and the instruction `STF R2,*AR0++` is executed a total of 40 times (repeated 39 times). Note that this `STF` instruction is also included within the block of code to be executed 40 times because it is in parallel with the preceding `LDF` instruction.

```
;ADAPTP.ASM-ADAPTIVE PREDICTOR
            .TITLE   "ADAPTP.ASM"        ;ADAPTIVE FIR FILTER
            .GLOBAL  RESET,BEGIN,FILT,ADAPT  ;REF/DEF SYMBOLS
            .SECT    "VECTORS"           ;ASSEMBLE INTO VECTOR SECTION
RESET       .WORD    BEGIN               ;RESET VECTOR
            .DATA                        ;ASSEMBLE INTO DATA SECTION
D_ADDR      .WORD    804000H             ;DESIRED SIGNAL PORT ADDRESS
DN_ADDR     .WORD    804001H             ;INPUT TO ADAPT FILTER PORT ADDR
OUT_ADDR    .WORD    804002H             ;OUTPUT PORT ADDRESS
DNS_ADDR    .WORD    DNS+LENGTH-1        ;LAST ADDR OF DN (HCOS312) SAMPLES
COEFF_ADDR  .WORD    .COEFF              ;ADDR OF COEFF H(N-1)
ERF_ADDR    .WORD    ERR_FUNC            ;ADDR OF ERROR FUNCTION
BETA        .FLOAT   5.0E-10             ;RATE OF ADAPTATION
LENGTH      .SET     41                  ;FILTER LENGTH N
```

FIGURE 7.16. Program for adaptive predictor using TMS320C30 code (`ADAPTP.ASM`)

258 Adaptive Filters

```
            .BSS    DNS,LENGTH           ;N SPACE FOR DN SAMPLES
            .BSS    COEFF,LENGTH         ;N SPACE FOR COEFFICIENTS
            .BSS    ERR_FUNC,1           ;1 SPACE FOR ERROR FUNTION
            .TEXT                        ;ASSEMBLE INTO TEXT SECTION
BEGIN       LDI     @D_ADDR,AR2          ;DESIRED SIGNAL ADDR -> AR2
            LDI     @DN_ADDR,AR3         ;DN ADDRESS           -> AR3
            LDI     @OUT_ADDR,AR4        ;OUTPUT ADDR          -> AR4
            LDI     @ERF_ADDR,AR6        ;ERROR FUNC ADDR      -> AR6
            LDI     LENGTH,BK            ;FILTER LENGTH N      -> BK
            LDI     @COEFF_ADDR,AR0      ;COEFF H(N-1) ADDRESS-> AR0
            LDI     @DNS_ADDR,AR1        ;LAST DN SAMPLE ADDR -> AR1
            LDF     0,R0                 ;INIT R0=0
            RPTS    LENGTH-1             ;PERFORM NEXT 2 INST. N TIMES
            STF     R0,*AR0++            ;INIT COEFF TO 0, IN // WITH
||          STF     R0,*AR1++%           ;INIT DN SAMPLES TO 0
LOOP        FLOAT   *AR3,R3              ;INPUT DN SAMPLE INTO R3
            STF     R3,*AR1++(1)%        ;STORE IN RAM
            FLOAT   *AR2,R4              ;INPUT D SAMPLE INTO R4
            LDI     @COEFF_ADDR,AR0      ;H(N-1) ADDR -> AR0
            CALL    FILT                 ;CALL FIR SUBROUTINE FILT
            FIX     R0,R1                ;CONVERT R0=Y TO INTEGER->R1
            STI     R1,*AR4              ;STORE R1 INTO OUTPUT PORT
            SUBF    R0,R4,R0             ;ERROR = D-Y -> R0
            MPYF    @BETA,R0             ;R0=ERR FUNC=BETA*ERROR
            STF     R0,*AR6              ;STORE ERROR FUNCTION
            LDI     LENGTH-2,RC          ;RESET REPEAT COUNTER
            LDI     @COEFF_ADDR,AR0      ;H(N-1) ADDR -> AR0
            CALL    ADAPT                ;CALL ADAPTING SUBROUTINE
            BR      LOOP                 ;REPEAT WITH NEXT SAMPLE
;FIR FILTER SUBROUTINE
FILT        LDF     0,R2                 ;R2 = 0
            RPTS    LENGTH-1             ;NEXT 2 INST.(LENGTH-1)TIMES
            MPYF    *AR0++,*AR1++%,R0    ;H(N-1-i)*x(n-(N-1-i))i=0,...,N-1
||          ADDF    R0,R2,R2             ;ACCUMULATE
            ADDF    R0,R2,R0             ;ADD LAST PRODUCT = y(n) -> R0
            RETSU                        ;RETURN FROM SUBROUTINE
;ADAPTATION SUBROUTINE
ADAPT       MPYF    *AR6,*AR1++%,R0      ;ERR FUNC * x(n-(N-1))    -> R0
            LDF     *AR0,R3              ;H(N-1) -> R3
            RPTB    LOOP_END             ;REPEAT(N-2 TIMES) UNTIL LOOP_END
            MPYF    *AR6,*AR1++%,R0      ;ERR FUNC * x(n-(N-1-i)) -> R0
||          ADDF    R3,R0,R2             ;H(N-1-i)+ERR FUNC * x(n-(N-1-i))
LOOP_END    LDF     *+AR0(1),R3          ;LOAD SUBSEQUENT H(k) -> R3
||          STF     R2,*AR0++            ;STORE/UPDATE COEFFICIENT
            ADDF    R3,R0,R2             ;H(n+1) = H(n) + ERR FUNC * x(n)
            STF     R2,*AR0              ;STORE/UPDATE THE LAST COEFF
            RETSU                        ;RETURN FROM SUBROUTINE
            .END                         ;END
```

FIGURE 7.16. (*Continued*)

Run this program. Configure the port map such that the desired signal $d(n)$ or SIN312 is at 80400h, the input to the adaptive filter HCOS312 is at 804001h, and the simulator output is at 804002h. Acquire enough output data points (execution should halt after 338 data points, the number of data points

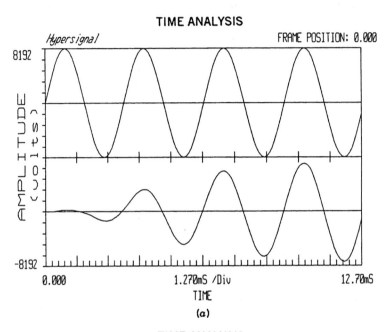

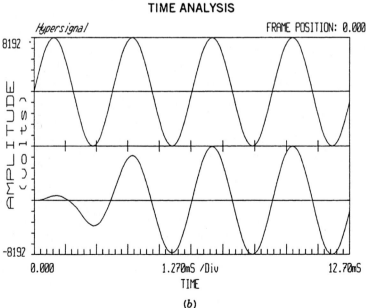

FIGURE 7.17. Output of adaptive filter converging to desired signal for different rate of adaptation: (*a*) $\beta = 1.5 \times 10^{-10}$; (*b*) $\beta = 5.0 \times 10^{-10}$

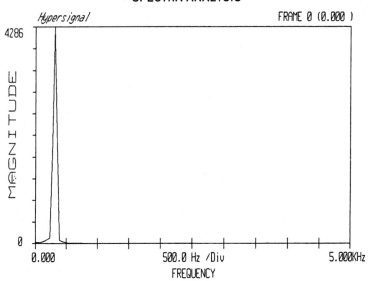

FIGURE 7.18. Frequency plot of output showing a delta function at the frequency of the desired signal

in the file HCOS312). Figure 7.17(a) shows the plot of the desired signal SIN312 and the output (lower graph) of the adaptive filter $y(n)$ converging to the desired signal, for an adaptation rate constant of $\beta = 1.5 \times 10^{-10}$. Figure 7.17(b) shows a faster adaptation rate with $\beta = 5.0 \times 10^{-10}$. The frequency plot of the adaptive filter's output $y(n)$ in Figure 7.18 shows a delta function at the frequency (312 Hz) of the desired signal. For a different adaptation constant or filter's length, change the set values of BETA and LENGTH in the program ADAPTP.ASM. A real-time project with this structure using C code is discussed in Chapter 8.

Example 7.5 Adaptive Filter for Noise Cancellation, Using TMS320C30 Code. The program ADAPTDNA.ASM in Figure 7.19 implements the adaptive filter structure in Figure 7.3 with 50 coefficients, using TMS320C30 code. The equivalent example implemented in C code was discussed in Example 7.2 (see also Examples 7.3 and 7.4). This implementation, unlike Examples 7.3 and 7.4, avoids separate filtering and adapting subroutines and uses a branching with delay instruction. This resulted in a faster execution time. Figure 4.9 associated with FIR filters provides much background for the program in Figure 7.19. The code section starting with ADAPT in Figure 7.19 updates the coefficients of the adaptive filter. Similar results are obtained as in Example 7.2. For a different number of filter coefficients, change the variable LENGTH.

7.4 Programming Examples using C and TMS320C30 Code

```
;ADAPTDNA.ASM-ADAPTIVE FILTER WITH SIGNAL+NOISE @ PRIMARY
            .TITLE    "ADAPTPDNA.ASM"   ;ADAPTIVE FIR FILTER
            .GLOBAL   RESET,BEGIN,FILT,ADAPT ;REF/DEF SYMBOLS
            .SECT     "VECTORS"         ;ASSEMBLE INTO VECTOR SECTION
RESET       .WORD     BEGIN             ;RESET VECTOR
            .DATA                       ;ASSEMBLE INTO DATA SECTION
D_ADDR      .WORD     804000H           ;DESIRED SIGNAL+NOISE PORT ADDR
DN_ADDR     .WORD     804001H           ;INPUT TO ADAPT FILTER PORT ADDR
OUT_ADDR    .WORD     804002H           ;OUTPUT PORT ADDRESS
DNS_ADDR    .WORD     DNS+LENGTH-1      ;LAST ADDR OF NOISE SAMPLES
COEFF_ADDR  .WORD     COEFF             ;ADDR OF COEFF H(N-1)
ERF_ADDR    .WORD     ERR_FUNC          ;ADDR OF ERROR FUNCTION
BETA        .FLOAT    2.5E-10           ;RATE OF ADAPTATION
LENGTH      .SET      50                ;FILTER LENGTH N
            .BSS      DNS,LENGTH        ;N SPACE FOR NOISE SAMPLES
            .BSS      COEFF,LENGTH      ;N SPACE FOR COEFFICIENTS
            .BSS      ERR_FUNC,1        ;1 SPACE FOR ERROR FUNTION
            .TEXT                       ;ASSEMBLE INTO TEXT SECTION
BEGIN       LDI       @D_ADDR,AR2       ;D+N ADDR           -> AR2
            LDI       @DN_ADDR,AR3      ;NOISE ADDRESS      -> AR3
            LDI       @OUT_ADDR,AR4     ;OUTPUT ADDR        -> AR4
            LDI       @ERF_ADDR,AR6     ;ERROR FUNC ADDR    -> AR6
            LDI       LENGTH,BK         ;FILTER LENGTH N    -> BK
            LDI       @COEFF_ADDR,AR0   ;COEFF H(N-1) ADDRESS-> AR0
            LDI       @DNS_ADDR,AR1     ;LAST N SAMPLE ADDR -> AR1
            LDF       0,R0              ;INIT R0=0
            RPTS      LENGTH-1          ;PERFORM NEXT 2 INST. N TIMES
            STF       R0,*AR0++         ;INIT COEFF TO 0, IN // WITH
||          STF       R0,*AR1++%        ;INIT DN SAMPLES TO 0
MAIN        LDI       LENGTH,AR5        ;AR5 = FILTER LENGTH
            SUBI      1,AR5             ;DECREMENT AR5
LOOP        FLOAT     *AR3,R3           ;INPUT NOISE SAMPLE INTO R3
            STF       R3,*AR1++(1)%     ;STORE IN RAM
            FLOAT     *AR2,R4           ;INPUT D+N SAMPLE INTO R4
            LDI       @COEFF_ADDR,AR0   ;H(N-1) ADDR -> AR0
FILT        LDF       0,R2              ;R2 = 0
            RPTS      LENGTH-1          ;NEXT 2 INST.(LENGTH-1)TIMES
            MPYF      *AR0++,*AR1++%,R0 ;H(N-1-i)*x(n-(N-1-i))i=0..N-1
||          ADDF      R0,R2,R2          ;ACCUMULATE
            ADDF      R0,R2,R0          ;ADD LAST PRODUCT = y(n) -> R0
            SUBF      R0,R4,R0          ;ERROR = (D+N)-Y -> R0
            FIX       R0,R1             ;CONVERT INTO INTEGER
            STI       R1,*AR4           ;OUTPUT
            MPYF      @BETA,R0          ;R0=ERR FUNC=BETA*ERROR
            STF       R0,*AR6           ;STORE ERROR FUNCTION
```

FIGURE 7.19. Adaptive filter program for noise cancellation using TMS320C30 code (ADAPTDNA.ASM)

262 Adaptive Filters

```
                LDI     LENGTH-2,RC         ;RESET REPEAT COUNTER
                LDI     @COEFF_ADDR,AR0     ;H(N-1) ADDR -> AR0
ADAPT           MPYF    *AR6,*AR1++%,R0     ;ERR FUNC*x(n-(N-1))->R0
                LDF     *AR0,R3             ;H(N-1) -> R3
                RPTB    LOOP_END            ;REPEAT(N-2 TIMES) UNTIL LOOP_END
                MPYF    *AR6,*AR1++%,R0     ;ERR FUNC * x(n-(N-1-i)) -> R0
||              ADDF    R3,R0,R2            ;H(N-1-i)+ERR FUNC * x(n-(N-1-i))
LOOP_END        LDF     *+AR0(1),R3         ;LOAD SUBSEQUENT H(k) -> R3
||              STF     R2,*AR0++           ;STORE/UPDATE COEFFICIENT
                DBNZD   AR5,LOOP            ;DELAYED BRANCH FOR NEXT n
                ADDF    R3,R0,R2            ;H(n+1) = H(n) + ERR FUNC * x(n)
                STF     R2,*AR0             ;STORE/UPDATE THE LAST COEFF
                NOP                         ;ADDED TO USE DELAYED BRANCH
                BR      MAIN                ;REPEAT FOR 50 OUTPUT POINTS
                .END                        ;END
```

FIGURE 7.19. *(Continued)*

To find the execution time for *one* sample, set a breakpoint (within the simulator) at the instruction

```
FLOAT *AR3,R3
```

for acquiring an input noise sample from port 804001h.

Execute and run the program until the set breakpoint and obtain a clock value of CLK = 47h. Execute and run the program again until the set breakpoint and obtain a new clock value of CLK = 12Ah. The main processing loop (starting at the label LOOP) for one sample executes in 13.62 μs, obtained from $(12A - 47) \times 60$ ns. The execution time for one sample using either C or assembly code follows:

C code: 44.34 μs (Example 7.2)

TMS320C20 code: 13.62 μs

illustrating that the assembly-code implementation executes 3.25 times faster than the equivalent C-code implementation. Note that the execution time of the C-code adaptive program depends on the C compiler version used. The next example illustrates a real-time implementation of this adaptive filter structure.

7.4 Programming Examples using C and TMS320C30 Code

Example 7.6 Real-Time Adaptive Filter for Noise Cancellation, Using TMS320C30 Code. This program example uses the basic adaptive structure in Figure 7.3 as an adaptive notch filter structure (see also Example 7.5). An AIC with two inputs is required for this example. Figure 3.16 provides the TLC32044 AIC interfacing diagram to the TMS320C30 serial port 1 through a 10-pin connector on the EVM. The primary (first) input IN to the AIC consist of a 2-Hz sinusoidal signal d added to a 60-Hz sinusoidal noise. Use a passive summer circuit or an OP AMP summer to add those two sinusoids.

```
;ADAPTER.ASM-ADAPTIVE STRUCTURE FOR NOISE CANCELLATION.OUTPUT AT E
            .TITLE  "ADAPTER.ASM"       ;ADAPTIVE FIR USING AIC
            .GLOBAL RESET,BEGIN,AICSET,AICSEC,IOPRI,IOAUX,SPSET
            .SECT   "vectors"           ;ASSEMBLE INTO VECT SECTION
RESET       .WORD   BEGIN               ;RESET VECTOR
            .DATA                       ;ASSEMBLE INTO DATA SECTION
AICSEC      .WORD   1428h,1h,5EBEh,73h  ;FOR AIC, Fs = 8 KHz
N_ADDR      .WORD   N+LENGTH-1          ;LAST ADDR OF NOISE SAMPLES
COEFF_ADDR  .WORD   COEFF               ;ADDR OF COEFF H(N-1)
ERF_ADDR    .WORD   ERR_FUNC            ;ADDR OF ERROR FUNCTION
BETA        .FLOAT  2.5E-12             ;RATE OF ADAPTATION CONSTANT
LENGTH      .SET    50                  ;FILTER LENGTH N
            .BSS    COEFF,LENGTH        ;N SPACE FOR COEFFICIENTS
            .BSS    ERR_FUNC,1          ;1 SPACE FOR ERROR FUNTION
N           .USECT  "XN_BUFF",LENGTH    ;NOISE SAMPLES BUFFER SIZE
            .TEXT                       ;ASSEMBLE INTO TEXT SECTION
BEGIN       LDP     SPSET               ;INIT DATA PAGE
            LDI     10h,R1              ;SERIAL PORT 1 OFFSET
            STI     R1,@SPSET           ;ENABLE SERIAL PORT 1
            CALL    AICSET              ;INIT AIC/TMS320C30
            LDI     @ERF_ADDR,AR6       ;ERROR FUNC ADDR       ->AR6
            LDI     LENGTH,BK           ;FILTER LENGTH N       ->BK
            LDI     @COEFF_ADDR.AR0     ;COEFF H(N-1) ADDRESS  ->AR0
            LDI     @N_ADDR,AR1         ;LAST NOISE SAMPLE ADDR->AR1
            LDF     0,R0                ;INIT R0=0
            RPTS    LENGTH-1            ;PERFORM NEXT 2 INST. N TIMES
            STF     R0,*AR0++           ;INIT COEFF TO 0, IN // WITH
 ||         STF     R0,*AR1++%          ;INIT NOISE SAMPLES TO 0
LOOP        CALL    IOAUX               ;OUTPUT,THEN GET N FROM AUX IN
            FLOAT   R6,R3               ;INPUT NOISE SAMPLE INTO R3
            STF     R3,*AR1++%          ;STORE IN RAM
            LDI     @COEFF_ADDR,AR0     ;H(N-1) ADDR -> AR0
            CALL    FILT                ;CALL FIR SUBROUTINE FILT
            CALL    IOPRI               ;OUTPUT,THEN GET D+N FROM PRI IN
```

FIGURE 7.20. Real-time adaptive filter program for noise cancellation using TMS320C30 code (ADAPTER.ASM)

264 Adaptive Filters

```
              FLOAT    R6,R4               ;R4=FLOATING POINT(D+N)
              SUBF3    R0,R4,R0            ;ERROR => R0 = (D+N)-Y
              FIX      R0,R7               ;R7=INTEGER(R0)
              MPYF     @BETA,R0            ;R0=ERR FUNC=BETA*ERROR
              STF      R0,*AR6             ;STORE ERROR FUNCTION
              LDI      LENGTH-2,RC         ;RESET REPEAT COUNTER
              LDI      @COEFF_ADDR,AR0     ;H(N-1) ADDR -> AR0
              CALL     ADAPT               ;CALL ADAPTING SUBROUTINE
              BR       LOOP                ;REPEAT WITH NEXT SAMPLE
FILT          LDF      0,R2                ;R2 = 0
              RPTS     LENGTH-1            ;NEXT 2 INST.(LENGTH-1)TIMES
              MPYF3    *AR0++,*AR1++%,R0   ;H(N-1-i)*x(n-(N-1-i)),i=0,...,N-1
||            ADDF3    R0,R2,R2            ;ACCUM
              ADDF3    R0,R2,R0            ;ADD LAST PRODUCT=y(n) -> R0
              RETSU                        ;RETURN FROM SUBROUTINE
ADAPT         MPYF3    *AR6,*AR1++%,R0     ;ERR FUNC * x(n-(N-1)) -> R0
              LDF      *AR0,R3             ;H(N-1) -> R3
              RPTB     LOOP_END            ;REPEAT (N-2 TIMES)UNTIL LOOP_END
              MPYF3    *AR6,*AR1++%,R0     ;ERR FUNC * x(n-(N-1-i)) -> R0
||            ADDF3    R3,R0,R2            ;H(N-1-i)+ERR FUNC * x(n-(N-1-i))
LOOP_END      LDF      *+AR0(1),R3         ;LOAD SUBSEQUENT H(k) -> R3
||            STF      R2,*AR0++           ;STORE/UPDATE COEFF
              ADDF3    R3,R0,R2            ;H(n+1) = H(n) + ERR FUNC * x(n)
              STF      R2,*AR0             ;STORE/UPDATE COEFF
              RETSU                        ;RETURN FROM SUBROUTINE
              .END                         ;END
```

FIGURE 7.20. *(Continued)*

The reference (second) input to the AIC (AUX IN) consists also of the 60-Hz sinusoidal noise. Example 3.10 discusses a loop program (`LOOPSW1.ASM`) that tests both inputs of an AIC connected through serial port 1 (see Chapter 3).

The main program `ADAPTER.ASM` is shown in Figure 7.20. Link this program with the AIC communication program `AICCOMA.ASM`, listed in Figure 3.21, which contains the subroutines `IOPRI` and `IOAUX` for I/O through the two AIC inputs (and one output). Consider the following:

1. The AIC configuration data is set for a sampling rate of 8 kHz. The rate of adaptation β is set at 2.5×10^{12}, and 50 coefficients are used for the adaptive filter. The fourth value, 73h, is `AICSEC` is used to delete the input bandpass filter within this AIC because very low frequencies are used. A value of 77h would be used if the input filter is to be inserted (if the third LSB is a 1 as in Figure 3.4).

2. The instructions

```
LDI 10h,R1
STI R1,@SPSET
```

enable serial port 1.

3. The input is in R6 (from AICCOMA.ASM) and the output in R7. When the subroutine IOAUX (in AICCOMA.ASM) is called, an output is performed (through R7) then a new 60-Hz noise sample is obtained through R6. The filter subroutine is then called for calculating the output of the filter at time n. Then the subroutine IOPRI (in AICCOMA.ASM) is called. An output is performed (through R7) and a new sample $(d + n)$ from the AIC primary input is obtained. The error signal $e(n)$, calculated as the difference between the AIC primary input $(d + n)$ and the adaptive filter's output y, is the overall output of the adaptive structure. After adaptation, the adaptive filter's output y is adapted to n and the error signal e approaches the desired signal d.

4. Initially, the 50 coefficients of the adaptive filter are set to zero. The adaptive subroutine ADAPT is similar to the adaptive predictor Example 7.4. In this case, 50 coefficients are being updated. The block of code (within ADAPT) between the repeat instruction RPTB and STF R2,*AR0++ (because it is in parallel with the preceding LDF instruction) is executed 49 times.

5. For each time n, both subroutines IOPRI and IOAUX are called in order to acquire a new sample $(d + n)$ from the AIC primary input and a new sample n from the AIC reference or auxiliary input, respectively.

Run the ADAPTER program. Figure 7.21(a) shows the primary input spectrum of the 2-Hz signal and 60-Hz noise before adaptation, and Figure 7.21(b) shows the spectrum of the output "error" signal after adaptation. Observe that after adaptation, the 60-Hz sinusoidal noise is reduced by approximately 41 dB.

For a different number of filter coefficients and adaptation rate, change LENGTH and BETA, respectively. A bandlimited noise signal can also be used in lieu of the 60-Hz sinusoidal noise. This example is further used in Chapter 8 to demonstrate the reduction of a 60-Hz artifact in an electrocardiogram (ECG) signal.

Exercise 7.1 Real-Time Adaptive Notch Filter. Run the adaptive filter of Example 7.6 using a desired signal of 1,000 Hz and sinusoidal noise of 400 Hz. Insert the input bandpass filter of the AIC. Change the program

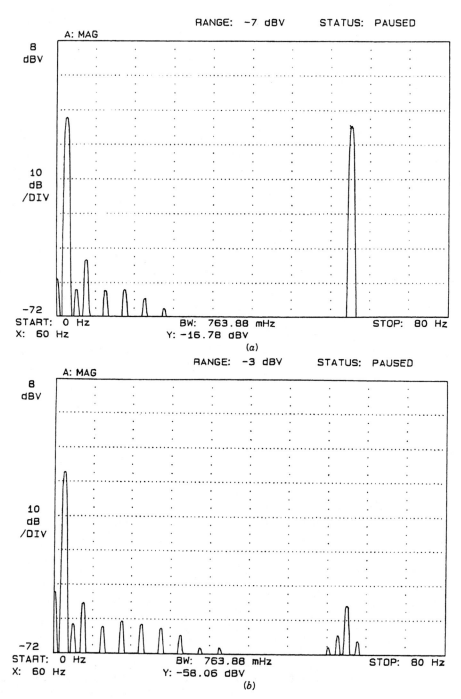

FIGURE 7.21. Spectrum of sinusoidal 2-Hz signal and 60-Hz noise: (*a*) before adaptation; (*b*) after adaptation

ADAPTER.ASM so that it becomes more efficient (without using separate filtering and adapting subroutines).

REFERENCES

[1] B. Widrow and S. D. Stearns, *Adaptive Signal Processing*, Prentice-Hall, Englewood Cliffs, N.J., 1985.

[2] B. Widrow, J. R. Glover, J. M. McCool, J. Kaunitz, C. S. Williams, R. H. Hearn, J. R. Zeidler, E. Dong, Jr., and R. C. Goodlin, Adaptive noise cancelling: Principles and applications, *Proceedings of the IEEE*, **63**, December 1975, pp. 1692–1716.

[3] B. Widrow and M. E. Hoff, Jr., Adaptive switching circuits, in *IRE WESCON*, pp. 96–104, 1960.

[4] J. R. Treichler, C. R. Johnson, Jr., and M. G. Larimore, *Theory and Design of Adaptive Filters*, Wiley, New York, 1987.

[5] R. Chassaing and D. W. Horning, *Digital Signal Processing with the TMS320C25*, Wiley, New York, 1990.

[6] D. G. Messerschmitt, Echo cancellation in speech and data transmission, *IEEE Journal on Selected Topics in Communications*, **SAC-2**, March 1984, pp. 283–303.

[7] M. L. Honig and D. G. Messerschmitt, *Adaptive Filters: Structures, Algorithms and Applications*, Kluwer Academic, Norwell, Mass., 1984.

[8] C. F. Cowan and P. F. Grant (editors), *Adaptive Filters*, Prentice-Hall, Englewood Cliffs, N.J., 1985.

[9] S. Haykin, *Adaptive Filter Theory*, Prentice-Hall, Englewood Cliffs, N.J., 1986.

[10] M. G. Bellanger, *Digital Filters and Signal Analysis*, Prentice-Hall, Englewood Cliffs, N.J., 1986.

[11] R. Chassaing, D. W. Horning, and P. Martin, Adaptive filtering with the TMS320C25, in *Proceedings of the 1989 ASEE Annual Conference*, June 1989.

[12] K. S. Lin (editor), *Digital Signal Processing Applications with the TMS320 Family–Theory, Algorithms, and Implementations*, Vol. 1, Texas Instruments, Inc., Dallas, Tex., 1987.

[13] P. Papamichalis (editor), *Digital Signal Processing Applications with the TMS320 Family–Theory, Algorithms, and Implementations*, Vol. 3, Texas Instruments, Inc., Dallas, Tex., 1990.

[14] D. J. DeFatta, J. G. Lucas, and W. S. Hodgkiss, *Digital Signal Processing: A System Approach*, Wiley, New York, 1988.

[15] J. G. Proakis and D. G. Manolakis, *Introduction to Signal Processing*, Macmillan, New York, 1988.

[16] L. R. Rabiner and B. Gold, *Theory and Application of Digital Signal Processing*, Prentice-Hall, Englewood Cliffs, N.J., 1975.

[17] S. Kuo, G. Ranganathan, P. Gupta, and C. Chen, Design and implementation of adaptive filters, *IEEE 1988 International Conference on Circuits and Systems*, June 1988.

[18] M. Bellanger, *Digital Processing of Signals: Theory and Practice*, Wiley & Sons Ltd., Paris, 1989.

[19] S. T. Alexander, *Adaptive Signal Processing Theory and Applications*, Springer-Verlag, New York, 1986.

8
Real-Time Digital Signal Processing Applications with C and the TMS320C30: Student Projects

This chapter can be used as a source of experiments, projects, and applications. Reference [1] contains a wide range of projects that have been implemented with the fixed-point TMS320. Many of these projects can be extended to the floating-point TMS320C30. Several projects developed by students are presented in this chapter, including applications in filtering and communications. These projects can be used as a source of ideas to implement other projects. The proceedings from the TMS320 Educators Conference published by Texas Instruments, Inc. [2], contains a number of TMS320-based articles and is also a good source of project ideas. A number of applications are also included in references [3]–[5]. The previous chapters on adaptive filtering, as well as the fast Fourier transform (FFT), can also be very useful.

I owe a special debt to all the students who have made this chapter possible— in particular, B. Bitler, R. Cabral, R. Cameron, E. Carpenter, J. Chopy, R. Cibelli, V. Dacosta, D. Fisher, G. Hollfelder, G. Letourneau, G. Monti, R. Oliver, B. Parillo, T. Pedchenko, W. Peterson, J. Rego, R. Siipola, R. Thayer, and B. Wilde.

8.1 PARAMETRIC EQUALIZER

This section discusses a parametric equalizer project using TMS320C30 code, with a PC host interactive program in C code. The goal of this project is to implement a software-activated and controlled parametric equalizer. The equalizer is to have control over the center frequency, bandwidth, and amplitude of a multiband filter, representing a bass range and a treble range.

Design Considerations

The ranges of the multiband filter should encompass the audio frequency range. The chosen bands are such that the bass range is from DC to

≃ 1 kHz, and the treble range is from ≃ 1 to 5 kHz. Each band should be controlled independently of the other, with amplitude gain or attenuation of 0 and ±6 dB, and with adjustable center frequency and bandwidth.

The Parametric Equalizer
The parametric equalizer is basically an audio frequency shaper similar to the graphic equalizer. The graphic equalizer contains several different bandpass filters that are used to increase or decrease the amplitude of a selected frequency. This provides control over as many different frequencies as there are filters. A commonly used graphic equalizer contains 10 different adjustable bandpass filters: five bass and five treble filters. The bass filters are typically at 30, 60, 120, 250, and 500 Hz. The treble filters are at 1, 2, 4, 8, and 16 kHz. There is a small "space" between each filter's passband, to provide separation between the filters' responses. Although this separation reduces the crosstalk between the filters, it causes losses of audio information that may exist within the separation space. In contrast to the graphic equalizer, the parametric equalizer consists of a multiband filter with only two bandpass regions, with all the filter parameters controllable. The use of the two broadband bandpass regions can reduce the lost audio information that was caused by the graphic equalizer's narrowband filters. Furthermore, the ability to control all the parameters of the multiband filter can give a simpler and more faithful reproduction of the audio signal.

The multiband filter provides the necessary conditions of the parametric equalizer. A number of bands, with bandwidth and gain, can be specified. To control these parameters with any change in the filter would require a new set of coefficients, because, even with a two-passband filter with two center frequencies, changing the sampling rate would change both (not just one) center frequencies. Hence, to produce the kind of variations that a parametric equalizer should have would require much memory and coding.

Implementation
The compromise that follows resulted in an acceptable (noncommercial) implementation. Sixteen filters are obtained with a set of coefficients representing the characteristics of each filter. These coefficients are stored in memory. A lookup table is used to obtain the user-selected filter. A sampling frequency of 16,234 Hz is chosen. The 16 filters are designed, each with 51 coefficients, such that the bass range is from 0 to 1.2 kHz and the treble range is from 1.2 to 5 kHz. Because each filter has the same number (51) of coefficients, this made the lookup table scheme much easier. Knowing the beginning address of the first set of coefficients, the address of subsequent sets of coefficients (offset from the first) can be readily obtained.

To run the parametric equalizer program, run the batch file PEQ.BAT, which contains the following:

```
PCPEQ
EVMLOAD  PEQ.OUT
```

```
/*PCPEQ.C-PC HOST/EVM COMMUNICATION PROGRAM*/
#include <stdio.h>
#include <dos.h>
#define IOBASE 0x280
#define PORT (IOBASE+0x800)
main()
    {
    int i;
    outp(PORT,0x00);
    clrscr();
    do
        {
        printf("\nEnter filter number (1-16) : ");
        scanf("%i",&i);
        }
    while ((i<1) || (i>16));
    outp(PORT,i);
    }
```

FIGURE 8.1. Interactive C program for parametric equalizer (PCPEQ.C)

This batch file executes the PC host interactive program PCPEQ in C code and downloads and executes the TMS320C30 assembly program PEQ.

A brief description of these two programs for the parametric equalizer project follows:

1. Consider the interactive C program PCPEQ.C, listed in Figure 8.1. When executed, the user is prompted to enter a number between 1 and 16, which corresponds to one of 16 filters that is to be selected and run. Section 3.5 describes the communication procedures between the PC host and the TMS320C30. Note that here, the PC I/O base address (IOBASE) is defined as 280h, not the default setting of 240h. A value of zero is initially transferred to the TMS320C30 memory addressed mapped to the PC I/O base address offset by 800h, until a valid value of i between 1 and 16 is entered.
2. Consider the main program PEQ.ASM. A partial version of this program, PEQPT, without the coefficients, is shown in Figure 8.2. The TMS320C30 address 804000h, with label OFSET, is mapped to the PC address (IOBASE + 0x800) for communication or transfer of the i value representing the selected filter. It is instructive to refer to the FIR filter programs (in Chapter 4) implemented with TMS320C30 code.
    (a) The starting address of each set of coefficients is stored in a table, specified with AR2. Each set of coefficients has a length of 51.
    (b) The offset input address 804000h, designated with OFSET, is used to retrieve the filter number (the i value) in order to select one of

```
;PEQPT-PARTIAL PARAMETRIC EQUALIZER WITHOUT 16 SETS OF COEF
         .TITLE   "PEQ"             ;2-PASSBANDS,51 COEFFS/FILTER
         .GLOBAL  RESET,BEGIN,AICSET,AICSEC,AICIO_P,OFSET,SPSET
         .SECT    "VECTORS"         ;ASSEMBLE INTO VECTOR SECTION
RESET    .WORD    BEGIN             ;RESET VECTOR
         .DATA                      ;ASSEMBLE INTO DATA SECT
OFSET    .WORD    804000H           ;OFFSET INPUT ADDRESS
AICSEC   .WORD    0E1Ch,1h,4286h,63H    ;Fs=16234 Hz
XN_ADDR  .WORD    XN+LENGTH-1       ;NEWEST (LAST) INPUT SAMPLE
HN_ADDR  .WORD    COEF1             ;COEFFICIENTS TABLE ADDR
STORE    .WORD    COF
XN       .USECT   "XN_BUFF",LENGTH  ;SAMPLES BUFFER SIZE
COF      .USECT   "COF_TABLE",16    ;COEFCNT SET ADDRS TABLE
         .TEXT                      ;ASSEMBLE INTO TEXT SECT
BEGIN    LDP      SPSET             ;INIT DATA PAGE
         CALL     AICSET            ;INIT AIC/TMS320C30
         LDI      @XN_ADDR,AR1      ;LAST SAMPLE ADDR ->AR1
         LDI      16,R5             ;SET TABLE LENGTH COUNTER ->R5
         LDI      LENGTH,R4         ;SET COEFCNT ADDRS OFFSET ->R4
         LDI      @HN_ADDR,AR3      ;FIRST COEFCNT ADDRS ->AR3
         LDI      @STORE,AR2        ;FIRST ADDR OF COEF TABLE->AR2
         LDI      @OFSET,AR4        ;LOAD ADDRESS OF OFFSET
LOOP1    STI      AR3,*AR2++(1)     ;STORE COEFCNT SET ADDRESSES
         ADDI     R4,AR3            ;INCREMENT TO NEXT COEFCNT SET
         SUBI     1,R5              ;DECREMENT TABLE LENGTH COUNTER
         BNZ      LOOP1             ;LOAD TABLE UNTIL FULL
         LDI      LENGTH,BK         ;BK = # COEFFS(FILTER LENGTH)
         LDF      0.0,R0            ;INIT R0=0
         RPTS     LENGTH-1          ;INIT ALL IN SAMPLES TO 0
         STF      R0,*AR1--(1)%     ;START & END AT LAST SAMPLE
         SUBI     16,AR2            ;RETURN TO FIRST COEFCNT SET
         LDI      *AR4,R1           ;LOAD ADDRESS OF OFFSET
         LDI      R1,IR0            ;SET THE OFFSET REGISTER
         AND      255,IR0           ;MASK OUT UNNECESSARY BITS
         SUBI     1,IR0             ;CORRECT OFFSET+1 TO OFFSET
LOOP2    CALL     AICIO_P           ;AIC I/O ROUTINE,IN->R6,OUT->R7
         FLOAT    R6,R3             ;INPUT NEW SAMPLE -> R3
         LDI      *+AR2(IR0),AR0    ;SELECTED COEF SET H(N-1)->AR0
         CALL     FILT              ;GO TO SUBROUTINE FILT
         FIX      R2,R7             ;R7=INTEGER(R2)
         BR       LOOP2             ;LOOP CONTINUOUSLY
FILT     STF      R3,*AR1++%        ;NEWEST SAMPLE TO MODEL DELAY
         LDF      0.0,R0            ;INIT R0=0
         LDF      0.0,R2            ;INIT R2=0
         RPTS     LENGTH-1          ;MULTIPLY LENGTH TIMES
```

FIGURE 8.2. Parametric equalizer program without coefficients (PEQPT)

```
            MPYF    *AR0++,*AR1++%,R0   ;H(N-1-i)*X(n-(N-1-i)
||          ADDF    R0,R2               ;// WITH ACC -> R2
            ADDF    R0,R2               ;LAST ACC -> R2
            RETS                        ;RETURN FROM SUBROUTINE
            .DATA
COEF1:      .FLOAT  -8.010325E-005
            .FLOAT  -2.127472E-005
              :
COEF16:     .FLOAT  4.269521E-004
              :
            .FLOAT  4.269521E-004       ;H0(16)
LENGTH      .SET    51                  ;LENGTH=51
            .END
```

FIGURE 8.2. (Continued)

the 16 filters. This address is mapped to the PC host communication address. The content of 804000h is stored as an index register IR0 that is used to select (with appropriate offset) one of the starting addresses of the filter's coefficients. The range of this index register is changed to 0–15 (from 1–16), because an offset value of IR0 = 0 selects the first set of coefficients (coefficient base address + 0), and so on.

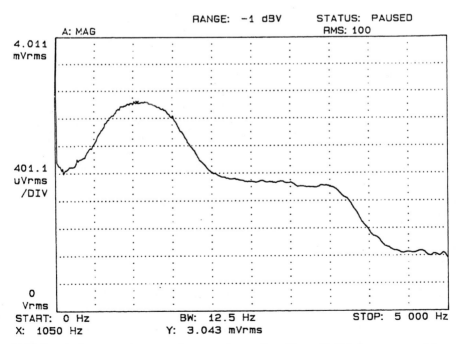

FIGURE 8.3. Frequency response of parametric equalizer filter with high bass range and low treble range

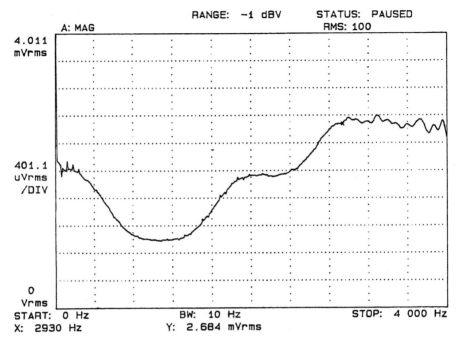

FIGURE 8.4. Frequency response of parametric equalizer filter with low bass range and high treble range

(c) This main program calls the AIC communication routines, with an output rate determined by polling.

Figure 8.3 shows the real-time magnitude response of one of the 16 filters, accentuating the bass range, and Figure 8.4 shows another filter with a low bass range and a high treble range. These filters were run using an impulse [obtained from the Hewlett Packard (HP) signal analyzer] as input. With music as input and the EVM output connected to a speaker, the different filters were run to demonstrate the accentuation and attenuation of different ranges of frequencies.

This project can be extended to include more than 16 filters, for better resolution. An additional gain value (in addition to 0 and ± 6 dB) can also improve this project. A faster interactive scheme between the PC host and the execution of the selected filter would be very useful.

8.2 ADAPTIVE NOTCH FILTER USING TMS320C30 CODE

This section discusses an adaptive filter project using TMS320C30 code. The goal of this project is to implement an adaptive filter capable of reducing the

60-Hz artifact in an electrocardiogram (ECG) signal. Three different types of filter are used to reduce the 60-Hz interference:

1. A 60-Hz notch filter
2. A 50-coefficient adaptive filter
3. A two-weight (two-coefficient) adaptive filter

Introduction

Sinusoidal 60-Hz interference is a frequent problem in electrocardiographic (ECG) monitoring, creating baseline artifacts that can obscure the true ECG waveform and hinder diagnostic interpretation of the ECG. The ECG is a representation of the electrical impulses that are associated with cardiac muscle contraction and relaxation. The heart muscle contracts in response to an electrical stimulus, which causes a depolarization wave of electrical activity that spreads from the upper heart chambers (atria) to the lower heart chambers (ventricles). After the ventricles contract, the ventricular muscle returns to an electrically neutral state during the repolarization stage. The ECG records the voltage magnitude generated by the heart as well as the time required for the voltage to travel throughout the heart. The waves of electrical impulses produced by the heart are conducted through the chest cavity to the surface of the skin. These small signals (in millivolts) can be detected by sensors (ECG leads and electrodes) that are attached to the chest. These sensors act as receiving antennas that detect skin voltage signals and feed them into a cardiac monitor for amplification and processing. In diagnostic cardiology, there are standard positions for electrodes on the body to monitor the electrical activity of the heart from a specific angle. These configurations are known as leads. The most commonly used one, lead II, records voltage down through the long axis of the heart, from the right arm (negative electrode) to the left leg (positive electrode). A third electrode is attached to the right leg and serves as an electrical reference, or ground, and does not provide cardiac information.

The ability to discern clearly the appearance of the various segments of the ECG is critical for proper diagnosis of cardiac abnormalities. The true ECG waveform can be distorted because the cardiac monitor will display any type of electrical voltage signal it detects, in addition to that produced by cardiac activity. Interference on the ECG due to 60-Hz noise is undesirable. The patient's body, the lead cables, the bed frame, and so on, all can act as antennas. Many situations, such as improperly grounded electrical devices, looping ECG cables across the patient's body, or the patient's limbs coming into contact with the metal bed frame, can allow 60-Hz noise to be superimposed on the baseline of the ECG waveform. Although this 60-Hz interference can be minimized by correcting problems such as grounding, in many cases the source of the problem cannot be determined.

Implementations

1. Frequency-Selective 60-Hz Notch Filter

As a warm-up for this project, a 60-Hz frequency-selective bandstop (notch) filter is implemented with 95 coefficients and a sampling frequency of 600 Hz. This filter was discussed in Chapter 4 (Exercises 4.1–4.4). The input highpass filter on the analog interface chip (AIC) was bypassed (deleted) to allow for the low-frequency signals used in this project. The frequency response of this 60-Hz notch filter is shown in Figure 8.5 (see also the frequency response of this filter in Figure 4.35 with a different input noise level). Two sinusoidal signals of frequencies 2 and 60 Hz are added and formed the primary input (IN) to the notch filter. An external analog interface chip connected through serial port 1 is used for I/O. The output shows a reduction of the 60-Hz noise by 46 dB.

2. 50-Coefficient Adaptive Filter

A patient simulator provided a 2-Hz (120 beats per minute) ECG signal corrupted (added) with 60-Hz "noise." The 2-Hz signal with added 60-Hz noise forms the primary input to a 50-coefficient adaptive filter structure shown in Figure 8.6. A 60-Hz sinusoid is used as the reference input (AUX IN), and the output is taken at $e(n)$. A sampling rate of 10 kHz is used.

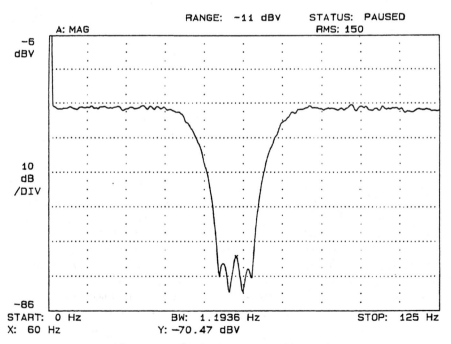

FIGURE 8.5. Frequency response of 60-Hz notch filter

8.2 Adaptive Notch Filter Using TMS320C30 Code

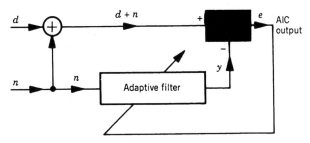

FIGURE 8.6. Adaptive filter structure with 50 coefficients for noise reduction

Chapter 7 provides background materials for this project, including a discussion on the least mean square (LMS) criterion [5]–[18]. Example 7.6 discusses the adaptive filter program ADAPTER.ASM, which can also be used for this project. The program is run and the output $e(n)$ of the adaptive filter structure [not $y(n)$, the adaptive filter's output] is displayed on the ECG monitor. Figure 8.7 shows the first part of the output waveform with the 60-Hz noise (before adaptation), and the latter part of the waveform with the 60-Hz noise considerably reduced (with adaptation). The negative spikes represent the 2-Hz signal (waveform is inverted). The rising and declining waveforms represent the atrial depolarization, the ventricular depolarization (higher negative spike), and the ventricular repolarization. The output $e(n)$ is also displayed on an HP signal analyzer, showing the 60-Hz noise reduced by 63 dB.

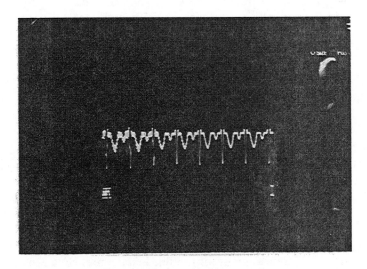

FIGURE 8.7. Adaptive filtering of 60-Hz noise displayed on ECG monitor

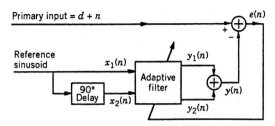

FIGURE 8.8. Two-weight adaptive notch filter structure

3. Two-Weight Adaptive Notch Filter

A faster adaptive scheme uses a two-weight (two-coefficient) adaptive notch filter, as shown in Figure 8.8 (see also Example 7.3). The same primary input (IN) as for the 50-coefficient filter is used, with the output also at $e(n)$. However, the reference input to the adaptive filter section consists of the following:

1. $x_1(n)$, a 60-Hz sinusoid as the AICs second input (AUX IN)
2. $x_2(n)$, the 60-Hz sinusoid shifted by 90°

The shift is obtained in software, because only two inputs on the AIC are available (the AIC on-board the EVM is not used). A program RSLOPE.ASM (on the accompanying disk) takes the slope of $x_1(n)$ [differentiates $x_1(n)$] to yield $x_2(n)$. This slope program (very sensitive to the sampling frequency) is to be linked with the main adaptive filter program. The main adaptive program is a real-time version of ADAPT2W.ASM, listed in Figure 7.11 and discussed in Example 7.3. See also the 50-coefficient adaptive program ADAPTER.ASM, listed in Figure 7.20. Similar output results are observed on the ECG monitor, with a faster adaptation process, because there are only two coefficients to update. However, only a 33-dB noise reduction is obtained (from the signal analyzer) as opposed to a 63-dB reduction obtained with the 50-coefficient adaptive filter structure.

Conclusions

These three methods demonstrate how a 60-Hz sinusoidal interference can be filtered out. These methods can also be used for the reduction of sinusoidal interference of a different frequency. The slope program can be improved to yield a more stable shifted sinusoid. This may in turn provide a greater noise reduction. An AIC (from serial port 1) with two inputs was used, one input (the primary) for the signal with added sinusoidal noise, and the second (the reference) input for the sinusoidal noise. Hence, a phase-shift circuit (in hardware) could not be used. The use of the AIC from serial port 0

8.3 ADAPTIVE FILTER FOR NOISE CANCELLATION USING C CODE

would provide a third input, allowing for the consideration of a phase-shift circuit.

8.3 ADAPTIVE FILTER FOR NOISE CANCELLATION USING C CODE

This project extends the adaptive predictor example (Example 7.4) to a real-time implementation, using C code only. Figure 7.15 shows the adaptive predictor structure. Two methods are used to obtain the delayed signal by shifting the primary input:

1. using an equation in terms of arcsine and arccosine
2. using a table lookup

The shifted or delayed signal becomes the input to a 21-coefficient adaptive filter.

A data file SIN312 used in Chapter 7 (on the accompanying disk) represents a sine function of frequency 312 Hz, generated using a sampling frequency of 10 kHz. The number of points per cycle is $F_s/312 = 32$ points per cycle. The file SIN312 was generated using a utility program available with the second-generation TMS320C25 simulator. It can also be generated with the Hypersignal-Plus DSP Software in the following fashion:

1. Select the DIFFERENCE EQUATIONS option, setting the equation as

   SIN(2*PI*312 / 10000)

   Save the generated file as SIN312.

2. Select the BINARY-ASCII-BINARY option. The resulting file SIN312.TIM created from step 1 needs to be converted to a decimal format. Use DEC for the ASCII format and B for the BIN-ASCII option. A file SIN312.TXT will then be created. Delete the first 10 values (which represent the number of sample points, the amplitude, and the sampling frequency) from SIN312.TXT, which will result in the file SIN312.

Shifting the Input Signal

1. The program SHIFT.C in Figure 8.9 can be used to shift or delay the input signal. This input (SIN312) is divided by 8,192, the maximum amplitude value of the sine function, in order to reduce its amplitude to +1 or −1. This shift is accomplished using either $\cos^{-1}(y)$ or $\sin^{-1}(y)$, depending on whether the input signal (y) is rising or falling. The equation

   X = X- (Pi / 2)

```
/*SHIFT.C-DEMONSTRATES 90 DEGREES DELAY IN PHASE */
#include <math.h>
#define Pi 3.1415926

main()
{
  int count;
  double Y,X,yo;
  volatile int *IO_INPUT = (volatile int*) 0x804000;
  volatile int *IO_OUTPUT = (volatile int*) 0x804001;
  for (count=0; count < 256; count++)
    {
    Y= *IO_INPUT;
    Y= Y/8192;                /*Y must be between 1 and -1 */
    if (yo>=Y)                /*is signal falling or rising*/
       X=acos(Y);             /*signal is falling          */
    else
       X=asin(Y)-(Pi/2);      /*signal is rising           */
    X=X-(Pi/2);               /*shift by 90 degrees        */
    *IO_OUTPUT=8192*cos(X);   /*shifted output Y value     */
    yo=Y;                     /*store Y value              */
    }
}
```

FIGURE 8.9. Program to shift an input (SHIFT.C)

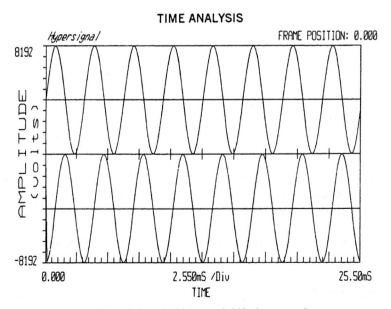

FIGURE 8.10. Plots of sinusoidal input and shifted output using SHIFT.C

8.3 Adaptive Filter for Noise Cancellation Using C Code

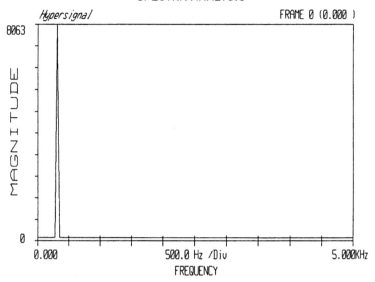

FIGURE 8.11. Frequency plot of shifted output using SHIFT.C

```
/*SHIFTTB.C-DEMONSTRATES -90 DEGREES SHIFT IN PHASE*/
#define Max 8192
#define Points 256
#define Pi   3.1415926
#include "scdat"         /*look-up table for acos, asin*/

main(void)
  {
  int count, Y1;
  double yo, Y, IN_data;
  volatile int *IO_INPUT = (volatile int*) 0x804000;
  volatile int *IO_OUTPUT = (volatile int*) 0x804001;
  for (count=0; count < Points; count++)
    {
    yo = IN_data;                  /*store signal value          */
    IN_data = *IO_INPUT;           /*input signal in             */
    Y1 = ((IN_data/Max)+1)*100;    /*step up array ( 0-> 200 )   */
    if (yo >= IN_data)             /*is signal falling or rising*/
       *IO_OUTPUT = yc[Y1]*Max;    /*signal falling              */
    else
       *IO_OUTPUT = ys[Y1]*Max;    /*signal rising               */
    }
  }
```

FIGURE 8.12. Program to shift an input using arccosine and arcsine table (SHIFTTB.C)

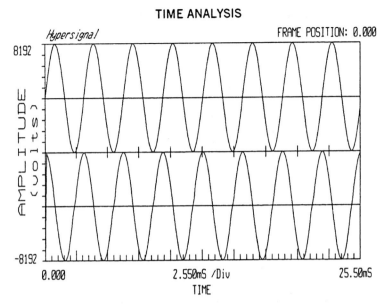

FIGURE 8.13. Plots of sinusoidal input and shifted output using SHIFTTB.C

represents a shift or delay by 90°. A delay such as 45° can be obtained by replacing (Pi / 2) by (Pi / 4) in the equation to find X. Figure 8.10 shows the input signal SIN312 (upper graph) and the resulting signal shifted by 90° (lower graph). The frequency plot of the shifted output sinusoidal signal is displayed in Figure 8.11.

2. The program SHIFTTB.C in Figure 8.12 can also produce a delayed signal. This faster (but not as clean) alternative can be used to obtain a delay of 270°. A lookup table file SCDAT, included on the accompanying disk, is used with SHIFTTB.C. The file SCDAT contains the arccosine and arcsine values, which are selected depending on whether the input signal is falling or rising. Figure 8.13 shows the input signal (upper graph) and the 270°-delayed output (lower graph).

Adaptation with Shifted Signal

The least mean square (LMS) algorithm discussed in Chapter 7 is used to illustrate the adaptation of the adaptive filter's output to the desired input signal. Figure 8.14 shows the program ADAPTSH.C, which incorporates the shifting program section of Figure 8.9 (SHIFT.C) and the LMS algorithm, discussed in Chapter 7. The input to the 21-coefficient adaptive filter is the input signal SIN312 delayed by 90°. A total of 256 samples are obtained. The adaptation process stops when the output of the adaptive filter is adapted to the desired signal SIN312, in which case, the error signal

```c
/*ADAPTSH.C-ADAPTIVE FILTER WITH SHIFTED INPUT*/
#include <math.h>
#define beta 0.004
#define N    20
#define NS   256
#define Pi   3.1415926
#define shift 90
#define max_input 8192

main()
{
  int I,T;
  float IN_data;
  double Y,Y1,X, E, D, yo;
  double W[N+1];
  double Delay[N+1];
  volatile int *IO_INPUT = (volatile int*) 0x804000;
  volatile int *IO_OUTPUT = (volatile int*) 0x804001;
  yo=0;
  for (T=0; T <= N; T++)
  {
    W[T] = 0.0;
    Delay[T] = 0.0;
  }
  for (T=0; T < NS; T++)
  {
    IN_data = *IO_INPUT;
    Y=IN_data/max_input;
    if (yo>=Y)
      X=acos(Y);
    else
      X=asin(Y)+(3*Pi/2);
    X=X-(shift);
    Delay[0]=cos(X);
    D=IN_data;
    Y1 = 0;
    yo=Y;
    for (I = 0; I <= N; I++)
      Y1 += (W[I] * Delay[I]);
    E = D - Y1;
    for (I = N; I >= 0; I--)
    {
      W[I] = W[I] + (beta*E*Delay[I]);
      if (I != 0)
      Delay[I] = Delay[I-1];
    }
    *IO_OUTPUT = Y1;
  }
}
```

FIGURE 8.14. Adaptive filter program incorporating shifting scheme used in SHIFT.C (ADAPTSH.C)

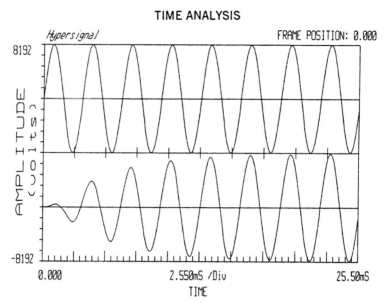

FIGURE 8.15. Plot of adaptive filter output (using ADAPTSH.C) converging to desired input signal

approaches zero. Figure 8.15 shows the output (lower graph) converging or adapting, to the desired input signal. A smaller constant beta can be used to obtain a slower convergence or adaptation rate.

The program ADAPTTB.C, listed in Figure 8.16, incorporates the table lookup procedure used in the program SHIFTTB.C for shifting an input signal. The input to the 21-coefficient adaptive filter is the input signal SIN312 delayed by 270°. Figure 8.17 shows the output converging to the desired input signal. The upper graph is obtained with an adaptation rate constant of beta = 0.004, and the lower graph with beta = 0.001.

Real-Time Implementation

The program ADAPTEVM.C (on the accompanying disk) implements in real time the adaptive filter for noise reduction, using the table (in SCDAT) lookup procedure for shifting. This program (ADAPTEVM.C) adjusts the input to the adaptive filter to a zero DC level with 2 V peak-to-peak. Figure 8.18 illustrates the adaptation in real-time. The upper graph represents the input which consists of a 2-kHz sinusoidal signal with added random noise. This noise is obtained from a source available with the Hewlett Packard signal/spectrum analyzer (3561A). The output (lower graph) shows the reduction of the noise, converging to the 2-kHz signal.

8.3 Adaptive Filter for Noise Cancellation Using C Code

```c
/*ADAPTTB.C-ADAPTIVE FILTER USING ASIN,ACOS       */
#define beta 0.004      /*rate of adaptation        */
#define N    20         /*order of filter (N+1)     */
#define NS   256        /*number of samples         */
#define Max 8192        /*max value of input signal*/
#include "scdat"        /*table for asin,acos       */

main()
{
  int I, J, T, Y;
  double E, yo, IN_data, OUT_data;
  double W[N+1];
  double Delay[N+1];
  volatile int *IO_INPUT  = (volatile int*) 0x804000;
  volatile int *IO_OUTPUT = (volatile int*) 0x804001;
  yo=0;
  for (T=0; T <= N; T++)
  {
    W[T] = 0.0;
    Delay[T] = 0.0;
  }
  for (T=0; T < NS; T++)
  {
    IN_data = *IO_INPUT;      /*input signal                          */
    IN_data = IN_data/Max;    /*scale for range between 1 and -1*/
    Y=((IN_data)+1)*100;      /*step up array between 0 and 200 */
    if (yo > IN_data)         /*is signal falling or rising       */
       Delay[0] = yc[Y];      /*signal is falling, acos domain   */
    else
       Delay[0] = ys[Y];      /*signal is rising, asin domain    */
    OUT_data = 0;             /*init filter output to zero            */
    yo=IN_data;               /*store input                             */
    for (I = 0; I <= N; I++)
       OUT_data += (W[I] * Delay[I]);   /*filter output          */
    E = IN_data - OUT_data;             /*error signal           */
    for (J=N; J >= 0; J--)
    {
      W[J] = W[J] + (beta*E*Delay[J]);  /*update coefficients */
      if (J != 0)
        Delay[J] = Delay[J-1];          /*update data samples */
    }
    *IO_OUTPUT = OUT_data*Max;          /* output signal        */
  }
}
```

FIGURE 8.16. Adaptive filter program incorporating shifting scheme with arccosine and arcsine table (ADAPTTB.C).

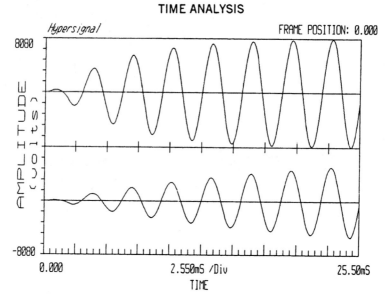

FIGURE 8.17. Plot of adaptive filter output converging to desired input signal for adaptation rate constant of $\beta = 0.004$ (upper graph) and $\beta = 0.001$ (lower graph)

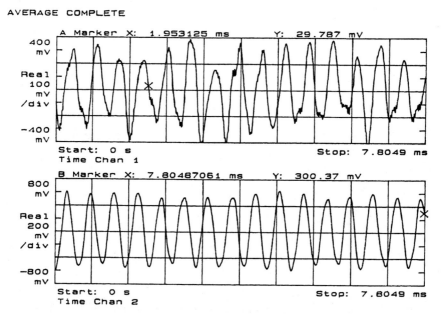

FIGURE 8.18. Real-time plot of adaptive filter output (lower graph) illustrating noise reduction

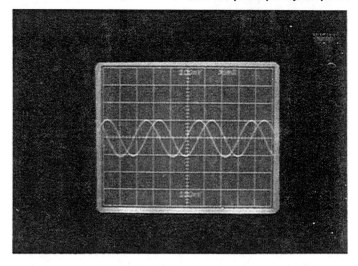

FIGURE 8.19. Input and shifted output signal using a phase-shifter circuit

Conclusions and Alternative Implementation

An adaptive filter can be implemented in real time using the LMS algorithm process. The program ADAPTEVM.C can be made more interactive as to allow for different rate of convergences, and so on. The sifted signal, which becomes the input to the adaptive filter, can also be produced using a phase-shifter circuit (such as the LM301 from Linear Technology). This feedback phase-shifter circuit includes an OP AMP and an R_i–C_i section. For a specific frequency, the resistance R_i can be varied for different values of desired phase shift between 0 and 180°. Figure 8.19 shows a photo of an input signal and the output of the LM301 shifted by 90°.

8.4 SWEPT FREQUENCY RESPONSE

The goal of this project is to find a circuit's magnitude response, which can then be displayed on the PC screen. It is implemented in C code.

Design Considerations and Implementations

The digital oscillator program of Chapter 5, based on a recursive difference equation, is extended to include a frequency-sweep feature. Figure 8.20 shows the program EVMSWEEP.C, which includes the function sinewave

```c
/*EVMSWEEP.C-GENERATES SINUSOID WITH FREQUENCY SWEEP        */
#include <math.h>              /*TI header of math functions */
#include "aiccom.c"             /*AIC comm routines           */
#define PI 3.14159              /*value for pi                */
#define SAMPLE_FREQ 10000       /*sampling frequency          */
volatile int *host = (volatile int *) 0x804000;
volatile int *dma  = (volatile int *) 0x808000;
void change_freq(float);        /*calculates diff. coeff    */
int average_peak(int *, int);   /*calculates average peaks*/
int AICSEC[4] = {0x1428,0x01,0x4A96,0x67}; /*AIC config data*/
float A,B,C;

void sinewave()
  {
  float Y[3]={0.0,0.0,0.0};    /*Y[N] array                 */
  float X[3]={0.0,0.0,1.0};    /*X[N] array                 */
  float current_freq;           /*latest frequency value     */
  int inval[200];               /*input sample array         */
  int peak;                     /*average peak value         */
  int counter=0;                /*sample counter             */
  int N=2;                      /*used for array indexing    */
  current_freq=100.0;           /*initial frequency          */
  change_freq(current_freq);
  while(1)
    {
    TWAIT;
    /* calculates and output sinewave value */
    Y[N]=A*Y[N-1] + B*Y[N-2] + C*X[N-1];
    PBASE[0x48]=-4 & (int)(Y[N]*2600);
    Y[N-2]=Y[N-1];             /*shift Y's back in array    */
    Y[N-1]=Y[N];
    X[N-2]=X[N-1];             /*shift X's back in array    */
    X[N-1]=X[N];
    X[N]=0.0;                  /*set future X's to 0        */
    inval[counter++]=(PBASE[0x4C]<<16)>>18; /*input sample*/
    if (counter==200)          /*if last sample             */
      {
      peak=average_peak(inval, counter);
      if (peak==0)             /*remove any 0 for host comm */
        peak=1;
      while(*host & 0x0000FFFF); /*any leftover value       */
      *host=peak;              /*send data                  */
      while(*host!=(peak & 0x0000FFFF));
      *host=0x0;               /*reset host register        */
      peak=0;                  /*initialize peak value      */
```

FIGURE 8.20. Frequency-sweep program (EVMSWEEP.C)

```
      counter=0;                    /*initialize counter      */
      Y[0]=Y[1]=Y[2]=0.0;           /*init difference arrays */
      X[0]=X[1]=0.0;
      X[2]=1.0;
      if (current_freq<2000.0)      /*if 2 kHz,reset to 100Hz*/
          current_freq+=100.0;
      else                          /*else increment by 100   */
          current_freq=100.0;
      change_freq(current_freq);    /*recalculate coeffs      */
      }
   } /* end of loop */
}

int average_peak(int *values, int count)
  {
  long int sum;                     /*sum of input samples    */
  int avg_peak;                     /*average input value     */
  int i;                            /*loop control variable   */
  sum=0;
  for (i=0;i<count;i++)
     {
     if (values[i]<0)               /*get absolute value      */
         values[i]=0-values[i];
     sum+=values[i];
     }
  avg_peak=sum/count;               /*determine average       */
  return(avg_peak);
  }

void change_freq(float newfreq)
  {
  float Fs;                         /*sampling frequency      */
  float T;                          /*sampling period         */
  float w;                          /*desired angular freq    */
  Fs=SAMPLE_FREQ;
  T=1/Fs;                           /*determine sample period*/
  w=2*PI*newfreq;                   /*determine angular freq.*/
  A=2 * cos((w * T));               /*determine coefficients */
  B=-1.0;
  C=sin((w * T));
  return;
  }

main()
```

FIGURE 8.20. *(Continued)*

```
{
AICSET();                    /*initialize AIC          */
INIT_EVM();                  /*initialize host comm    */
*host=0x0;                   /*reset host comm reg     */
sinewave();                  /*call sweep function     */
}
```

FIGURE 8.20. *(Continued)*

(which generates a sinusoidal waveform), and performs the host data communication and frequency sweep. A range of frequencies between 100 and 2,000 Hz is selected. A brief description of this program follows:

1. From the program segment (within the function sinewave)

```
if (current_freq<2000.0)
   current_freq += 100.0;
else
   current_freq = 100.0;
change_freq(current_freq);
```

the current frequency is incremented by 100, until it reaches 2,000 Hz. This frequency is then initialized back to 100 Hz. The function change_freq is called next, passing to it the current frequency value. This new frequency value is used to recalculate a new set of coefficients A and C ($B = -1.0$) to generate a sinusoid at that (new) frequency.

2. For each frequency $100, 200, \ldots, 2,000$ Hz, 200 samples are obtained, incrementing the variable counter in the program. After 200 samples are obtained, the average_peak function is called. This function determines the average peak value of the input sinewave samples, first taking the absolute value of each sample. This value is downloaded to the PC host computer. The sweep frequency is then incremented and the previous steps repeated.

The program HSTSWEEP.C (on the accompanying disk) is used in conjunction with the EVMSWEEP.C program (listed in Figure 8.20). Each average peak value obtained from step 2 is downloaded (transferred) from the EVM to the host PC. A total of 20 average values are transferred to the host PC, each corresponding to the 20 frequency values ranging from 100 to 2,000 Hz. The program HSTSWEEP.C includes a function that converts these values into decibels (dB), with reference to a 1-V peak-to-peak signal output by the EVM. These values in decibels are then plotted with respect to frequency. This program includes functions to initialize the graphics driver and print screen and to graph the array of values.

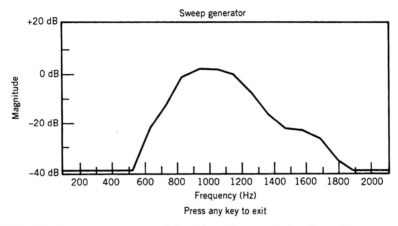

FIGURE 8.21. Frequency response of bandpass filter as displayed on PC screen using frequency-sweep technique

The function init_graph () in HSTSWEEP.C sets up the EGA–VGA graphics handler (the Turbo C file EGAVGA.BGI must be included). The init_graph () function draws the rectangular outline that is used to display the frequency plot and generates the screen title, labels and values, and the message to the user for terminating the program. A viewport is also set with the function setviewport. A (0, 0) reference coordinate is set at the upper left-hand corner of the viewport.

To test this project, connect the output of the EVM (outputting the sweep frequency) to the input of an analog filter (such as a second-order analog bandpass filter). The analog filter's output becomes the input to the EVM. Because the frequency range is limited to 2,000 Hz, the roll-off region of the filter should be less than 2,000 Hz. Figure 8.21 shows the frequency response of a second-order analog bandpass filter, as displayed on a PC screen. Note that if no key is pressed, the process continues: reading 20 values, converting them to decibels, then plotting them on the PC screen.

This project can be improved by extending the magnitude and frequency ranges for a better display (without clipping). This clipping can be noted in Figure 8.21 when the magnitude response is below -40 dB.

8.5 MULTIRATE FILTER

The goal of this project is to implement an efficient programmable noise generator using multirate techniques. With multirate processing, a filter can be realized with fewer coefficients than an equivalent single-rate approach. Possible applications include a controlled noise source and background noise synthesis.

Introduction

Multirate processing uses more than one sampling frequency to perform a desired processing operation. The two basic operations are *decimation*, which is a sampling-rate reduction, and *interpolation*, which is a sampling-rate increase [19]–[26]. Decimation techniques have been used in filtering. Multirate decimators can reduce the computational requirements of the filter [19]. Interpolation can be used to obtain a sampling-rate increase. For example, a sampling-rate increase by a factor of K can be achieved by padding $K - 1$ zeros between pairs of consecutive input samples x_i and x_{i+1}. We can also obtain a noninteger sampling-rate increase or decrease by cascading the decimation process with the interpolation process. For example, if a net sampling-rate increase of 1.5 is desired, we would interpolate by a factor of 3, padding (adding) two zeros between each input sample, and then decimate with the interpolated input samples shifted by 2 before each calculation. Decimating or interpolating over several stages generally results in better efficiency.

Design Considerations

A binary random signal is fed into a bank of filters that are used to shape the output spectrum. The functional block diagram of the multirate filter is shown in Figure 8.22. The frequency range is divided into 10 octave bands, with each band $\frac{1}{3}$-octave controllable. The control of each octave band is achieved with three filters. The coefficients of these filters are combined to yield a composite filter with one set of coefficients for each octave. Only three unique sets of filter coefficients (low, middle, and high) are required, because the center frequency and the bandwidth are proportional to the sampling frequency. Each of the $\frac{1}{3}$-octave filters has a bandwidth of approximately 23% of its center frequency, a stopband rejection of greater than 45 dB, with an amplitude that can be controlled individually. This control provides the capability of shaping an output pseudorandom noise spectrum. The sampling rate of the output is chosen to be 16,384 Hz. Forty-one (41) coefficients are used for the highest $\frac{1}{3}$-octave filter to achieve these requirements.

In order to meet the filter specifications in each region with a *constant* sampling rate, the number of filter coefficients must be doubled from one octave filter to the next-lower one. As a result, the lowest-octave filter would require 41×2^9 coefficients. With 10 filters ranging from 41 coefficients to 41×2^9 coefficients, the computational requirements would be considerable. To overcome these computational requirements, a multirate approach is used, as shown in Figure 8.22.

The noise generator is a software-based implementation of a maximal length sequence technique used for generating pseudorandom numbers, as illustrated in Figure 8.23. The output of the noise generator provides uncor-

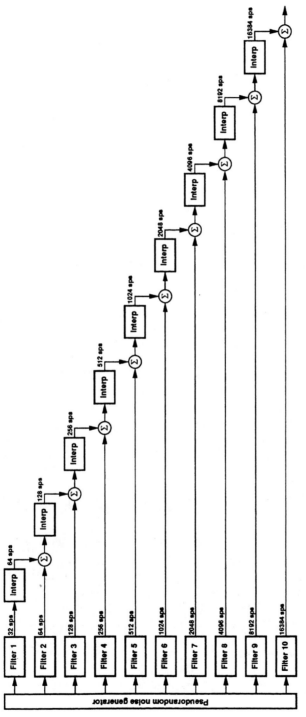

FIGURE 8.22. Functional block diagram of 10-band multirate filter

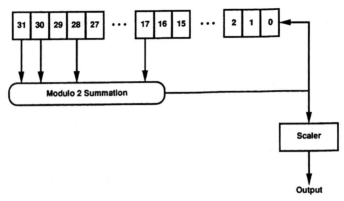

FIGURE 8.23. 32-Bit noise generator

related noise input to each of the 10 sets of bandpass filters. The two noise generation examples in Chapter 3 use the process shown in Figure 8.23.

Because each $\frac{1}{3}$-octave filter can be scaled individually, a total of 30 levels can be controlled. The output of each octave bandpass filter (except the last one) becomes the input to an interpolation lowpass filter, using a $2:1$ interpolation factor. The ripple in the output spectrum is minimized by having each adjacent $\frac{1}{3}$-octave filter with crossover frequencies at the 3-dB points.

The center frequency and bandwidth of each filter are determined by the sampling rate. The sampling rate of the highest-octave filter is processed at 16,384 samples per second, and each successively lower-octave band is processed at half the rate of the next-higher band.

Only three separate sets of 41 coefficients are used for the lower, middle, and higher $\frac{1}{3}$-octave bands. For each octave band, the coefficients are combined as follows:

$$H_{ij} = (H_{lj})(L_{3i-2}) + (H_{mj})(L_{3i-1}) + (H_{hj})(L_{3i})$$

where $i = 1, 2, \ldots, 10$ bands and $j = 0, 1, \ldots, 40$ coefficients; $L_1, L_2, \ldots, L_{30}$ represent the level of each $\frac{1}{3}$-octave band filter, and H_{lj}, H_{mj}, H_{hj} represent the jth coefficient of the lower, middle, and higher $\frac{1}{3}$-octave band FIR filter. For example, for the first band ($i = 1$),

$$H_0 = (H_{l0})(L_1) + (H_{m0})(L_2) + (H_{h0})(L_3)$$
$$H_1 = (H_{l1})(L_1) + (H_{m1})(L_2) + (H_{h1})(L_3)$$
$$\vdots$$
$$H_{40} = (H_{l40})(L_1) + (H_{m40})(L_2) + (H_{h40})(L_3)$$

8.5 Multirate Filter

and, for band 10 ($i = 10$),

$$H_0 = (H_{l0})(L_{28}) + (H_{m0})(L_{29}) + (H_{h0})(L_{30})$$
$$H_1 = (H_{l1})(L_{28}) + (H_{m1})(L_{29}) + (H_{h1})(L_{30})$$
$$\vdots$$
$$H_{40} = (H_{l40})(L_{28}) + (H_{m40})(L_{29}) + (H_{h40})(L_{30})$$

For an efficient design with the multirate technique, lower-octave bands are processed at a lower sampling rate, then interpolated up to a higher sampling rate, by a factor of 2, to be summed with the next-higher octave band filter output, as shown in Figure 8.22. Each interpolation filter is a 21-coefficient FIR lowpass filter, with a cutoff frequency of approximately one-fourth of the sampling rate. For each input, the interpolation filter provides two outputs, or

$$y_1 = x_0 I_0 + 0 I_1 + x_1 I_2 + 0 I_3 + \cdots + x_{10} I_{20}$$
$$y_2 = 0 I_0 + x_0 I_1 + 0 I_2 + x_1 I_3 + \cdots + x_9 I_{19}$$

where y_1 and y_2 are the first and second interpolated outputs, respectively, x_n are the filter inputs, and I_n are the interpolation filter coefficients. The interpolator is processed in two sections to provide the data-rate increase by a factor of 2.

For the multirate filter, the approximate number of multiplication operations (with accumulation) per second is

$$\text{MA} = (41 + 21)(32 + 64 + 128 + 256 + 512 + 1{,}024 + 2{,}048 + 4{,}096 + 8{,}192)$$
$$+ (41)(16{,}384)$$
$$\approx 1.686 \times 10^6$$

Note that no interpolation is required for the last stage. To find the approximate equivalent number of multiplications for the single-rate filter to yield the same impulse response duration, let

$$N_s T_s = N_m T_m$$

where N_s and N_m are, respectively, the number of single-rate and multirate coefficients, and T_s and T_m are, respectively, the single-rate and multirate sampling periods. Then,

$$N_s = N_m \frac{F_s}{F_m}$$

where F_s and F_m are, respectively, the single-rate and multirate sampling frequencies. Then, for band 1,

$$N_s = (41)(16{,}384/32) = 20{,}992$$

where F_s is the sampling rate of the highest band, and $F_m = 32$ for the first band. For band 2,

$$N_s = (41)(16{,}384/64) = 10{,}496$$

For band 3, $N_s = 5{,}248$; for band 10, $N_s = 41$. The total number of coefficients for the single-rate filter would then be

$$N_s = 20{,}992 + 10{,}496 + \cdots + 41 = 41{,}943$$

The approximate number of multiplications (with accumulation) per second for an equivalent single-rate filter is then

$$\text{MA} = N_s F_s = 687 \times 10^6$$

which would considerably increase the processing time and data storage requirements.

Implementation

The TMS320C30 program `MR10AIC.ASM` (on the accompanying disk) is used to implement the multirate filter in real time.

Run the multirate program `MR10AIC.ASM`. The sampling frequency is set at Fs = 16,234 Hz (in lieu of 16,384) and is specified in `AICSEC`. All the levels or "scale" values are initially set to zero. These levels (L_1–L_{30}) are specified in the program by `SCALE_1L, SCALE_1M, SCALE_1U, SCALE_2L, SCALE_2M, . . . , SCALE_10U` (which represent the lower, middle, and upper $\frac{1}{3}$-octave scales for the 10 bands). Set L_{29} (`SCALE_10M`) to 1 in order to turn *on* the middle $\frac{1}{3}$-octave filter of band 10. Figure 8.24 shows the frequency response of the middle $\frac{1}{3}$-octave filter of band 10. The center frequency is at approximately 4,000 Hz ($\frac{1}{4}F_s$). Turn on the middle $\frac{1}{3}$-octave filter of band 1 by setting L_{29} to 0 and L_2 (`SCALE_1M`) to 1. Figure 8.25 shows the frequency response of the middle $\frac{1}{3}$-octave filter of band 1, with a center frequency of approximately 8 Hz. Note that the effective sampling rate for band 1 is approximately 32 Hz, as shown in the functional block diagram in Figure 8.22. Figure 8.26 shows the three $\frac{1}{3}$-octave band filters of band 9, illustrating the crossover frequencies at the 3-dB points.

8.5 Multirate Filter 297

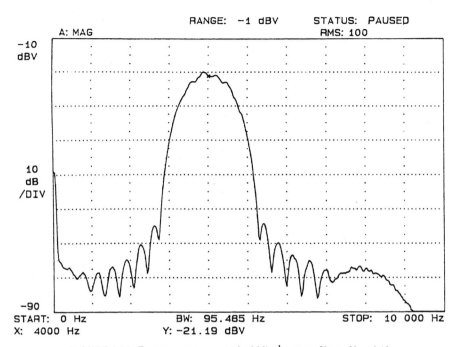

FIGURE 8.24. Frequency response of middle $\frac{1}{3}$-octave filter of band 10

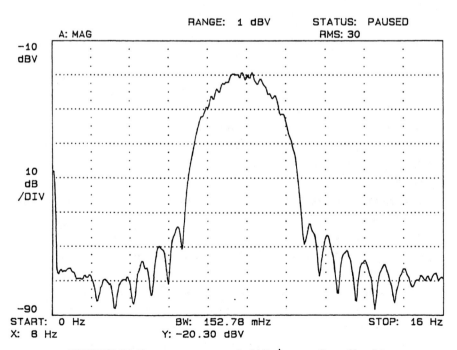

FIGURE 8.25. Frequency response of middle $\frac{1}{3}$-octave filter of band 1

298 Real-Time DSP Applications: Student Projects

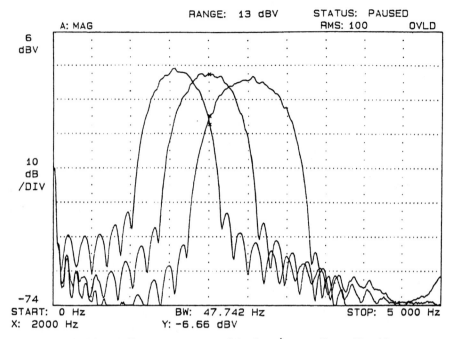

FIGURE 8.26. Frequency response of the three $\frac{1}{3}$-octave filters of band 9

A brief description of the main processing follows, for the first time through:

Band 1

1. Run bandpass filter and obtain one output sample.
2. Run lowpass interpolation filter twice and obtain two outputs. The interpolator provides two sample outputs for each input sample.
3. Store in buffer B_2, size 512, at locations 1 and 2 (in memory).

Band 2

1. Run bandpass filter two times and sum with the two previous outputs stored in B_2, from band 1.
2. Store summed values in B_2, at the same locations 1 and 2 (again).
3. Pass sample in B_2 at location 1 to interpolation filter twice and obtain two outputs.
4. Store these two outputs in buffer B_3, size 256, at locations 1 and 2.
5. Pass sample in B_2 at location 2 to interpolation filter twice and obtain two outputs.
6. Store these two outputs in buffer B_3, at locations 3 and 4.

Band 3

1. Run bandpass filter four times and sum with the previous four outputs stored in B_3, from band 2.
2. Store summed values in B_3 at locations 1 through 4.
3. Pass sample in B_3 at location 1 to interpolation filter twice and obtain two outputs.
4. Store these two outputs in buffer B_2, at locations 1 and 2.
5. Pass sample in B_3 at location 2 to interpolation filter twice and obtain two outputs.
6. Store these two outputs in buffer B_2, at locations 3 and 4.
7. Repeat steps 3 and 4 for the other two samples at locations 3 and 4 in B_3. For each of these samples, obtain two outputs, and store each set of two outputs in buffer B_2 at locations 5 through 8.

⋮

Band 10

1. Run bandpass filter 512 times and sum with the previous 512 outputs stored in B_2, from band 9.
2. Store summed values in B_2 at locations 1 through 512.

No interpolation is required for band 10. After all the bands are processed, wait for the output buffer B_1, size 512, to be empty. Then switch the buffers B_1 and B_2—the last working buffer with the last output buffer. The main processing is then repeated again.

A processing time of $\simeq 8.8$ ms was measured. This value was obtained using the following procedure:

1. Output to an oscilloscope a positive value set at the beginning of the main processing loop.
2. At the end of the main processing loop, negate the value set in the previous step. Output this negative level to the oscilloscope.
3. The processing time, which is the duration of the positive level set in step 1 can be measured readily with an oscilloscope.

The highest sampling rate (in kilosamples per second) is approximately 58 ksps, because

$$F_s(\max) = \frac{\text{number of samples}}{\text{processing time}}$$

$$= \frac{512}{8.8 \text{ ms}} = 58 \text{ ksps}$$

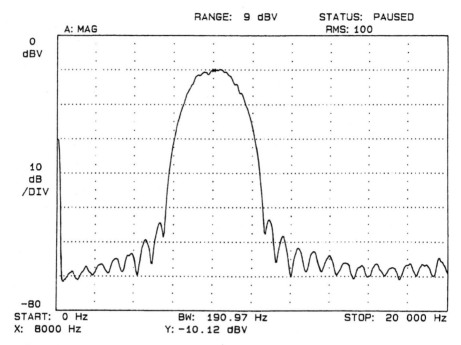

FIGURE 8.27. Frequency response of middle $\frac{1}{3}$-octave filter of band 10 using the XDS1000/AIB ($F_s = 32{,}786$ Hz).

The main processing time also can be measured using the program MR10.ASM with the simulator. Instruction cycle times can be obtained with the simulator clock (CLK).

Another (shorter) version of the multirate program using macros (MR10MCR.ASM) is listed in Figure F.1 (Appendix F). The multirate filter has also been implemented with the XDS1000 emulator in conjunction with the AIB (see Chapter 3 and Appendix E). The program used for simulation MR10.ASM (on the accompanying disk) can run on the XDS1000/AIB without any changes. The sampling rate in MR10.ASM is set at Fs = 32,786 Hz. The frequency response of the middle $\frac{1}{3}$-octave filter of band 10 (with a center frequency at $\frac{1}{4}F_s \simeq 8$ kHz) using the XDS1000/AIB is shown in Figure 8.27. Note that the maximum sampling rate of the AIC on-board the EVM is 19.2 kHz (25 kHz with the TLC32046). If the sampling rate Fs = 50 kHz, the center frequency of the middle $\frac{1}{3}$-octave filter of band 10 would be at approximately 12,500 Hz, band 9 at $\simeq 6{,}250$ Hz, and so on. This can also be verified through simulation with MR10.ASM. Obtain a large number of output sample points when using MR10.ASM.

The Burr–Brown analog evaluation fixture (through serial port 1), discussed in Chapter 3 and Appendix D, also has been used to implement the multirate filter, yielding similar frequency responses.

8.6 INTRODUCTION TO IMAGE PROCESSING: VIDEO LINE RATE ANALYSIS

The goal of this project, implemented in C code, is to analyze a video signal at the horizontal (line) rate. Interactive algorithms commonly used in image processing for filtering, averaging, and edge enhancement are utilized for this analysis.

To minimize the hardware requirements, a single line of video is sampled out of each television field. The results of this line of video are processed by the TMS320C30 and displayed on the PC monitor in a format similar to a video waveform monitor at the line rate. The user can select several image processing options using the function keys and can see the results immediately on the monitor. The algorithms available include edge enhancement, digital filtering, and image averaging. The particular line of video can be selected with the up- and down-arrow keys. A module was designed to sample the video signal. The circuitry for this module (shown in Figure 8.28) can be built from readily available and inexpensive parts. A top-view diagram of the module is shown in Figure 8.29.

Video Line Rate Analysis

The video signal is the electrical representation of a complete television picture [27]. For this project, the source of the video signal is a charge coupled device (CCD) camera. The television picture as viewed on a television receiver or monitor comprises approximately 30 complete images displayed per second, generally referred to as a *frame*. Each frame of video information comprises two fields of video information that are combined to form a complete television picture. The two fields of video information are combined in an interlaced method to conserve system bandwidth and eliminate any objectionable flicker that would otherwise be present at a frame rate of 30 Hz. The bandwidth of a broadcast video signal is 4.2 MHz.

The timing characteristics of an NTSC video signal are derived as follows. The color subcarrier frequency is 3.579545 MHz and was chosen as the 455th harmonic of one-half the horizontal scanning frequency. This allows the luminance and chrominance information to be interleaved in the same sideband for transmission. The horizontal frequency is

$$\text{Horizontal frequency} = 2\left(\frac{3.579545 \text{ MHz}}{455}\right) = 15{,}734.26 \text{ Hz}$$

Because there are 525 horizontal intervals in a complete television picture (generally referred to as a frame), the frame frequency can be obtained as

$$\text{Frame frequency} = \frac{15{,}734.26}{525} = 29.97 \text{ Hz}$$

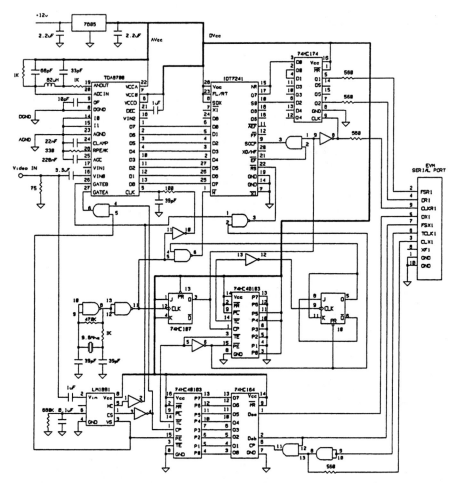

FIGURE 8.28. Circuit diagram for video line rate analysis

Because the human eye can perceive a flicker in the television picture at 29.97 Hz, the television picture (frame) is divided into two fields of information, with each field containing every other horizontal line of information. Subsequently, the field frequency is

$$\text{Field frequency} = \frac{15{,}734.26}{525/2} = 59.94 \text{ Hz}$$

This scheme, referred to as interlace scanning, effectively reduces the flicker without increasing the bandwidth of the video signal, which is limited to 4.2 MHz for commercial TV broadcasting.

8.6 Introduction to Image Processing: Video Line Rate Analysis

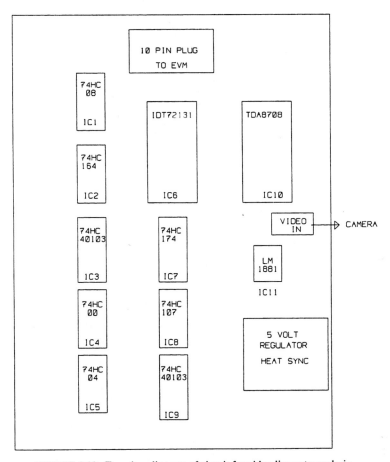

FIGURE 8.29. Top-view diagram of circuit for video line rate analysis

Each field of video information consists of 262.5 horizontal scanning lines occurring at a frequency of 15,734.26 Hz. Subsequently, the horizontal rate is $(1/15{,}734) = 63.5$ μs and consists of synchronization information occupying 11.1 μs (generally referred to as horizontal blanking). This leaves approximately $(63.5 - 11.1) = 52.4$ μs of video information. To sample the video information contained in one horizontal line, a sampling rate of 9.77 MHz is required to obtain 512 samples, because

$$F_s = \frac{512 \text{ samples}}{52.4 \ \mu\text{s}} = 9.77 \text{ MHz}$$

The vertical synchronization information occurs in every field and occupies the time of approximately 21 horizontal lines. The 21-line interval is used by

304 Real-Time DSP Applications: Student Projects

various pieces of television equipment to obtain synchronization. It is also used by the television receiver to retrace and begin scanning the next field, and is used for the transmission of test signals used by the television receiver or the TV broadcasting station. For example, videotape recorders accomplish head switching during this interval, and video production switchers change video sources in order to obtain a smooth transition from one video camera to another.

The desired line to be sampled is downloaded into the binary down-counter on the video module during this interval. The rate of the vertical interval is $262.5/15{,}734.26 = 16.68$ ms, thus requiring the processing of the signal to occur within this period of time. The processing of the signal consists of sampling the desired line of video, transferring the eight-bit digital representation of the signal to the EVM, and performing the signal processing algorithm. The results are then displayed on the screen of the PC. The line number to be sampled is transferred from the TMS320C30 to the timing circuitry on the video module.

Hardware Considerations

The majority of the circuitry for this project (Figure 8.28) is the timing circuit required to achieve proper timing between the TMS320C30 and the video signal. Figure 8.30 displays a timing diagram associated with several of the components. A parts list of the components used in Figure 8.28 is included in Appendix F. The incoming composite video signal is stripped of its synchro-

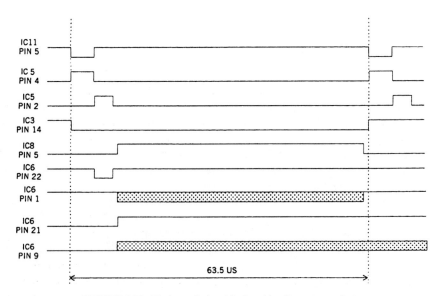

FIGURE 8.30. Timing relationship for video line rate analysis

nization pulses by a sync separator (LM1881). The composite and vertical sync are used by the timing circuitry to select the desired line of video to be sampled. The video signal is also connected to the input of an eight-bit video analog-to-digital convertor (ADC). The high sampling rate ADC is well suited for this type of application because it contains many of the preprocessing requirements of a video signal, such as the automatic gain control (AGC) and the clamp circuitry required to process the video signal before conversion. The horizontal sync and clamp pulses required by the ADC are provided by the sync separator (LM1881).

The line number to be sampled is transmitted by the TMS320C30 to a serial-to-parallel shift register (74HC164), where the eight-bit parallel output is loaded into the binary down-counter (74HC40103) during the vertical interval. The horizontal sync pulse provides the clock to the binary down-counter. When the desired line of video occurs, its output (pin 14) goes low.

A second binary counter, IC9 in Figure 8.29, is enabled during the desired line of video to be sampled and counts down to zero during the blanking interval; thus signalling the start of the video information. A JK flip-flop connected to the second binary down-counter outputs (from pin 5) an "enable pulse" to a NAND gate. This enable pulse is asserted for approximately 52.4 μs (for 512 samples). The serial receive clock frequency (CLKR1) is derived from the sampling clock as follows:

$$\text{CLKR1} = \frac{\text{sampling frequency}}{2} = \frac{9.77 \text{ MHz}}{2} = 4.885 \text{ MHz}$$

Because the sample to be transferred is eight bits wide, the transfer rate is determined by the following:

$$\text{Transfer rate} = \frac{\text{CLKR1}}{8} = \frac{4.885 \text{ MHz}}{8} = 610.625 \text{ kHz}$$

During the video portion of the desired line to be converted, the sampling clock is used to write the output of the ADC to the parallel-to-serial FIFO buffer. The FIFO serves as a rate buffer between the 9.77-MHz sampling rate and the transfer rate of 610.625 kHz to the EVM. After the first sample is written into the FIFO, the empty flag goes high and is used to initiate the transfer of the serial data to the EVM. The receive frame sync pulse to the EVM (FSR1) is derived from the control pulse width of the FIFO (Q7) used to set the length of the word to be transmitted from the FIFO. The frame sync pulse is delayed by one serial clock pulse with a D flip-flop (74HC174) and occurs while the first bit (MSB) is being transferred to the EVM. The serial data and the frame sync pulse (second time through the D flip-flop) are clocked through the D flip-flop prior to the transfer to the EVM. When the internal read pointer and the write pointer of the FIFO are at the same memory location (the 511th location), the empty flag is asserted and the serial data transfer to the TMS320C30 is completed until the next field. The time required to transfer the sampled line of video to the EVM is derived as follows:

$$\text{Time of serial clock} = \frac{1}{\text{serial receive clock frequency}}$$

$$= \frac{1}{4.885 \text{ MHz}} = 0.2047 \, \mu s$$

The transfer of 512 (8-bit) samples then takes $(512 \times 8)(0.2047 \, \mu s) = 0.8385$ ms.

Because the vertical interval rate is 16.68 ms, the digital signal processor then has (16.68 ms − 0.8385 ms) = 15.84 ms to complete any image processing algorithm and transfer the results to the host computer.

Image Processing Algorithms

Several commonly used image processing algorithms, including image averaging, edge enhancement, and digital filtering, are demonstrated in this project.

1. The image averaging algorithm is a simple averaging technique that is commonly used to reduce random noise in an image. The technique is accomplished by averaging each of the 512 samples with the samples from previous lines of video. Subsequently, an image lag occurs in areas where motion is present. When motion is kept to a minimum, random noise is effectively reduced without a loss of high-frequency information.
2. The edge enhancement algorithm is used to increase the risetime of the transitions in the video signal. The effects of this technique when viewed on a waveform monitor can be seen as an increase in the amplitude of the high-frequency components of the video signal and as overshoots on lower-frequency transitions. Two edge enhancement algorithms are available, derived from
   (a) Edge1 = 3**b** − (**a** + **b**)
   (b) Edge2 = 2**b** − (**a** + **c**)/2
   where **a** represents the previous sample, **b** the current sample, and **c** the subsequent sample. The difference between these algorithms can be seen in the increase in amplitude of the high-frequency components of the transitions of the video signal.
3. The digital filtering algorithm implements three sixth-order lowpass IIR filters. Each filter consists of three cascaded second-order sections. The three filters have the following cutoff frequencies: 500 kHz, 1 MHz, and 3 MHz.

8.6 Introduction to Image Processing: Video Line Rate Analysis 307

Demonstration — Running the Video Programs

1. Connect the module to a +12-V power supply. Connect the video input of the module to a camera and the serial 1 port on the EVM to the 10-pin connector on the module.
2. Connect the camera to a video monitor.
3. Create and run a batch file to download and execute the program VIDEO.OUT and execute the PC host communication program for the video project VIDEOPC.EXE. A listing of these source files are included in Appendix F. Figure 8.31 shows the display on the PC screen, with the camera pointing at a "test chart." The test chart is focused on the image sensor of the camera by the lens. The resulting video signal contains specific frequency bursts and patterns that are used to analyze the performance of the system. The resolution of the sensor and video circuitry can be measured as well as the image processing algorithms used in this project.
4. The desired line of video can be selected by pressing the up and down arrow keys. Figure 8.31 shows a horizontal line selected from the center of the picture (line number 128 of 256). Because approximately 21 horizontal lines contain synchronization information, the remaining 241.5 lines (262.5 − 21) can be easily selected by an eight-bit word. The function key F6 is used to specify the number of lines to step through when selecting the desired line to be sampled.

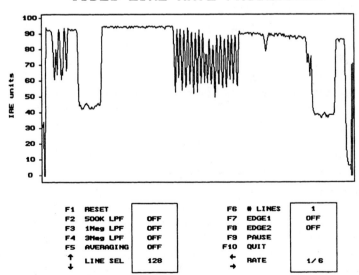

FIGURE 8.31. PC screen display of one horizontal video line signal (line number 128 selected)

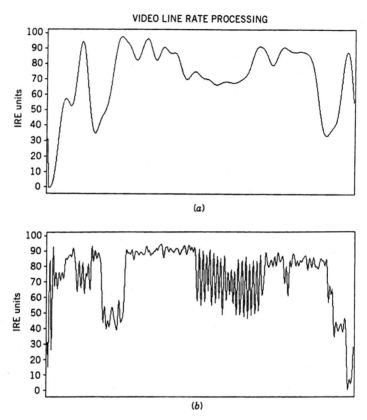

FIGURE 8.32. PC screen display of horizontal video line: (*a*) with 500-kHz lowpass filter *on*; (*b*) with 3-MHz lowpass filter *on*

5. Press F2 to assert the 500-kHz lowpass filter, resulting in the display shown in Figure 8.32(*a*). Although this filter reduces the noise considerably, the information at higher frequencies of the video signal is lost. Press F4 to turn on the 3-MHz lowpass filter, and observe from Figure 8.32(*b*) that more of the higher-frequency information is retained, although the noise reduction is not as pronounced as with the lower-bandwidth filter.

6. Press F5 and obtain the display in Figure 8.33(*a*). This averaging function provides a significant noise reduction without sacrificing the high-frequency video information. This improvement is possible because the noise is random. Figure 8.33(*a*) is obtained after first accentuating the noise embedded in the signal. This is accomplished by reducing the amount of light illuminating the sensor (by closing the iris). The automatic gain control (AGC) increases the gain of the video

8.6 Introduction to Image Processing: Video Line Rate Analysis

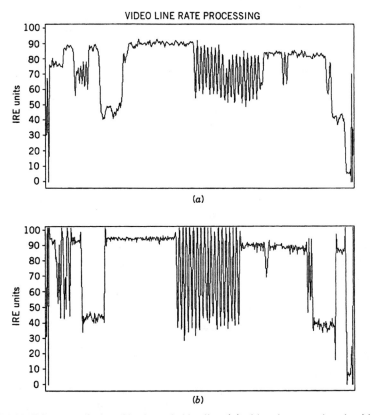

FIGURE 8.33. PC screen display of horizontal video line: (*a*) with noise averaging algorithm *on*; (*b*) with edge enhancement algorithm *on*

signal with a significant sacrifice in the signal-to-noise ratio of the camera.

Furthermore, the left- and right-arrow keys specify the averaging rate, used in conjunction with F5. For example, a rate of $\frac{1}{6}$ averages 10 fields ($\frac{60}{6}$), whereas a rate of $\frac{1}{3}$ averages 20 fields ($\frac{60}{3}$), resulting in greater noise reduction.

7. Press F7 and obtain the display in Figure 8.33(*b*). Note the edge enhancement (see also Example C.2 in Appendix C), or accentuation of the signal at the edges. F8 provides a lesser degree of edge enhancement.

Appendix F contains the listing of the supporting programs (also on the accompanying disk), all in C code.

8.7 PID CONTROLLER

The goal of this project is to design a digital speed control system for a servomotor. A well-known proportional, integral, and derivative (PID) control algorithm is used and is written in C code. The *error* is the difference between a speed command (reference) and the speed from a tachometer. The PID controller is utilized to nullify this error. A graphical user interface allows one to view the results on a PC monitor.

PID Controller

The structure of the control system is shown in Figure 8.34, which illustrates a PID (proportional, integral, derivative) controller [28–30]. A proportional (P) term is obtained by multiplying the error by a proportional constant, an integral (I) term is obtained by multiplying the integral of the error by an integral constant (and is added with the previous integral term), and a derivative (D) term is obtained by multiplying the slope (derivative) of the error (between the previous sample and the current sample) by a derivative constant. These three terms are then summed to form the controller output [30]. The integral term can increase the low-frequency gain to reduce the steady-state errors. However, this term can create an undesirable effect called windup that requires a long period of time after the error changes its sign to bring the system back to normal. This is the result of the integral term becoming extremely large, thus requiring a large number of samples (of opposite sign) to reduce the magnitude of the integral term. The derivative term can add phase lead to the system, to improve stability. A disadvantage of the derivative term is that it can amplify the high-frequency noise in the system.

Implementation

A block diagram of the control system is shown in Figure 8.35, which includes a DC motor used as a servomotor and a tachgenerator to translate speed into voltage. A separate tachgenerator DC motor is used because this alternative is less expensive than using a servomotor with a built-in tachometer. A flywheel, used as a load, couples the shafts of the motor and generator directly.

Because the output voltage of the particular tachgenerator employed is small, adjustments are made to allow for the analysis of the small feedback values (a good argument for a floating-point processor). The driver circuit in Figure 8.36 is used to convert the output of the EVM into a voltage range with a suitable current to run the servomotor and to provide protection from transients generated by the motor. The driver circuit is needed because the output of the EVM is not capable of driving the (available) servomotor

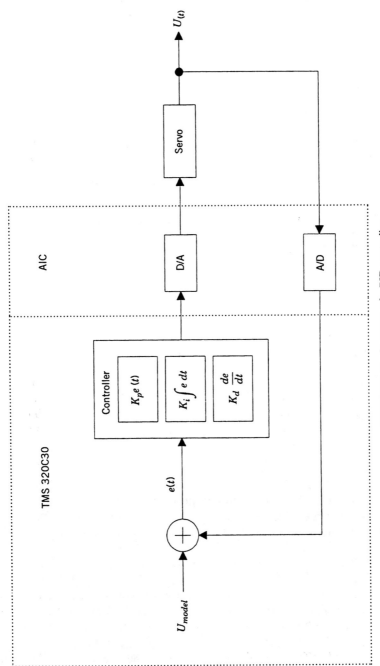

FIGURE 8.34. Structure of a PID controller

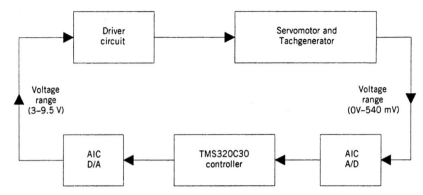

FIGURE 8.35. Block diagram of control system

directly. A linear relationship is obtained between the input to the servomotor and the output of the tachgenerator as shown in the following figure.

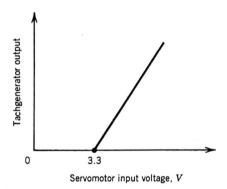

The output of the tachgenerator becomes the input to the EVM. The input voltage to the driver circuit is limited to a range between 3.0 and 9.5 V—chosen because the motor requires 3.3 V at the input of the driver circuit to overcome friction, whereas 9.5 V limits the maximum speed of the motor to insure that the output of the tachgenerator is within the input voltage range of the AIC.

To demonstrate the PID controller, the resulting effects of the proportional, integral, and derivative constants are viewed immediately on the PC monitor. The program CONTPID.C (on the accompanying disk) implements the PID controller. The program CONTPCD.C enables the PC host communication (through DMA) and the graphics support for plotting the results on the PC monitor. The AIC communication program AICCOM01.C is "included", allowing the use of serial port 1 (see Chapter 3). The output capacitor in the AIC–TMS320C30 interface diagram (Figure 3.16) is bypassed because the control system bandwidth is from DC to 300 Hz. The

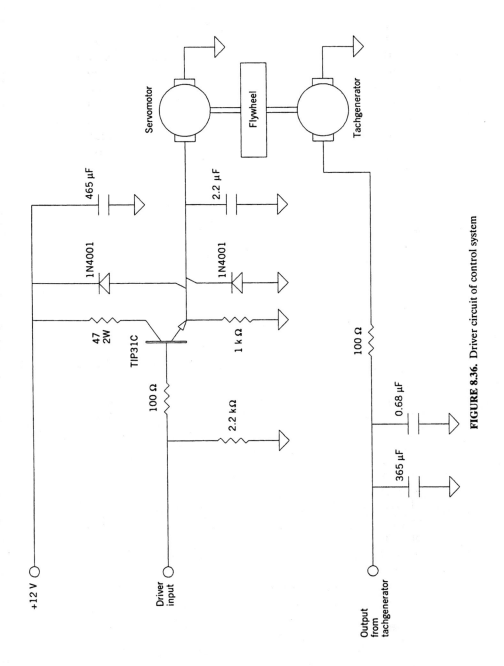

FIGURE 8.36. Driver circuit of control system

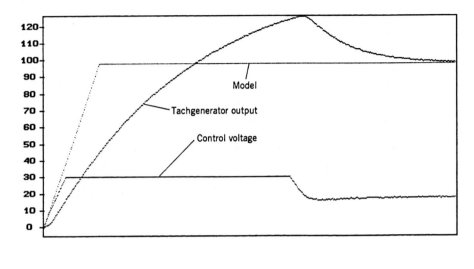

FIGURE 8.37. Effects on motor speed with different values of integral gain constants (PGAIN = 10, DGAIN = 0): (a) IGAIN = 0.01; (b) IGAIN = 0.05; (c) IGAIN = 0.001

8.7 PID Controller   315

sampling rate of the AIC is set at 600 Hz, and the AIC input highpass filter is disabled.

In the program CONTPID.C, PCNTRL represents the proportional term (PGAIN is the proportional constant), ICNTRL is the integral term (IGAIN is the integral constant), and DCNTRL is the derivative term (DGAIN is the derivative constant). The control voltage is derived from the sum of the proportional, integral, and derivative terms, minus an offset value. The offset value ($-2,000$), when placed in the transmit register of serial port 1, provides an input voltage of 3 V at the driver circuit before the system is started. Once the system starts, the control voltage may reach its maximum value of 9.5 V at the driver circuit before the desired speed is obtained. This maximum value is obtained by limiting the control value to 4,000 before the offset is applied. The maximum control voltage of 9.5 V is equivalent to a value of 2,000 in the transmit register of port 1.

The maximum speed of the servo with an input voltage of 9.5 V results in a tachgenerator output voltage of 0.54 V, or -1.08 V at the input of the AIC (the AIC has an input gain of -2). This voltage is within the allowable AIC voltage range of ± 1.5 V. The speed of the servomotor (output of the tachgenerator) is sampled and the results stored (in variable OMEGA). A small offset is added (to OMEGA) to obtain a better PC display. The error is the

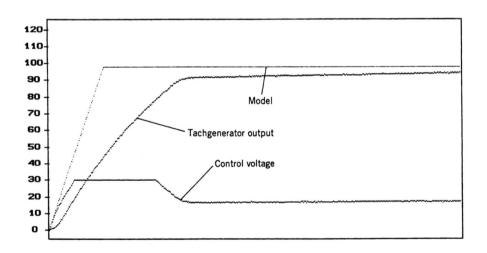

FIGURE 8.37. *(Continued)*

316   Real-Time DSP Applications: Student Projects

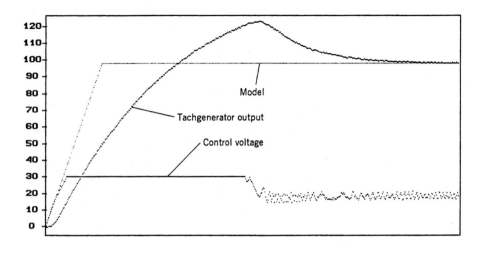

PGAIN = 10
IGAIN = 0.01
DGAIN = 50

(a)

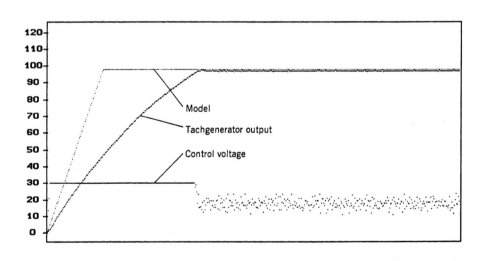

PGAIN = 100
IGAIN = 0.001
DGAIN = 100

(b)

FIGURE 8.38. Effects of PID gain constants on motor speed: (a) PGAIN = 10, IGAIN = 0.01, DGAIN = 50; (b) PGAIN = 100, IGAIN = 0.001, DGAIN = 100; (c) PGAIN = 10, IGAIN = 0.002, DGAIN = 0

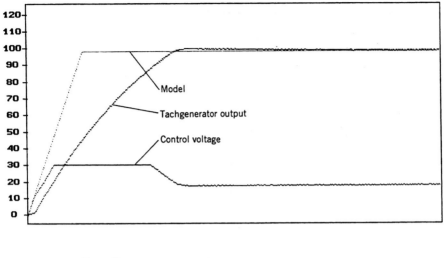

PGAIN = 10
IGAIN = 0.002
DGAIN = 0

(c)

FIGURE 8.38. *(Continued)*

difference between the servo speed as measured by the tachgenerator and a reference (goal) speed.

To run the control system, execute CONTPID.BAT. The user is then requested to enter the gain constants PGAIN, IGAIN, and DGAIN. A discussion on how to determine these gain constants can be found in references [29] and [30]. When powered, the motor turns on, but stops when the batch file CONTPID.BAT is run. After the user enters the constant gain values, the motor resumes its speed. When the desired speed is obtained, the output of the tachgenerator remains constant until the system is disturbed. For example, if the flywheel is touched (applying friction), the algorithm will increase the output voltage so that the desired (goal) speed is maintained. Figures 8.37 and 8.38 (from the PC monitor) illustrate the effects of the gain constants PGAIN, IGAIN, and DGAIN on the responses versus time. A model with a maximum value of 100 and a control voltage with a maximum value of 30 are also plotted. The lower graph is the control voltage (output of AIC) that drives the motor. This maximum value is limited to 30 so as not to interfere with the goal value of 100. In order to capture better the effects of the gain constants, only the 20th sample point is plotted. With a sampling frequency of 600 Hz, this produces 30 samples per second. Because a total of 512 samples can be displayed on the monitor, after approximately 17 s ($\frac{512}{30}$), the PC screen is cleared (incorporated in CONTPCD.C) and the plotting process resumes.

Figure 8.37 is essentially a PI controller (derivative gain set to 0). Figure 8.37(a) shows that the output of the tachgenerator continues to overshoot because of the windup effect of the integral term. Then, as the integral term becomes smaller, the control voltage decreases, slowing the motor speed toward the goal speed. An "antiwindup" algorithm can also be devised to reduce this windup effect [29]. Figure 8.37(b) shows that a larger value of IGAIN (0.05) can cause the speed of the motor to oscillate, resulting in an unstable system. A smaller value of IGAIN (0.001) can eliminate the overshoots, as displayed in Figure 8.37(c). However, a longer time is required for the motor to reach its desired (goal) speed. Figure 8.38(a) shows the effects of the derivative gain constant. After the overshoot occurs, the control voltage decreases sooner than in Figure 8.37(a), illustrating the phase lead aspect and a faster rate of convergence toward the goal. The high-frequency components (noise) of the control voltage also can be seen. Increasing the derivative constant gain DGAIN would further increase the high-frequency components of the control voltage. Figures 8.38(b) and (c) illustrate two different sets of gain constants that produce minimum overshoots and a fast convergence toward the desired speed.

This project can be extended to examine the effects of the sampling frequency on the tachgenerator output. An alternative system with a servomotor and a built-in tachometer can be modeled and simulated. Simulation results can then be compared with experimental ones.

8.8 WIREGUIDED SUBMERSIBLE

This project illustrates the control of a submersible. The 512-point FFT of Example 6.8, and Example 3.13 illustrating the communication technique between the PC host and the TMS320C30 through DMA, provide the background for this project. The two main programs are written in C, and the FFT function is in assembly.

Design and Implementation

A wireguided submersible system is controlled via a wire, enabling the submersible to be stirred in the water. The control sends signals of different frequencies to the submersible. The submersible is maneuvered based on the specific frequency of the received signal. A frequency generator is used as the control element (input to the EVM). The received signal is transformed using a 512-point real-valued FFT (see FFT_RL.ASM of Example 6.8). The program SUBEVM.C (which can be found on the accompanying disk) calls the FFT function FFT_RL.ASM. The sampling frequency is set at 10kHz. The frequency representation of the input control signal is sent through the PC host through DMA (as in Example 3.13). The program SUBHOST.C (also found on the accompanying disk) determines the frequency of the received signal by finding where the maximum magnitude occurs. This host program also provides the necessary graphics for plotting.

With a 512-point ($N = 512$) FFT, the Nyquist frequency corresponds to $N/2$, the 256th point. A 1-kHz signal would correspond to $N/10$, a 2-kHz signal to $N/5$, and so on. Four windows of width ± 20 from the corresponding points are set in the host program. Following is the response of the submersible based on the received frequency:

Ideal Frequency (kHz)	Frequency range (Hz)	Window Width	Response
1	605–1387	31–71	Turn right
2	1602–2383	82–122	Turn left
3	2617–3398	134–174	Go shallow
4	3613–4395	185–225	Go deep

Figure 8.39 shows a display of the simulated maneuvers on the PC screen. A signal of frequency between 605 and 1387 Hz would stir the submersible to

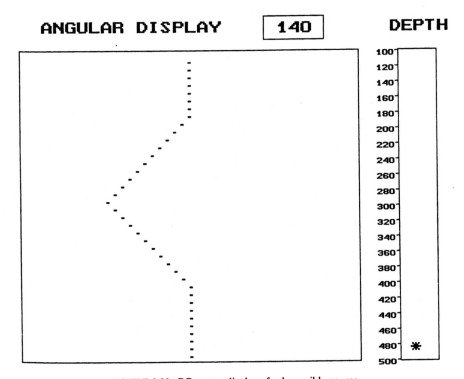

FIGURE 8.39. PC screen display of submersible course

the right, and a signal of frequency between 1602 and 2383 Hz would stir the submersible to the left. The angular display represents the direction of the submersible—the turn status. When the submersible is not changing course, (e.g., with a control signal frequency below 605 Hz or between 1387 and 1602 Hz), the dots on the PC screen will track straight in the vertical direction, as shown in the lower and upper regions in Figure 8.39.

Unlike the angular display, the depth display indicates the "actual" depth of the submersible. The program initializes the submersible with a depth of 200, using a range between 120 and 480. A 4-kHz signal (control signal between 3613 and 4395 Hz) will cause the submersible to dive until it reaches a depth of 480 or until the input frequency signal is changed. A 3-kHz control signal will cause the submersible to rise until it reaches a depth of 120.

This project can be extended in a number of ways. A microphone could be used as the input control. The submersible could filter out any noise received with a control signal. Multiple frequencies can be sent down the wire to the submersible and the submersible would acknowledge the received control signal and specify the response—the course being followed. The duration of the control signal could also specify the course to be taken (while using a single frequency).

8.9 FREQUENCY SHIFT USING MODULATION

The goal of this project is to obtain a 1, 2, 3, or 4 kHz signal using modulation.

Implementation

This project is implemented in C code. A 1-kHz sinusoidal input is used to generate either a filtered 1.001, 2, 3, or 4 kHz modulated output signal. The program MODHOST.C provides the interaction between the TMS320C30 and the PC host. See Examples 5.11 and 3.13. The program MOD.C includes the code for filtering. The coefficients of four 6th-order IIR bandpass filters, centered at 1, 2, 3, and 4 kHz, respectively, are included in the file MODCOEF.H. The frequency is shifted using

$$y(n) = x \cos(2\pi n f/F_s)$$

where x is a 1-kHz input signal used as a carrier frequency signal, f is the desired shifted frequency sent by the PC host to the TMS320C30 via DMA, and F_s is the sampling frequency. The modulated frequency signal $y(n)$ becomes the input to the selected IIR filter.

To verify this project, download the COFF file MOD.OUT into the EVM and execute MODHOST. The source programs MOD.C and MODHOST.C are included on the accompanying disk. Select 1, 2, 3, or 4 to create modulated signals of frequencies 1.001, 2, 3, or 4 kHz, respectively.

This project was also tested using the output modulated signal as the input control in the previous submersible project (in lieu of a function generator). It can also be extended to include a wider range of modulated signals and can be readily modified to select a different carrier frequency signal.

8.10 FOUR-CHANNEL MULTIPLEXER FOR FAST DATA ACQUISITION

The objective of this project is to design and construct a four-channel multiplexer which can be used for fast data acquisition. An input signal, acquired through one of the four channels, is to be transformed, and the result displayed on the PC screen.

Design and Implementation

The circuit diagram of the four-channel multiplexer module is shown in Figure 8.40. This module is connected to the serial port 1 available on the EVM. It includes an 8-bit flash analog-to-digital converter ADC304 by Datel, an IDT7202 FIFO by Integrated Devices Technology, and an IH6108 MUX by Intersil. Since the ADC can run faster than the serial port can accommodate, a FIFO buffer is utilized for storing the acquired data. When the FIFO is full, FSR is asserted, signaling the TMS320C30 processor. Communication between the EVM and the PC host is via DMA.

The program MAINEVM.C (found on the accompanying disk) contains the appropriate data for communication through serial port 1. This program calls the FFT function FFT_RL.ASM to perform a 128-point FFT on the input signal. (See also Examples 6.8 and 3.13).

The C program MAINPC.C (also found on the accompanying disk) contains the code for selecting a desired channel. The program is run by typing MAINPC. Pressing the function keys F1 through F4 selects channels 1 through 4, respectively. An LED, located next to each channel BNC connector, indicates which of the four channels is selected. A special plot program allows for the plotting of the FFT data.

Figure 8.41(a) shows the PC screen display of the FFT of a 350-kHz sinusoidal input signal, acquired through channel 2 of the four-channel multiplexer. A 2-MHz crystal is used in this implementation. The Nyquist frequency of 1 MHz is at $F_s/2$. Figure 8.41(b) displays the FFT of a 995-kHz sinusoidal input signal.

A 20-MHz crystal oscillator was also used in this project, in lieu of the 2-MHz crystal producing a noisier signal. Note that an input line driver (buffer) would have helped to reduce the effects of the length of the cable which connects the EVM to the multiplexer board.

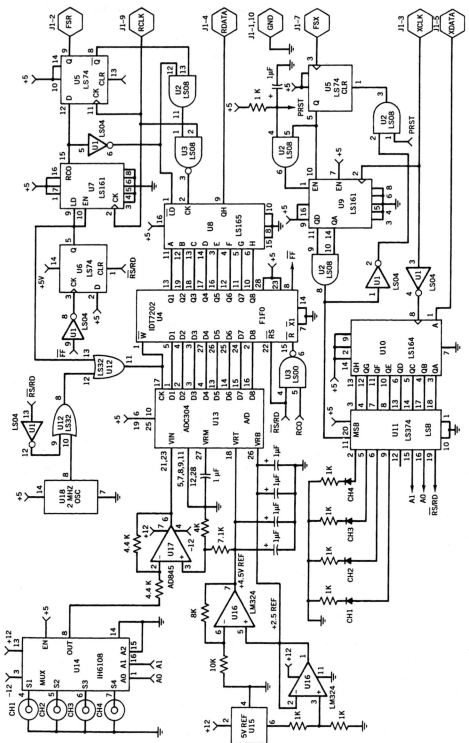

FIGURE 8.40. Circuit diagram of 4-channel multiplexer

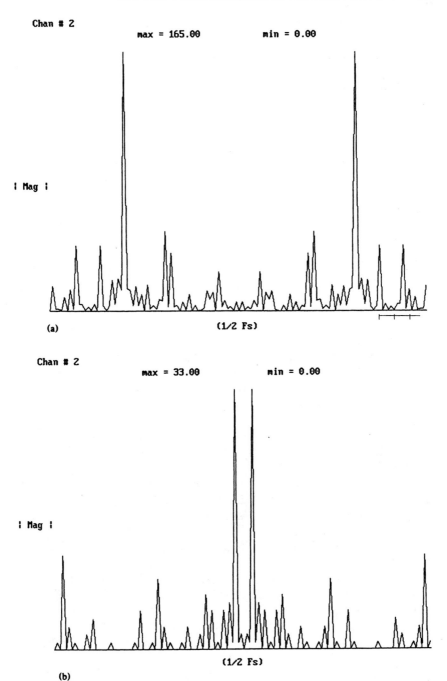

FIGURE 8.41. PC screen display of the FFT of a sinusoidal input signal acquired through channel 2 of the four-channel multiplexer: (*a*) 350 kHz signal; (*b*) 995 kHz signal.

8.11 NEURAL NETWORK FOR SIGNAL RECOGNITION

The goal of this project is to recognize a signal. The FFT of a signal becomes the input to a neural network which is trained to recognize the signal using the back-propagation learning rule.

Design and Implementation

The neural network consists of three layers with a total of 90 nodes: 64 input nodes in the first layer, 24 nodes in the middle or hidden layer, and two output nodes in the third layer. The 64 points as input to the neural network are obtained by retaining half of 128 points resulting from a 128-point FFT of the signal to be recognized. Communication between the TMS320C30 and the PC host is via DMA. Examples 6.8 and 3.13 provide the background for this project. Much research in the area of neural network is currently taking place. While such research was very extensive in the 1950s and 1960s, much of it diminished from 1969 until 1982, when renewed interest took place [31–33]. In recent years, a large number of books and articles on neural networks have been published. Neural network products are now available from many vendors.

Many different rules have been described in the literature for training a neural network. The back-error propagation is probably the most widely used for a wide range of applications. Given a set of input, the network is trained to give a desired response. If the network gives the wrong answer, then the network is corrected by adjusting its parameters so that the error is reduced. During this correction process, one starts with the output nodes and propagation is backward to the input nodes (back propagation). Then, the propagation process is repeated.

To illustrate the procedure for training a neural network using the back-propagation rule, consider a simple three-layer network with 7 nodes, as shown in Figure 8.42. The input layer consists of 3 nodes, and the hidden layer and output layer, each consists of 2 nodes. Given the following set of inputs: input No. 1 = 1 into node 0, input No. 2 = 1 into node 1, and input No. 3 = 0 into node 2, the network is to be trained in order to yield the desired output 0 at node 0 and 1 at node 1. Let the subscripts i, j, k be associated with the 1st, 2nd, and 3rd layer, respectively. A set of random weights are initially chosen as shown in Figure 8.42. For example, the weight $w_{11} = 0.9$ represents the weight value associated with node 1 in layer 1 and node 1 in the middle or hidden layer 2. The weighted sum of the input value is

$$s_j = \sum_{i=0}^{2} w_{ji} x_i \qquad (8.1)$$

8.11 Neural Network for Signal Recognition

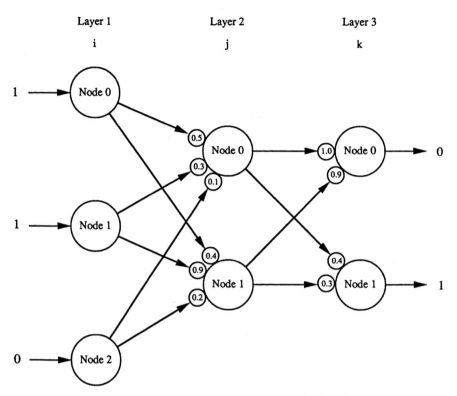

FIGURE 8.42. Three-layer neural network with 7 nodes

where $j = 0, 1$ and $i = 0, 1, 2$. Then.

$$s_0 = w_{00}x_0 + w_{01}x_1 + w_{02}x_2 = (0.5)(1) + (0.3)(1) + (0.1)(0) = 0.8$$

Similarly, $s_1 = 1.3$. A function of the resulting weighted sum $f(s_j)$ is next computed. This transfer function f of a processing element must be differentiable. For this project, f is chosen as the hyperbolic tangent function tanh. Other functions such as the unit step function or the smoother sigmoid function also can be used. The output of the transfer function associated with the nodes in the middle layer is

$$x_j = f(s_j) = \tanh(s_j), \quad j = 0, 1 \qquad (8.2)$$

The output of node 0 in the hidden layer then becomes

$$x_0 = \tanh(0.8) = 0.664$$

Similarly, $x_1 = 0.862$. The weighted sum at each node in layer 3 is

$$s_k = \sum_{j=0}^{1} w_{kj} x_j, \quad k = 0, 1 \quad (8.3)$$

Then, using (8.3),

$$s_0 = w_{00} x_0 + w_{01} x_1 = (1.0)(0.664) + (0.9)(0.862) = 1.44$$

Similarly, $s_1 = 0.524$. Using (8.2), associated with the output layer, and replacing j by k,

$$x_k = f(s_k), \quad k = 0, 1 \quad (8.4)$$

Then $x_0 = \tanh(1.44) = 0.894$, and $x_1 = \tanh(0.524) = 0.481$. The error in the output layer can now be found using

$$e_k = (d_k - x_k) f'(s_k) \quad (8.5)$$

where $d_k - x_k$ reflects the amount of error, and $f'(s)$ represents the derivative of $\tanh(s)$, or

$$f'(x) = (1 + f(s))(1 - f(s)) \quad (8.6)$$

Then,

$$e_0 = (0 - 0.894)(1 + \tanh(1.44))(1 - \tanh(1.44)) = -0.18$$

Similarly, $e_1 = 0.399$. Based on this output error, the contribution to the error by each hidden layer node is to be found. The weights are then adjusted [33] based on this error using,

$$\Delta w_{kj} = \eta e_k x_j \quad (8.7)$$

where η is the network learning rate constant, chosen as 0.3. A large value of η can cause instability, and a very small one can make the learning process much too slow. Using (8.7),

$$\Delta w_{00} = (0.3)(-0.18)(0.664) = -0.036$$

Similarly, $\Delta w_{01} = -0.046$, $\Delta w_{10} = 0.08$, and $\Delta w_{11} = 0.103$. The error associated with the hidden layer is

$$e_j = f'(s_j) \sum_{k=0}^{1} e_k w_{kj} \quad (8.8)$$

Then,

$$e_0 = (1 + \tanh(0.8))(1 - \tanh(0.8))\{(-0.18)(1.0) + (0.399)(0.4)\}$$
$$= -0.011$$

Similarly, $e_1 = -0.011$. Using again (8.7), in order to change the weights between layers i and j,

$$\Delta w_{ji} = \eta e_j x_i$$

Then,

$$\Delta w_{00} = (0.3)(-0.011)(1) = -0.0033$$

Similarly, $\Delta w_{01} = -0.0033$, $\Delta w_{02} = 0$, $\Delta w_{10} = -0.0033$, $\Delta w_{11} = -0.0033$, and $\Delta w_{12} = 0$. This gives an indication of by how much to change the original set of weights chosen. For example, the new set of coefficients becomes

$$w_{00} = w_{00} + \Delta w_{00} = 0.5 - 0.0033 = 0.4967$$

and $w_{01} = 0.2967$, $w_{02} = 0.1$, and so on.

This new set of weights represents only the values after one complete cycle. These weight values have been verified using a training program for this project. For this procedure of training the network, re-adjusting the weights is continuously repeated, until the output values converge to the set desired output values. For this project, the training program is such that the training process can be halted by the user who can still use the resulting weights.

To implement this project, two sets of inputs were chosen: a sinusoidal and a square wave input. The FFT (128-point) of each input signal is captured and stored in a file, with a total of 4800 points: 200 vectors, each with 64 features (retaining one-half of the 128 points). Another program scales each set of data (sine and square wave) so that the values are between 0 and 1.

To demonstrate this project, two output values for each node are displayed on the PC screen. A value of $+1$ for node 0 and -1 for node 1 indicate that a sinusoidal input is recognized, and a value of -1 for node 0 and $+1$ for node 1 indicate that a square wave input is recognized.

This project was successful, but implemented for only the 2 sets of chosen data. Much work remains to be done, such as training more complex sets of data, and examining the effects of different training rules based on the different signals to be recognized.

REFERENCES

[1] R. Chassaing and D. W. Horning, *Digital Signal Processing with the TMS320C25*, Wiley, New York, 1990.

[2] *Proceedings of the 1991 TMS320 Educators Conference*, Texas Instruments, Inc., Dallas, Tex., July 1991.

[3] K. S. Lin (editor), *Digital Signal Processing Applications with the TMS320 Family—Theory, Algorithms, and Implementations*, Vol. 1, Texas Instruments, Inc., Dallas, Tex., 1987.

[4] P. Papamichalis (editor), *Digital Signal Processing Applications with the TMS320 Family—Theory, Algorithms, and Implementations*, Vol. 2, Texas Instruments, Inc., Dallas, Tex., 1989.

[5] P. Papamichalis (editor), *Digital Signal Processing Applications with the TMS320 Family—Theory, Algorithms, and Implementations*, Vol. 3, Texas Instruments, Inc., Dallas, Tex., 1990.

[6] B. Widrow and S. D. Stearns, *Adaptive Signal Processing*, Prentice-Hall, Englewood Cliffs, N.J., 1985.

[7] B. Widrow, J. R. Glover, J. M. McCool, J. Kaunitz, C. S. Williams, R. H. Hearn, J. R. Zeidler, E. Dong, Jr., and R. C. Goodlin, Adaptive noise cancelling: Principles and applications, *Proceedings of the IEEE*, **63**, December 1975, pp. 1692–1716.

[8] J. R. Treichler, C. R. Johnson, Jr., and M. G. Larimore, *Theory and Design of Adaptive Filters*, Wiley, New York, 1987.

[9] S. T. Alexander, *Adaptive Signal Processing Theory and Applications*, Springer-Verlag, New York, 1986.

[10] B. Widrow, J. McCool, M. Larimore, and R. Johnson, Stationary and nonstationary learning characteristics of the LMS adaptive filter, *Proceedings of the IEEE*, **64**, 1976, pp. 1151–1162.

[11] M. L. Honig and D. G. Messerschmitt, *Adaptive Filters: Structures, Algorithms and Applications*, Kluwer Academic, Norwell, Mass., 1984.

[12] C. F. Cowan and P. F. Grant (editors), *Adaptive Filters*, Prentice-Hall, Englewood Cliffs, N.J., 1985.

[13] S. Haykin, *Adaptive Filter Theory*, Prentice-Hall, Englewood Cliffs, N.J., 1986.

[14] R. Chassaing, D. W. Horning, and P. Martin, Adaptive filtering with the TMS320C25, in *Proceedings of the 1989 ASEE Annual Conference*, June, 1989.

[15] R. Chassaing and P. Martin, Digital filtering with the floating-point TMS320C30 digital signal processor, in *Proceedings of the 21st Annual Pittsburgh Conference on Modeling and Simulation*, May 1990.

[16] S. Kuo, G. Ranganathan, P. Gupta, and C. Chen, Design and implementation of adaptive filters, *IEEE 1988 International Conference on Circuits and Systems*, June 1988.

[17] M. Bellanger, *Digital Processing of Signals Theory and Practice*, Wiley & Sons Ltd., Paris, 1989.

[18] D. G. Messerschmitt, Echo cancellation in speech and data transmission, *IEEE Journal on Selected Areas in Communications*, **SAC-2**, March 1984, pp. 283–297.

[19] R. E. Crochiere and L. R. Rabiner, *Multirate Digital Signal Processing*, Prentice-Hall, Englewood Cliffs, N.J., 1983.

[20] R. W. Schafer and L. R. Rabiner, A digital signal processing approach to interpolation, *Proceedings of the IEEE*, **61**, June 1973, pp. 692–702.

[21] R. E. Crochiere and L. R. Rabiner, Optimum FIR digital filter implementations for decimation, interpolation and narrow-band filtering, *IEEE Transactions on Acoustics, Speech, and Signal Processing*, **ASSP-23**, October 1975, pp. 444–456.

[22] R. E. Crochiere and L. R. Rabiner, Further considerations in the design of decimators and interpolators, *IEEE Transactions on Acoustics, Speech, and Signal Processing*, **ASSP-24**, August 1976, pp. 296–311.

[23] M. G. Bellanger, J. L. Daguet, and G. P. Lepagnol, Interpolation, extrapolation, and reduction of computation speed in digital filters, *IEEE Transactions on Acoustics, Speech, and Signal Processing*, **ASSP-22**, August 1974, pp. 231–235.

[24] R. Chassaing, P. Martin, and R. Thayer, Multirate filtering using the TMS320C30 floating-point digital signal processor, in *Proceedings of the 1991 ASEE Annual Conference*, June 1991.

[25] R. Chassaing, Digital broadband noise synthesis by multirate filtering using the TMS320C25, in *Proceedings of the 1988 ASEE Annual Conference*, Vol. 1, June 1988.

[26] R. Chassaing, W. A. Peterson, and D. W. Horning, A TMS320C25-based multirate filter, *IEEE Micro*, October 1990, pp. 54–62.

[27] G. P. McGinty, *Video Cameras: Theory and Servicing*, H. W. Sams, Indianapolis, Ind., 1984.

[28] Y. Dote, *Servo Motor and Motion Control Using Digital Signal Processors*, Prentice-Hall, Englewood Cliffs, N.J., 1990.

[29] I. Ahmed (editor), *Digital Control Applications with the TMS320 Family*, Texas Instruments, Inc., Dallas, Tex., 1991.

[30] K. Åström and B. Wittenmark, *Computer Controlled Systems—Theory and Design*, Prentice-Hall, Englewood Cliffs, N.J., 1984.

[31] B. Widrow and R. Winter, Neural nets for adaptive filtering and adaptive pattern recognition, *Computer Magazine*, Computer Society of the IEEE, March 1988, pp. 25–39.

[32] R. P. Lippmann, An introduction to computing with neural nets, *IEEE ASSP Magazine*, April 1987, pp. 4–22.

[33] D. E. Rumelhart, J. L. McClelland, and the PDP Research Group, *Parallel Distributed Processing: Explorations in the Microstructure of Cognition*, Vol. 1, MIT Press, Cambridge, Mass., 1986.

[34] A. V. Oppenheim and R. Schafer, *Discrete-Time Signal Processing*, Prentice-Hall, Englewood Cliffs, N.J., 1989.

[35] T. W. Parks and C. S. Burrus, *Digital Filter Design*, Wiley, New York, 1987.

[36] D. J. DeFatta, J. G. Lucas, and W. S. Hodgkiss, *Digital Signal Processing: A System Approach*, Wiley, New York, 1988.

[37] L. B. Jackson, *Digital Filters and Signal Processing*, Kluwer Academic, Norwell, Mass., 1986.

[38] J. G. Proakis and D. G. Manolakis, *Introduction to Signal Processing*, Macmillan, New York, 1988.

[39] L. R. Rabiner and B. Gold, *Theory and Application of Digital Signal Processing*, Prentice-Hall, Englewood Cliffs, N.J., 1975.

[40] N. Ahmed and T. Natarajan, *Discrete-Time Signals and Systems*, Reston, Reston, Va., 1983.

A
Introduction to C Programming

BILL BITLER

Programming in C has several distinct advantages over assembly language programming for digital signal processing (DSP) applications. As a high-level language, C is generally easier to program compared with assembly language and is portable to a variety of processors, including the **TMS320C30** *floating-point digital signal processor. This enables applications to be developed more quickly when a DSP chip with a different instruction set is introduced to the market. When an application requires minimum execution time and memory requirements, portions of the code can be developed in assembly functions that are called by a C program. In this case, only the C-callable assembly functions need to be revised for a different processor.*

The C programming language is a structured language offering "low-level" capabilities (such as address manipulation with pointers) that are used often in DSP applications. A C program comprises one or more functions used to perform specific operations in a program. Every C program contains at least one function called main(), *which is the main function where program execution begins. The parentheses following the word* main *are used to pass an argument list to a function.*

To introduce the C programming language, several algorithms in DSP will be implemented on an IBM PC using the Borland Turbo C compiler. These examples will introduce the reader to programming techniques required in more-advanced applications in DSP. The following example illustrates a simple C program that displays a message on the monitor.

Example A.1 *Monitor Display.* This example demonstrates output to a PC screen. Consider the following program:

```
#include <stdio.h>
main()
{
printf("C Programming for Digital Signal Processing \n");
}
```

The result of the program on the computer display is

```
C Programming for Digital Signal Processing
```

The line that starts with the symbol # is not a C statement and is referred to as a compiler directive. This directive instructs the complier to include the contents of the file stdio.h when compiling the source code. This file is commonly referred to as a header file and includes information about the functions included in the Turbo C library. Because the printf function is found in the Turbo C library, the corresponding header file needs to be compiled with the source code.

This program has a function called main() that passes no arguments to the function. The fourth and sixth lines of the program contain the braces {} that enclose a statement for printing that belongs to the main function. The printf statement calls a function from the Turbo C library that outputs the information contained within the parentheses. The double quotes " " are used to instruct the compiler that the information contained between the quotes is a character string. The sequence \n is called an escape sequence, which instructs the compiler to issue a new-line character to terminate the current line and move the cursor to the left margin of the next line. The semicolon at the end of the printf statement is referred to as a statement terminator.

Example A.2 *Use of Variables and Arithmetic Operations.* Consider the program in Figure A.1. Comments are inserted between the /* and */

```
/*ARITHM.C - USE OF VARIABLES AND ARITHMETIC OPERATIONS*/
#include <stdio.h>

main()
{
   float x = 5.0;    /*declare and initialize x to 5.0*/
   float y = 2.0;    /*declare and initialize y to 2.0*/
   float z;          /*declare z*/
   int A,B;
   A = 5;
   B = 2;
   z = x/y;
   printf ("x/y = %f\n", z);                  /*output z*/
   printf ("A/B = %i\n", A/B);                /*output int part of A / B*/
   printf ("A mod B = %i\n", A%B);            /*output modulus of A / B */
   printf ("A/B = %f\n", (float)A/(float)B);  /*floating output of A / B*/
}
```

FIGURE A.1. C program to illustrate the use of variables and arithmetic operations (ARITHM.C)

characters and are ignored by the compiler. They are freely used to document the source code. The first statement in the main program declares and initializes a floating-point variable x to 5.0. The third statement declares a floating-point variable z without initializing it to a value. The C compiler does not initialize variables automatically. When a variable is declared, whatever information resides in the associated memory location becomes the contents of the declared variable. The variable z should be initialized before it is used, or unpredictable results may occur. The fourth statement declares two integer variables A and B. Each variable must be of the same type and separated by a comma. When more than one variable is declared in a statement, any number of the variables could be initialized on one line. For example, from

```
int A = 5, B = 2;
```

the integer variables A and B could have been initialized and declared at the same time. This technique is often preferred since the resulting source code requires fewer statements.

The statement z = x / y; initializes the variable z with the value of variable x divided by y. This satisfies the requirement of initializing a variable before it is used.

The first printf statement outputs the character string specified between the double quotes and the value of the variable z on the computer monitor. The format specifier %f instructs the compiler to format the variable z as a floating-point value and replaces the format specifier with this result.

The second printf statement is similar to the first, with the exception of the argument and format specifier. The argument is the result of variable A divided by variable B, whereas the format specifier indicates that the result of the argument is to be formatted as an integer value (for example, in integer division, $\frac{5}{2} = 2$ with a remainder of 1). Also, the ability to perform an arithmetic operation as an argument allows the programmer to write code that is compact and easy to read. The next printf statement demonstrates the modulo (remainder) operation. This operation provides a way to create a form of circular buffering used to implement digital filters and is discussed in Chapters 2 and 4.

The last printf statement uses a floating-point format specifier and a cast operator to perform a floating-point division, (float)A / (float)B. Instead of performing an integer division, the compiler is forced to perform a floating-point division. The cast operator (float) preceding the variable does not necessarily need to be of the floating type. This is an important concept for DSP programming because most I/O operations are of the integer type and DSP algorithms often perform their calculations using floating-point operations.

The results of the program on the computer display are

```
x / y = 2.500000
A / B = 2
A mod B = 1
A / B = 2.500000
```

Example A.3 Concept of Arrays and Loops. Consider the multiplication matrix program in Figure A.2. The variable A is a two-dimensional array containing three rows and three columns of floating-point data. An important point to remember is that the row and column subscripts always start at zero, resulting in the subscript being one less than the desired row or column to be accessed. Therefore, the following notation A[0][0] accesses the first row and the first column. Likewise, the second row and third column is accessed using A[1][2]. The variable B is a one-dimensional array containing three columns of floating-point data. In this case, array B requires only one subscript to access a particular column of data. The integer variables i and j (which represent row and column) are used to access the appropriate positions of array A, and variable j is used in conjunction with array B. The statement

```
for (i = 0; i < 3; i++)
```

is used to define a for loop in C. The expressions contained within the parentheses of the for statement control the execution of the loop. Execution continues as long as i is less than 3 (starting with i = 0). The variable i is then postincremented by one. The statements contained within the scope of the for loop are located between the braces {}. The braces are not necessary if only one statement is to be executed.

A second for loop is nested within the first and consists of the following code:

```
for (j = 0; j < 3; j++)
   result += A[i][j]*B[j];
```

This loop is used to accumulate the results of the matrix algorithm. The statement

```
result += A[i][j]*B[j];
```

is equivalent to

```
result = result + A[i][j]*B[j];
```

```
/*MAT.C - CONCEPTS OF ARRAYS AND LOOPS*/

main()
{
  float A[3][3] = {{1.0,2.0,3.0},
                   {4.0,5.0,6.0},
                   {7.0,8.0,9.0}};
  float B[3] =    {1.0,2.0,3.0};
  int i, j;
  float result;
  for (i = 0; i < 3; i++)            /*first for loop*/
  {
    result = 0;
    for (j = 0; j < 3; j++)          /*second for loop*/
      result += A[i][j]*B[j];
    printf("%f\n", result);
  }
}
```

FIGURE A.2. C program to illustrate arrays and loops (MAT.C)

The printf statement is associated with the first loop and is used to output the overall result.

The results of the program on the computer display are

14.000000
32.000000
50.000000

A useful loop for real-time signal processing is an endless loop. The following segment implements an endless for loop:

```
for (;;)
{
 ...
}
```

Example A.4 Volatile Variable. Consider the program in Figure A.3. The declaration of a volatile variable is an important concept in DSP. In real-time applications, data is often modified by external events, such as an interrupt routine servicing an I/O port. An optimizing compiler, such as the Texas Instruments, Inc., TMS320C30 C compiler, often will reduce an expression used in conjunction with an interrupt routine or an I/O port access. The

```
/*VOLATILE.C - USE OF VOLATILE VARIABLE*/
#include <dos.h>
#include <stdio.h>
void interrupt (*old_int_routine)(void);/*declare interrupt function pointer*/
volatile int clock = 0;                 /*declare a volatile variable*/

void interrupt timer()
{
  clock++;
  if (clock % 18 == 0)                  /*approximately 18.2 clock ticks/second*/
    printf("The following seconds have elapsed %i\n", clock/18);
  old_int_routine();                    /*call old interrupt routine*/
}

main()
{
  old_int_routine = getvect(0x8);       /*save original interrupt vector*/
  setvect(0x8, timer);                  /*install new interrupt handler*/
  while (clock < 200);                  /*idle for 200 clock ticks */
  setvect(0x8, old_int_routine);        /*install orginal interrupt vector*/
}
```

FIGURE A.3. C program to illustrate the volatile variable (VOLATILE.C)

volatile variable declaration warns the compiler not to make any assumptions about its value. This also prevents the compiler from assigning it as a register variable.

The interrupt function timer is used to output the elapsed time in seconds to the display. In the PC, the timer interrupt occurs approximately 18.2 times per second. The clock variable is postincremented by 1 every time an interrupt occurs. The modulo operator in the if statement is used to determine when one second has passed (approximately when 18 timer interrupts have occurred) and updates the new elapsed time on the display. The original timer routine, old_int_routine(), is executed to satisfy the requirements of the PC environment.

The first two statements in the main function are used to save the address of the original interrupt routine and to install the new interrupt routine. The while statement defines another type of loop commonly used in C. In this example, the while loop contains no statement within its scope and causes the program to idle until 200 timer interrupts occur. Depending on the optimization level of the compiler, this statement could be reduced to a single execution of the loop when the variable clock is not defined as volatile. This may occur because the variable is not changed within the scope of the while loop and the compiler is not aware that the variable is modified externally by the timer interrupt routine. The last statement in the main function is used to reinstall the original timer interrupt routine.

Example A.5 *Recursive IIR Filter.* Consider the program IIRA.C listed in Figure A.4. The first line of code, #define STAGES 3, is a preprocessor

```c
/*IIRA.C - IIR FILTER*/
#define STAGES 3            /* number of 2nd-order stages in filter */
#include <stdio.h>
double A[STAGES][3] = { /* numerator coefficients */
                { 5.3324E-02, 0.0000E+00, -5.3324E-02},
                { 5.3324E-02, 0.0000E+00, -5.3324E-02},
                { 5.3324E-02, 0.0000E+00, -5.3324E-02}};
double B[STAGES][2] = { /* denominator coefficients */
                { -1.4435E+00, 9.4879E-01},
                { -1.3427E+00, 8.9514E-01},
                { -1.3082E+00, 9.4377E-01}};
double DLY[STAGES][2] = {0};        /*delay storage elements*/
FILE *IO_IN, *IO_OUT;

void IIR(int n, int len)            /*IIR function*/
{
  int i, loop = 0;
  double dly, yn, input;
  long temp;
  while (loop < len)                /*while loop*/
  {
    ++loop;
    fscanf(IO_IN, "%D\n", &temp);
    input = temp;
    for (i = 0; i < n; i++)         /*for loop*/
    {
      dly = input - B[i][0]*DLY[i][0] - B[i][1]*DLY[i][1];
      yn = A[i][2]*DLY[i][1] + A[i][1]*DLY[i][0] + A[i][0]*dly;
      DLY[i][1] = DLY[i][0];
      DLY[i][0] = dly;
      input = yn;
    }
    fprintf (IO_OUT, "%d\n", (int)yn);   /*output data to file*/
  }
}

main()                                   /*main function*/
{
  #define length 345
  IO_IN = fopen ("IMPULSE.DAT", "rt");   /*open IMPULSE data file*/
  IO_OUT = fopen ("IIRA.DAT", "wt");     /*open OUTPUT data file */
  IIR(STAGES, length);                   /*execute IIR function  */
  fclose (IO_IN);                        /*close IMPULSE.DAT file*/
  fclose (IO_OUT);                       /*close IIROUT.DAT file */
}
```

FIGURE A.4. C program to illustrate IIR bandpass filter (IIRA.C)

directive that defines a macro substitution. The compiler defines the character string STAGES with 3.

The declaration of a FILE pointer is used in the main and IIR functions, the FILE pointer *IO_IN will point to the DOS file IMPULSE.DAT while *IO_OUT will point to the DOS file IIROUT.DAT. These pointers are used by the program to input and output data to and from the appropriate DOS files.

Consider the function IIR: The word void preceding IIR instructs the compiler that no data is returned by the function IIR. The statements contained within the pair of braces {} are executed when this function is called. The arguments within the parentheses allow information to be passed to the function when it is called. The first three lines declare variables that are used. Because these variables are defined within the scope of the function IIR, they cannot be accessed by the main function. The statements contained within the pair of braces {} associated with the while loop will execute as long as the expression (loop < len) is true. The fscanf statement inside the while loop is used to input data from a DOS file. The first argument in the fscanf statement is the FILE pointer, the second argument contains formatting instruction, and the third argument instructs the compiler to place the formatted results at the memory location specified with the variable temp. In this example, a for loop is nested within the while loop. The fprintf statement inside the while loop is used to output formatted data to a file. With the exception of a FILE pointer passed as the first argument, the fprintf statement is similar to the printf statement described earlier. Because the output data is written to a file, no data is returned from the IIR function back to the main function. This fact is reflected by the void declaration in the IIR function statement.

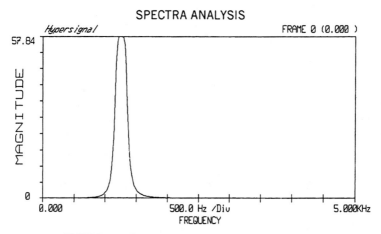

FIGURE A.5. Frequency response of IIR bandpass filter

Within the `main` function, the `fopen` statements open the appropriate DOS files and assign them to `FILE` pointers `IO_IN` and `IO_OUT`. After the files are opened, the `IIR` function is called and the digital filter algorithm is performed on the impulse data from the DOS file (`IMPULSE.DAT` on the accompanying disk) and the results are written back to the DOS file called `IIRA.DAT`. After the `IIR` function is executed, the `fclose` statements close the appropriate DOS files and terminate the execution of the program.

The results of the IIR filter (bandpass with a center frequency of 1,250 Hz) shown in Figure A.5 were obtained using a commercially available DSP utility package from Hyperception, Inc. This package performed a fast Fourier transform (FFT) on the data file from the `IIR` filter program (`IIRA.DAT`) and plotted the resulting magnitude.

Chapter 5 also discusses this IIR filter example. For a more comprehensive study of the C programming language, see references [1–3].

REFERENCES

[1] B. W. Kernighan and D. M. Ritchie, *The C Programming Language*, Prentice-Hall, Englewood Cliffs, N.J., 1978.

[2] *Turbo C++ Getting Started*, Borland International, Inc., Scotts Valley, Calif., 1990.

[3] S. G. Kochan, *Programming in C*, Hayden Books, 1991.

[4] P. M. Embree and B. Kimble, *C Language Algorithms for Digital Signal Processing*, Prentice-Hall, Englewood Cliffs, N.J., 1990.

[5] *TMS320 Floating-Point DSP Optimizing C Compiler User's Guide*, Texas Instruments, Inc., Dallas, Tex., 1991.

[6] R. Chassaing, B. Bitler, and D. W. Horning, "Real-time digital filters in C, in *Proceedings of the 1991 ASEE Annual Conference, June, 1991*.

B
Instruction Set, Registers, and Memory Maps

B.1 TMS320C30 INSTRUCTION SET

Table B.1 shows a summary of the TMS320C30 instruction set, and Table B.2 contains a list of parallel instructions.

B.2 REGISTER FORMATS AND MEMORY MAPS

The following register formats and memory maps can be very useful during program development:

1. *Status (ST) register*. The format of this register is shown in Figure B.1. It provides information about the state of the CPU. The condition flags of the ST register are set based on resulting operations.
2. *Interrupt enable (IE) register*. The format of this register is shown in Figure B.2. A 1 (or 0) enables (or disables) the corresponding interrupt.
3. *Interrupt flag (IF) register*. The format of the IF register is shown in Figure B.3. An interrupt is set with a 1 in a flag register bit. An interrupt is cleared with a 0 into the corresponding flag register.
4. *I / O flag (IOF) register*. The format of the IOF register is shown in Figure B.4. It controls the function of the external pins XF0 and FX1 for I/O. This register is zero at reset.

B.2 Register Formats and Memory Maps

TABLE B.1 TMS320C30 Instruction Set Summary

Mnemonic	Description	Operation
ABSF	Absolute value of a floating-point number	$|src| \rightarrow Rn$
ABSI	Absolute value of an integer	$|src| \rightarrow Dreg$
ADDC	Add integers with carry	$src + Dreg + C \rightarrow Dreg$
ADDC3	Add integers with carry (3-operand)	$src1 + src2 + C \rightarrow Dreg$
ADDF	Add floating-point values	$src + Rn \rightarrow Rn$
ADDF3	Add floating-point values (3-operand)	$src1 + src2 \rightarrow Rn$
ADDI	Add integers	$src + Dreg \rightarrow Dreg$
ADDI3	Add integers (3-operand)	$src1 + src2 + \rightarrow Dreg$
AND	Bitwise logical-AND	Dreg AND src $\rightarrow$ Dreg
AND3	Bitwise logical-AND (3-operand)	src1 AND src2 $\rightarrow$ Dreg
ANDN	Bitwise logical-AND with complement	Dreg AND $\overline{src}$ $\rightarrow$ Dreg
ANDN3	Bitwise logical-ANDN (3-operand)	src1 AND $\overline{src2}$ $\rightarrow$ Dreg
ASH	Arithmetic shift	If count $\geq$ 0: (Shifted Dreg left by count) $\rightarrow$ Dreg Else: (Shifted Dreg right by $|count|$) $\rightarrow$ Dreg
ASH3	Arithmetic shift (3-operand)	If count $\geq$ 0: (Shifted src left by count) $\rightarrow$ Dreg Else: (Shifted src right by $|count|$) $\rightarrow$ Dreg
B*cond*	Branch conditionally (standard)	If cond = true: If Csrc is a register, Csrc $\rightarrow$ PC If Csrc is a value, Csrc + PC $\rightarrow$ PC Else, PC + 1 $\rightarrow$ PC

Mnemonic	Description	Operation
B*cond*D	Branch conditionally (delayed)	If cond = true: If Csrc is a register, Csrc $\rightarrow$ PC If Csrc is a value, Csrc + PC + 3 $\rightarrow$ PC Else, PC + 1 $\rightarrow$ PC
BR	Branch unconditionally (standard)	Value $\rightarrow$ PC
BRD	Branch unconditionally (delayed)	Value $\rightarrow$ PC
CALL	Call subroutine	PC + 1 $\rightarrow$ TOS Value $\rightarrow$ PC
CALL*cond*	Call subroutine conditionally	If cond = true: PC + 1 $\rightarrow$ TOS If Csrc is a register, Csrc $\rightarrow$ PC If Csrc is a value, Csrc + PC $\rightarrow$ PC Else, PC + 1 $\rightarrow$ PC
CMPF	Compare floating-point values	Set flags on Rn $-$ src
CMPF3	Compare floating-point values (3-operand)	Set flags on src1 $-$ src2
CMPI	Compare integers	Set flags on Dreg $-$ src
CMPI3	Compare integers (3-operand)	Set flags on src1 $-$ src2
DB*cond*	Decrement and branch conditionally (standard)	ARn $-$ 1 $\rightarrow$ ARn If cond = true and ARn $\geq$ 0: If Csrc is a register, Csrc $\rightarrow$ PC If Csrc is a value, Csrc + PC + 1 $\rightarrow$ PC Else, PC + 1 $\rightarrow$ PC
DB*cond*D	Decrement and branch conditionally (delayed)	ARn $-$ 1 $\rightarrow$ ARn If cond = true and ARn $\geq$ 0: If Csrc is a register, Csrc $\rightarrow$ PC If Csrc is a value, Csrc + PC + 3 $\rightarrow$ PC Else, PC + 1 $\rightarrow$ PC

TABLE B.1 (Continued)

Mnemonic	Description	Operation
FIX	Convert floating-point value to integer	Fix (src) → Dreg
FLOAT	Convert integer to floating-point value	Float(src) → Rn
IDLE	Idle until interrupt	PC + 1 → PC Idle until next interrupt
LDE	Load floating-point exponent	src(exponent) → Rn(exponent)
LDF	Load floating-point value	src → Rn
LDF*cond*	Load floating-point value conditionally	If cond = true, src → Rn Else, Rn is not changed
LDFI	Load floating-point value, interlocked	Signal interlocked operation src → Rn
LDI	Load integer	src → Dreg
LDI*cond*	Load integer conditionally	If cond = true, src → Dreg Else, Dreg is not changed
LDII	Load integer, interlocked	Signal interlocked operation src → Dreg
LDM	Load floating-point mantissa	src (mantissa) → Rn (mantissa)
LSH	Logical shift	If count ≥ 0: (Dreg left-shifted by count) → Dreg Else: (Dreg right-shifted by |count|) → Dreg
LSH3	Logical shift (3-operand)	If count ≥ 0: (src left-shifted by count) → Dreg Else: (src right-shifted by |count|) → Dreg
MPYF	Multiply floating-point values	src × Rn → Rn
MPYF3	Multiply floating-point value (3-operand)	src1 × src2 → Rn
MPYI	Multiply integers	src × Dreg → Dreg
MPYI3	Multiply integers (3-operand)	src1 × src2 → Dreg
NEGB	Negate integer with borrow	0 − src − C → Dreg
NEGF	Negate floating-point value	0 − src → Rn
NEGI	Negate integer	0 − src → Dreg
NOP	No operation	Modify ARn if specified
NORM	Normalize floating-point value	Normalize (src) → Rn
NOT	Bitwise logical-complement	$\overline{\text{src}}$ → Dreg
OR	Bitwise logical-OR	Dreg OR src → Dreg
OR3	Bitwise logical-OR (3-operand)	src1 OR src2 → Dreg
POP	Pop integer from stack	*SP−− → Dreg
POPF	Pop floating-point value from stack	*SP−− → Rn
PUSH	Push integer on stack	Sreg → *++ SP
PUSHF	Push floating-point value on stack	Rn → *++ SP
RETI*cond*	Return from interrupt conditionally	If cond = true or missing: *SP−− → PC 1 → ST (GIE) Else, continue
RETS*cond*	Return from subroutine conditionally	If cond = true or missing: *SP−− → PC Else, continue
RND	Round floating-point value	Round (src) → Rn
ROL	Rotate left	Dreg rotated left 1 bit → Dreg
ROLC	Rotate left through carry	Dreg rotated left 1 bit through carry → Dreg

B.2 Register Formats and Memory Maps

TABLE B.1 *(Continued)*

Mnemonic	Description	Operation
ROR	Rotate right	Dreg rotated right 1 bit → Dreg
RORC	Rotate right through carry	Dreg rotated right 1 bit through carry → Dreg
RPTB	Repeat block of instructions	src → RE 1 → ST (RM) Next PC → RS
RPTS	Repeat single instruction	src → RC 1 → ST (RM) Next PC → RS Next PC → RE
SIGI	Signal, interlocked	Signal interlocked operation Wait for interlock acknowledge Clear interlock
STF	Store floating-point value	Rn → Daddr
STFI	Store floating-point value, interlocked	Rn → Daddr Signal end of interlocked operation
STI	Store integer	Sreg → Daddr
STII	Store integer, interlocked	Sreg → Daddr Signal end of interlocked operation
SUBB	Subtract integers with borrow	Dreg − src − C → Dreg
SUBB3	Subtract integers with borrow (3-operand)	src1 − src2 − C → Dreg
SUBC	Subtract integers conditionally	If Dreg − src ≥ 0: [(Dreg − src) << 1] OR 1 → Dreg Else, Dreg << 1 → Dreg
SUBF	Subtract floating-point values	Rn − src → Rn
SUBF3	Subtract floating-point values (3-operand)	src1 − src2 → Rn
SUBI	Subtract integers	Dreg − src → Dreg
SUBI3	Subtract integers (3-operand)	src1 − src2 → Dreg
SUBRB	Subtract reverse integer with borrow	src − Dreg − C → Dreg
SUBRF	Subtract reverse floating-point value	src − Rn → Rn
SUBRI	Subtract reverse integer	src − Dreg → Dreg
SWI	Software interrupt	Perform emulator interrupt sequence
TRAP*cond*	Trap conditionally	If cond = true or missing: Next PC → * ++ SP Trap vector N → PC 0 → ST (GIE) Else, continue
TSTB	Test bit fields	Dreg AND src
TSTB3	Test bit fields (3-operand)	src1 AND src2
XOR	Bitwise exclusive-OR	Dreg XOR src → Dreg
XOR3	Bitwise exclusive-OR (3-operand)	src1 XOR src2 → Dreg

LEGEND:

src	general addressing modes	Dreg	register address (any register)
src1	three-operand addressing modes	Rn	register address (R0 — R7)
src2	three-operand addressing modes	Daddr	destination memory address
Csrc	conditional-branch addressing modes	ARn	auxiliary register n (AR0 — AR7)
Sreg	register address (any register)	addr	24-bit immediate address (label)
count	shift value (general addressing modes)	cond	condition code (see Chapter 11)
SP	stack pointer	ST	status register
GIE	global interrupt enable register	RE	repeat interrupt register
RM	repeat mode bit	RS	repeat start register
TOS	top of stack	PC	program counter
		C	carry bit

(Reprinted by permission of Texas Instruments)

TABLE B.2 TMS320C30 Parallel Instructions

Mnemonic	Description	Operation
Parallel Arithmetic With Store Instructions		
ABSF ‖ STF	Absolute value of a floating-point	|src2| → dst1 ‖ src3 → dst2
ABSI ‖ STI	Absolute value of an integer	|src2| → dst1 ‖ src3 → dst2
ADDF3 ‖ STF	Add floating-point	src1 + src2 → dst1 ‖ src3 → dst2
ADDI3 ‖ STI	Add integer	src1 + src2 → dst1 ‖ src3 → dst2
AND3 ‖ STI	Bitwise logical-AND	src1 AND src2 → dst1 ‖ src3 → dst2

ASH3 ‖ STI	Arithmetic shift	If count ≥ 0:
		  src2 << count → dst1
		  ‖ src3 → dst2
		Else:
		  src2 >> |count| → dst1
		  ‖ src3 → dst2

FIX ‖ STI	Convert floating-point to integer	Fix(src2) → dst1 ‖ src3 → dst2
FLOAT ‖ STF	Convert integer to floating-point	Float(src2) → dst1 ‖ src3 → dst2
LDF ‖ STF	Load floating-point	src2 → dst1 ‖ src3 → dst2
LDI ‖ STI	Load integer	src2 → dst1 ‖ src3 → dst2

LSH3 ‖ STI	Logical shift	If count ≥ 0:
		  src2 << count → dst1
		  ‖ src3 → dst2
		Else:
		  src2 >> |count| → dst1
		  ‖ src3 → dst2

MPYF3 ‖ STF	Multiply floating-point	src1 x src2 → dst1 ‖ src3 → dst2
MPYI3 ‖ STI	Multiply integer	src1 x src2 → dst1 ‖ src3 → dst2
NEGF ‖ STF	Negate floating-point	0 − src2 → dst1 ‖ src3 → dst2

(Reprinted by Permission of Texas Instruments)

TABLE B.2 *(Continued)*

Mnemonic	Description	Operation
\multicolumn{3}{c}{Parallel Arithmetic With Store Instructions (Concluded)}		
NEGI ‖ STI	Negate integer	0 − src2 → dst1 ‖ src3 → dst2
NOT3 ‖ STI	Complement	$\overline{src1}$ → dst1 ‖ src3 → dst2
OR3 ‖ STI	Bitwise logical-OR	src1 OR src2 → dst1 ‖ src3 → dst2
STF ‖ STF	Store floating-point	src1 → dst1 ‖ src3 → dst2
STI ‖ STI	Store integer	src1 → dst1 ‖ src3 → dst2
SUBF3 ‖ STF	Subtract floating-point	src1 − src2 → dst1 ‖ src3 → dst2
SUBI3 ‖ STI	Subtract integer	src1 − src2 → dst1 ‖ src3 → dst2
XOR3 ‖ STI	Bitwise exclusive-OR	src1 XOR src2 → dst1 ‖ src3 → dst2
\multicolumn{3}{c}{Parallel Load Instructions}		
LDF ‖ LDF	Load floating-point	src2 → dst1 ‖ src4 → dst2
LDI ‖ LDI	Load integer	src2 → dst1 ‖ src4 → dst2
\multicolumn{3}{c}{Parallel Multiply And Add/Subtract Instructions}		
MPYF3 ‖ ADDF3	Multiply and add floating-point	op1 × op2 → op3 ‖ op4 + op5 → op6
MPYF3 ‖ SUBF3	Multiply and subtract floating-point	op1 × op2 → op3 ‖ op4 − op5 → op6
MPYI3 ‖ ADDI3	Multiply and add integer	op1 × op2 → op3 ‖ op4 + op5 → op6
MPYI3 ‖ SUBI3	Multiply and subtract integer	op1 × op2 → op3 ‖ op4 − op5 → op6

LEGEND:

src1	register addr (R0 — R7)	src2	indirect addr (disp = 0, 1, IR0, IR1)
src3	register addr (R0 — R7)	src4	indirect addr (disp = 0, 1, IR0, IR1)
dst1	register addr (R0 — R7)	dst2	indirect addr (disp = 0, 1, IR0, IR1)
op3	− register addr (R0 or R1)	op6	register addr (R2 or R3)

op1,op2,op4,op5 − Two of these operands must be specified using register addr, and two must be specified using indirect.

31	30	29	28	27	26	25	24	23	22	21	20	19	18	17	16
xx	xx	xx	xx	xx	xx	xx	xx	xx	xx	xx	xx	xx	xx	xx	xx

15	14	13	12	11	10	9	8	7	6	5	4	3	2	1	0
xx	xx	GIE	CC	CE	CF	xx	RM	OVM	LUF	LV	UF	N	Z	V	C
		R/W	R/W	R/W			R/W	R/W	R/W	R/W	R/W	R/W	R/W	R/W	R/W

NOTE: xx = reserved bit.
R = read, W = write.

Status Register Bits Summary

Bit	Name	Function
0	C	Carry flag.
1	V	Overflow flag.
2	Z	Zero flag.
3	N	Negative flag.
4	UF	Floating-point underflow flag.
5	LV	Latched overflow flag.
6	LUF	Latched floating-point underflow flag.
7	OVM	Overflow mode flag. This flag affects only the integer operations. If OVM = 0, the overflow mode is turned off; integer results that overflow are treated in no special way. If OVM = 1, a) integer results overflowing in the positive direction are set to the most positive 32-bit two's-complement number (7FFFFFFFh) b) integer results overflowing in the negative direction are set to the most negative 32-bit two's-complement number (80000000h). Note that the function of V and LV is independent of the setting of OVM.
8	RM	Repeat mode flag. If RM = 1, the PC is being modified in either the repeat block or repeat-single mode.
9	Reserved	Read as 0.
10	CF	Cache freeze. When CF = 1, the cache is frozen. If the cache is enabled (CE = 1), fetches from the cache are allowed, but no modification of the state of the cache is performed. This function can be used to save frequently used code resident in the cache. At reset, 0 is written to this bit. Cache clearing (CC=1) is allowed when CF=0.
11	CE	Cache enable. CE = 1 enables the cache, allowing the cache to be used according to the least recently used (LRU) cache algorithm. CE = 0 disables the cache; no update or modification of the cache can be performed. No fetches are made from the cache. This function is useful for system debug. At system reset, 0 is written to this bit. Cache clearing (CC = 1) is allowed when CE=0.
12	CC	Cache clear. CC = 1 invalidates all entries in the cache. This bit is always cleared after it is written to and thus always read as 0. At reset, 0 is written to this bit.
13	GIE	Global interrupt enable. If GIE = 1, the CPU responds to an enabled interrupt. If GIE = 0, the CPU does not respond to an enabled interrupt.
14—15	Reserved	Read as 0.
16—31	Reserved	Value undefined.

FIGURE B.1. Status (ST) register format (Reprinted by permission of Texas Instruments)

B.2 Register Formats and Memory Maps

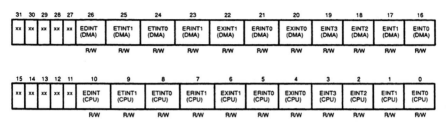

IE Register Bits Summary

Bit	Name	Reset Value	Function
0	EINT0	0	Enable external interrupt 0 (CPU)
1	EINT1	0	Enable external interrupt 1 (CPU)
2	EINT2	0	Enable external interrupt 2 (CPU)
3	EINT3	0	Enable external interrupt 3 (CPU)
4	EXINT0	0	Enable serial-port 0 transmit interrupt (CPU)
5	ERINT0	0	Enable serial-port 0 receive interrupt (CPU)
6	EXINT1	0	Enable serial-port 1 transmit interrupt (CPU)
7	ERINT1	0	Enable serial-port 1 receive interrupt (CPU)
8	ETINT0	0	Enable timer 0 interrupt (CPU)
9	ETINT1	0	Enable timer 1 interrupt (CPU)
10	EDINT	0	Enable DMA controller interrupt (CPU)
15 — 11	Reserved	0	Value undefined
16	EINT0	0	Enable external interrupt 0 (DMA)
17	EINT1	0	Enable external interrupt 1 (DMA)
18	EINT2	0	Enable external interrupt 2 (DMA)
19	EINT3	0	Enable external interrupt 3 (DMA)
20	EXINT0	0	Enable serial-port 0 transmit interrupt (DMA)
21	ERINT0	0	Enable serial-port 0 receive interrupt (DMA)
22	EXINT1	0	Enable serial-port 1 transmit interrupt (DMA)
23	ERINT1	0	Enable serial-port 1 receive interrupt (DMA)
24	ETINT0	0	Enable timer 0 interrupt (DMA)
25	ETINT1	0	Enable timer 1 interrupt (DMA)
26	EDINT	0	Enable DMA controller interrupt (DMA)
31 — 27	Reserved	0–0	Value undefined

FIGURE B.2. Interrupt enable (IE) register format (Reprinted by permission of Texas Instruments)

Instruction Set, Registers, and Memory Maps

31	30	29	28	27	26	25	24	23	22	21	20	19	18	17	16
xx	xx	xx	xx	xx	xx	xx	xx	xx	xx	xx	xx	xx	xx	xx	xx

15	14	13	12	11	10	9	8	7	6	5	4	3	2	1	0
xx	xx	xx	xx	xx	xx	DINT	TINT1	TINT0	RINT1	XINT1	RINT0	XINT0	INT3	INT2	INT1	INT0
						R/W	R/W	R/W	R/W	R/W	R/W	R/W	R/W	R/W	R/W	R/W

NOTE: xx = reserved bit, read as 0.
R = read, W = write.

CPU Interrupt Flag (IF) Register Bits Summary

Bit	Name	Function
0	INT0	External interrupt 0 flag
1	INT1	External interrupt 1 flag
2	INT2	External interrupt 2 flag
3	INT3	External interrupt 3 flag
4	XINT0	Serial port 0 transmit interrupt flag
5	RINT0	Serial port 0 receive interrupt flag
6	XINT1	Serial port 1 transmit interrupt flag
7	RINT1	Serial port 1 receive interrupt flag
8	TINT0	Timer 0 interrupt flag
9	TINT1	Timer 1 interrupt flag
10	DINT0	DMA channel interrupt flag
11 — 31	Reserved	Value undefined

FIGURE B.3. Interrupt flag (IF) register format (Reprinted by permission of Texas Instruments)

31	29	29	28	27	26	25	24	23	22	21	20	19	18	17	16
xx	xx	xx	xx	xx	xx	xx	xx	xx	xx	xx	xx	xx	xx	xx	xx

15	14	13	12	11	10	9	8	7	6	5	4	3	2	1	0
xx	xx	xx	xx	xx	xx	xx	xx	INXF1	OUTXF1	I/OXF1	xx	INXF0	OUTXF0	I/OXF0	xx
								R	R/W	R/W		R	R/W	R/W	

NOTE: xx = reserved bit, read as 0.
R = read, W = write.

IOF Register Bits Summary

Bit	Name	Function
0	Reserved	Read as 0.
1	I/OXF0	If I/OXF0 = 0, XF0 is configured as a general-purpose input pin. If I/OXF0 = 1, XF0 is configured as a general-purpose output pin.
2	OUTXF0	Data output on XF0.
3	INXF0	Data input on XF0. A write has no effect..
4	Reserved	Read as 0.
5	I/OXF1	If I/OXF1 = 0, XF1 is configured as a general-purpose input pin. If I/OXF1 = 1, XF1 is configured as a general-purpose output pin.
6	OUTXF1	Data output on XF1.
7	INXF1	Data input on XF1. A write has no effect.
8 — 31	Reserved	Read as 0.

FIGURE B.4. I/O flag (IOF) register format (Reprinted by permission of Texas Instruments)

B.2 Register Formats and Memory Maps

The following memory maps also can be very useful during program development:

1. *Reset and interrupt vector locations.* Figure B.5 shows the memory addresses for the reset and interrupt vectors. When the TMS320C30 is reset, the content of memory location 0h is loaded into the program counter. This reset vector specifies the address where execution is to begin.
2. *Peripheral bus registers.* The addresses of the peripheral registers are shown in Figure B.6. Note the timer register location (when using interrupt) and the serial port register (when using external I/O).
3. *Peripheral timers.* Two timers are available on the TMS320C30. The peripheral addresses of these timers are shown in Figure B.7. They can be used to specify interrupt and I/O rate. The mode of the timer is specified by the timer global control register, and the signal frequency by the timer period register. The timer counter register is for incrementing the count from zero to the period register value.
4. *External interface control registers.* Two external interfaces are available, as shown in Figure B.8: the primary bus and the expansion bus. These buses are very useful for I/O and support wait states.

31	24	23	0
x	x	address	

Reset and Interrupt Vector Locations

Reset or Interrupt	Vector Location	Priority	Function
RESET	0h	0	External reset signal input on the RESET pin.
INT0	1h	1	External interrupt input on the INT0 pin.
INT1	2h	2	External interrupt input on the INT1 pin.
INT2	3h	3	External interrupt input on the INT2 pin.
INT3	4h	4	External interrupt input on the INT3 pin.
XINT0	5h	5	Internal interrupt generated when serial port 0 transmit buffer is empty.
RINT0	6h	6	Internal interrupt generated when serial port 0 receive buffer is full.
XINT1	7h	7	Internal interrupt generated when serial port 1 transmit buffer is empty.
RINT1	8h	8	Internal interrupt generated when serial port 1 receive buffer is full.
TINT0	9h	9	Internal interrupt generated by timer 0.
TINT1	0Ah	10	Internal interrupt generated by timer 1.
DINT	0Bh	11	Internal interrupt generated by DMA controller 0.

FIGURE B.5. Reset and interrupt vector memory maps (Reprinted by permission of Texas Instruments)

Instruction Set, Registers, and Memory Maps

Peripheral-Bus Memory-Map Registers

Address	Register
808000h – 80800Fh	DMA Controller Registers (16)
808010h – 80801Fh	Reserved (16)
808020h – 80802Fh	Timer 0 Registers (16)
808030h – 80803Fh	Timer 1 Registers (16)
808040h – 80804Fh	Serial Port 0 Registers (16)
808050h – 80805Fh	Serial Port 1 Registers (16)
808060h – 80806Fh	Primary and Expansion Port Registers (16)
808070h – 8097FFh	Reserved

FIGURE B.6. Peripheral bus registers (Reprinted by permission of Texas Instruments)

Peripheral Address		Register
Timer 0	Timer 1	
808020h	808030h	Timer Global Control (See Figure D–18)
808021h	808031h	Reserved
808022h	808032h	Reserved
808023h	808033h	Reserved
808024h	808034h	Timer Counter
808025h	808035h	Reserved
808026h	808036h	Reserved
808027h	808037h	Reserved
808028h	808038h	Timer Period
808029h	808039h	Reserved
80802Ah	80803Ah	Reserved
80802Bh	80803Bh	Reserved
80802Ch	80803Ch	Reserved
80802Dh	80803Dh	Reserved
80802Eh	80803Eh	Reserved
80802Fh	80803Fh	Reserved

FIGURE B.7. Peripheral timers (Reprinted by permission of Texas Instruments)

5. *Serial port peripheral addresses.* Two serial ports are available on the TMS320C30, as shown in Figure B.9. The functions of these various registers are described in the *TMS320C3X User's Guide* [1]. Several examples, associated with the use of both serial ports 0 and 1, are described in Chapter 3 in conjunction with the AIC communication routines.

B.2 Register Formats and Memory Maps

Peripheral Address	Register
808060h	Expansion-Bus Control (see Figure D–25)
808061h	Reserved
808062h	Reserved
808063h	Reserved
808064h	Primary-Bus Control (see Figure D–26)
808065h	Reserved
808066h	Reserved
808067h	Reserved
808068h	Reserved
808069h	Reserved
80806Ah	Reserved
80806Bh	Reserved
80806Ch	Reserved
80806Dh	Reserved
80806Eh	Reserved
80806Fh	Reserved

FIGURE B.8. External interface control registers (Reprinted by permission of Texas Instruments)

Peripheral Address		Register
Serial Port 0	Serial Port 1	
808040h	808050h	Serial-Port Global-Control (See Figure D–20)
808041h	808051h	Reserved
808042h	808052h	FSX/DX/CLKX Port Control (See Figure D–21)
808043h	808053h	FSR/DR/CLKR Port Control (See Figure D–22)
808044h	808054h	R/X Timer Control (See Figure D–23)
808045h	808055h	R/X Timer Counter
808046h	808056h	R/X Timer Period
808047h	808057h	Reserved
808048h	808058h	Data Transmit
808049h	808059h	Reserved
80804Ah	80805Ah	Reserved
80804Bh	80805Bh	Reserved
80804Ch	80805Ch	Data Receive
80804Dh	80805Dh	Reserved
80804Eh	80805Eh	Reserved
80804Fh	80805Fh	Reserved

FIGURE B.9. Serial port peripheral addresses (Reprinted by permission of Texas Instruments)

6. *Direct memory access (DMA) registers.* The peripheral addresses used for DMA are shown in Figure B.10. These registers can be used to alleviate the CPU from I/O operations. The function of these registers are described in [1].

Peripheral Address	Register
808000h	DMA Global Control (See Figure D–16)
808001h	Reserved
808002h	Reserved
808003h	Reserved
808004h	DMA Source Address
808005h	Reserved
808006h	DMA Destination Address
808007h	Reserved
808008h	DMA Transfer Counter
808009h	Reserved
80800Ah	Reserved
80800Bh	Reserved
80800Ch	Reserved
80800Dh	Reserved
80800Eh	Reserved
80800Fh	Reserved

FIGURE B.10. Direct memory access (DMA) registers (Reprinted by permission of Texas Instruments)

REFERENCE

[1] *TMS320C3X User's Guide*, Texas Instruments, Inc., Dallas, Tex., 1991.

C
Digital Signal Processing Tools

C.1 INTRODUCTION

A wide range of digital signal processing (DSP) tools is commercially available from a number of vendors, such as Hyperception, Inc. [1], and Atlanta Signal Processors, Inc. [2].

The software package MATLAB [3] includes utilities for finding the FFT and for the design of filters. For example, using MATLAB, the coefficients of an FIR filter can be found using the Parks–McClellan algorithm, then the frequency response of the filter can be plotted.

These DSP tools include filter design packages with processor-specific code generators, and utilities for spectral analysis, correlation, and so forth. The Hypersignal-Plus DSP Software from Hyperception, Inc. contains several menu-driven utilities for time and frequency analysis. A digital filter design package is available for the design of both FIR and IIR filters, with a code generator included for implementing digital filters in both C and TMS320C30 code. This filter design package allows for the design of frequency-selective filters. It is a powerful design tool that allows a person to design and implement a digital filter in a matter of minutes. The code generator supports the TMS320 family of processors (as well as several other processors). The utilities include the fast Fourier transform (FFT), the generation of a sinusoid, and the design of finite impulse response (FIR) and infinite impulse response (IIR) filters. Plots in both time and frequency domains can be readily obtained.

To access the Hypersignal package, type

354 Digital Signal Processing Tools

There are three main menus to choose from:

1. Time domain
2. Frequency domain
3. Utilities

Within each of the three menus, several options are available, such as the `magnitude display` option within the frequency domain main menu. For each of the options, a `HELP` menu is available with `F1`. Use the `ESC` and arrow keys to access the various options. For example, from the utilities main menu, choose the `USER SETUP` option, to set up the drive and subdirectory used. Do not worry about the time and frequency filenames at this time.

The following examples will illustrate the use of this package.

C.2 TIME AND FREQUENCY DOMAIN UTILITIES

Example C.1 *Plot of a Sine Sequence.* In Examples 3.4 and 3.5, a sinusoidal sequence with four points is generated. The simulator output file `SINE4.DAT` (or `SINE4C.DAT`) contains the output sequence, which must first be converted before being plotted.

1. From the utilities menu, select the file acquisition option, to convert the hex file `SINE4.DAT` to the file `SINE4`, as shown in Figure C.1. Note

```
                    HYPERSIGNAL-PLUS DSP SOFTWARE, V3.01

TIME: 23:40:27   SESSION: 0:06:17         DRIVE: C:
DATE: 13 DEC 1991                         SUBDIRECTORY: \HSP\
                                          TIME FILENAME: SINE4
SELECTION: FILE ACQUISITION               FREQ FILENAME: SINE4

#89460  Roger Williams College, Bristol, RI. Single-owner lic. #1.
        HYPERSIGNAL TIME-DOMAIN FILE ACQUISITION
                                                UTILITIES
INPUT FILENAME(or SP or SEG:OFS): SINE4.DAT
       OUTPUT TIME-DOMAIN FILENAME: SINE4       • FILE ACQUISITION
       FORMAT (DSP/BIN/SWP/DEC/HEX/FP): HEX     • analog conversion
             SAMPLE WIDTH (in nibbles): 8       • binary-ASCII-binary
    INITIAL BYTES/SAMPLES TO DISCARD: 0         • user setup
    NUMBER OF INPUT SAMPLES (0=ALL): 0          • directory
              SAMPLING FREQUENCY (Hz): 10000    • ext module config
                           FRAMESIZE: 1024      • graphics/mouse config
                  SUBTRACT MEAN (Y/N): N        • serial port config
     CHANNEL STRIP, MASK (0-SS,0-±32): 0,0      • system config
                                                • code generate/quantize
                                                • user-defined
           Copyright (C) 1985 through 1990 by Hyperception

      help: F1    elete: Backspc    delete: Del    move: ↕→    menu: Esc
```

FIGURE C.1. `File acquisition` screen from utilities menu

C.2 Time and Frequency Domain Utilities

```
                    HYPERSIGNAL-PLUS DSP SOFTWARE, V3.01

TIME: 23:41:04   SESSION: 0:06:54          DRIVE: C:
DATE: 13 DEC 1991                    SUBDIRECTORY: \HSP\
                                    TIME FILENAME: SINE4
SELECTION: WAVEFORM DISPLAY/EDIT    FREQ FILENAME: SINE4

#89460  Roger Williams College, Bristol, RI. Single-owner lic. #1.

    TIME DOMAIN              GRAPHICAL TIME WAVEFORM DISPLAY/EDIT

  • WAVEFORM DISPLAY/EDIT              FILENAME1:  SINE4
  • FFT generation                     FILENAME2:
  • FIR filter constructio             FRAMESIZE:  64
  • convolution/correlatio   Y-LABELS (SCALE,UNITS): 1.0,(volts)
  • LPC autocorrelation     MAX SCALE (0=AUTOSCALE): 0
  • IIR filter constructio          PLOT (B/C/D):  C
  • recursive filtering             GRID (N/S/D/I): N
  • difference equations
  • digital oscilloscope
  • userdef:sinewavegen
  • user-defined
              Copyright (C) 1985 through 1990 by Hyperception

help: F1    elete: Backspc    delete: Del    move: ↕→    menu: Esc
```

FIGURE C.2. Waveform display / edit screen from time domain menu

that the file SINE4.DAT is in hex and contains 32 bits (8 nibbles). A sampling frequency of 10 kHz is chosen.

2. From the time domain menu, select the waveform display option, as shown in Figure C.2. A frame size of 64 is chosen. A much larger size would make the sine waveform plot too condensed. Plot and grid options are available with this menu. The plot of the sine sequence is shown in Figure C.3.

3. To obtain an equivalent plot in the frequency domain, select the FFT generation from the time domain menu as in Figure C.4. An FFT order of 10, which corresponds to 1,024 points, is chosen (order 8 is for 256 points, order 9 for 512 points), using a Hamming window. Other window functions such as Hanning and Blackman, also can be chosen.

4. The magnitude of the sine function can be plotted, using the specifications listed in Figure C.5 (with the magnitude display option within the frequency menu). Figure C.6 displays the FFT of the simulator output sequence SINE4.DAT, showing a delta function at the frequency of the sine function. This frequency is

$$f = \frac{F_s}{\text{number of points}} = \frac{10 \text{ kHz}}{4} = 2.5 \text{ kHz}$$

5. Repeat the preceding steps, and choose a sampling frequency of 8 kHz (from Figure C.1) to obtain a plot similar to Figure C.6, but showing the delta function at 2 kHz.

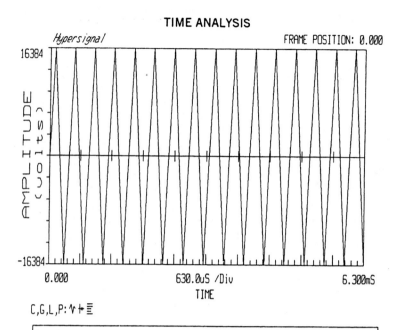

FIGURE C.3. Time domain plot of sine sequence

HYPERSIGNAL-PLUS DSP SOFTWARE, V3.01

```
TIME: 23:37:06    SESSION: 0:02:56        DRIVE: C:
DATE: 13 DEC 1991                         SUBDIRECTORY: \HSP\
                                          TIME FILENAME: SINE4
SELECTION: FFT GENERATION                 FREQ FILENAME: SINE4
```

#89460   Roger Williams College, Bristol, RI. Single-owner lic. #1.

```
       TIME DOMAIN                     FAST FOURIER TRANSFORM

  ▪ waveform display/edit                 FILENAME: SINE4
  ▪ FFT GENERATION                       FRAMESIZE: 694
  ▪ FIR filter constructio               FFT ORDER: 10
  ▪ convolution/correlatio           FRAME OVERLAP: 0
  ▪ LPC autocorrelation          WINDOW (R/Hm/Hn/Bl/Bt): HM
  ▪ IIR filter constructio       REAL/IMAG OUTPUT (Y/N): n
  ▪ recursive filtering          PREC: INTEGER/REAL (I/R): R
  ▪ difference equations         PHASE/MAG APPROX. (Y/N): Y
  ▪ digital oscilloscope
  ▪ userdef:sinewavegen
  ▪ user-defined
           Copyright (C) 1985 through 1990 by Hyperception

   help: F1     elete: Backspc    delete: Del    move: ↕→    menu: Esc
```

FIGURE C.4. Fast Fourier transform screen from time domain menu

```
                    HYPERSIGNAL-PLUS DSP SOFTWARE, V3.01
  Number of samples: 513
     Type/complex: mag/real
    Max |amplitude|: 23955              DRIVE: C:
 Analysis framesize: 694             SUBDIRECTORY: \HSP\
        FFT order: 10              TIME FILENAME: SINE4
 Sampling frequency: 10000Hz       FREQ FILENAME: SINE4
 waveform summary for SINE4.FRQ    stol, RI. Single-owner lic. #1.
                                           MAGNITUDE DISPLAY
       time domain              FREQUEN
                                                    FILENAME1: SINE4
                               • MAGNITU            FILENAME2:
                               • phase d      POWER SPEC (P/M): M
                               • 3-D spe      AMP SQUARE (Y/N): N
                               • iFFT ge    LOG (N/+/-, N/base^NN): N
                               • 2-D spe    Y-LABELS (SCALE, UNITS): 1
                               • pole-ze    MAX SCALE (0=AUTOSCALE): 0
                               • power s           PLOT (B/C/D): C
                               • spectru          GRID (N/S/D/I): N
                               • user-de
                               • user-de
                               • user-de
                  Copyright (C) 1985 through 1990 by Hyperception

    help: F1    elete: Backspc    delete: Del    move: ↕→    menu: Esc
```

FIGURE C.5. Magnitude display screen from frequency domain menu

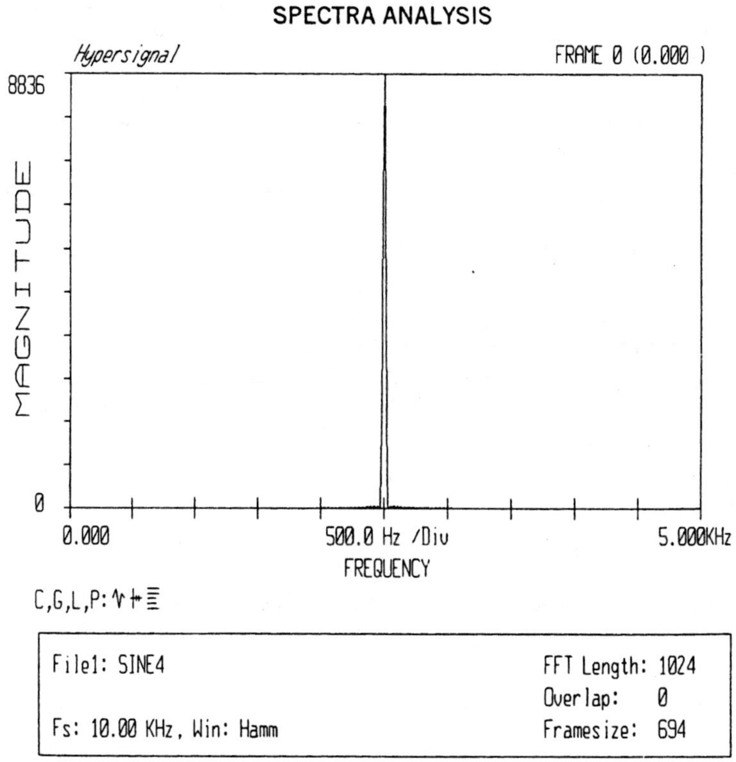

SPECTRA ANALYSIS

Hypersignal   FRAME 0 (0.000)

8836

MAGNITUDE

0
  0.000   500.0 Hz /Div   5.000KHz
                            FREQUENCY

C,G,L,P: ⋁ ╫ ≡

```
 File1: SINE4                        FFT Length: 1024
                                     Overlap:    0
 Fs: 10.00 KHz, Win: Hamm            Framesize:  694
```

FIGURE C.6. Frequency domain plot of sine sequence

```c
/*EDGE.C-EDGE ENHANCEMENT PROGRAM*/
main()
{
 volatile int *IO_INPUT = (volatile int *) 0x804000;
 volatile int *IO_OUTPUT = (volatile int *) 0x804001;
 int A[3] = {0,0,0};
 int i, result;
  for (i = 0; i < 75; ++i)
    {
    A[2] = *IO_INPUT;
    *IO_OUTPUT = 3*A[1] - A[2] - A[0];
    A[0] = A[1];
    A[1] = A[2];
    }
}
```

FIGURE C.7. Edge enhancement program using C code (EDGE.C)

Example C.2 Edge Enhancement Using C Code. This program example implements a technique called edge enhancement commonly used in image processing. This horizontal edge enhancement technique is routinely used in solid-state imaging cameras employing charged coupled devices (CCD) as the image sensor. These CCD devices typically offer lower resolution compared to that of an analog imaging device like the Vidicon camera tube. When viewing a camera with edge enhancement, a crisper image can be noticed, compared to the same camera without this technique. This example illustrates a significant increase in the amplitude of the higher harmonics in the frequency spectrum.

Figure C.7 shows the program EDGE.C for the edge enhancement technique. In an image processing application, the sampling rate is typically 10 MHz. However, the same effect can be obtained with a lower sampling rate of 10 kHz for simulation purposes. A square-wave input file EDGEIN.DAT (on the accompanying disk), which contains the values

0x00000000
0x00000000
0x00000000
0x00000000
0x00000020
0x00000040
0x00000040
    .
    .
    .

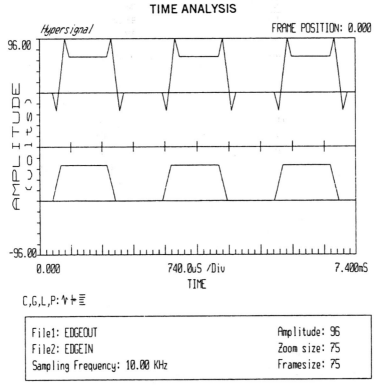

FIGURE C.8. Plots of output and input waveforms for edge enhancement

is used as input to the simulator. Compile and run the program `EDGE.C`. Figure C.8 shows a time plot of both the simulator output file `EDGEOUT.DAT` (upper graph) and the input file `EDGEIN.DAT`. Figure C.9 shows the frequency plot of the simulator output (upper graph) and the input square wave (lower graph), displaying an increase in the amplitude of the output waveform higher harmonics.

C.3 FIR FILTER DESIGN AND CODE GENERATION

This section illustrates the design of a finite impulse response (FIR) filter. Access the Hypersignal-Plus DSP Software as in the previous section by typing `HSP`. From the time domain main menu, select the option `FIR Filter Construction`. Frequency-selective filters using the Kaiser window function and the Parks–McClellan algorithm can be designed.

360 Digital Signal Processing Tools

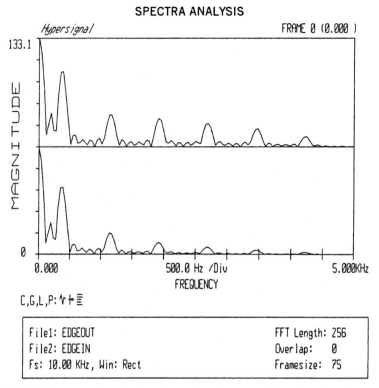

FIGURE C.9. Frequency domain plots of output and input waveforms for edge enhancement

***Example C.3** FIR Bandpass Filter with 45 Coefficients.* This example illustrates how to design an FIR filter with 45 coefficients, centered at 1 kHz. The coefficients generated are further used in Chapter 4. A stopband attenuation of 26 dB and a passband ripple of 6 dB are chosen. To generate the coefficients proceed as follows:

1. Type HSP to access Hypersignal.
2. From the time domain menu, select the option FIR Filter Construction. The characteristics of the desired filter are shown in Figure C.10. From the FIR filter design plot, press C to construct this filter. Note that increasing the stopband attenuation, or decreasing the passband ripple, will require more coefficients for this design (a stopband, or notch, filter is designed using a negative value for the stopband attenuation). Construct this filter using 45 coefficients (not the estimated length of 43).
3. Select the waveform display option, as previously done in Example C.1 (Figure C.2) to produce the time response shown in Figure C.11.

```
                    HYPERSIGNAL-PLUS DSP SOFTWARE, V3.01
        Text file type: filter spec
           Filter type: BP (FIR)           DRIVE: C:
    Design method used: KW (class)         SUBDIRECTORY: \HSP\
   Filter order/length: 45                 TIME FILENAME: BP45
    Impulse resp length: 45                FREQ FILENAME: BP45
    Cascade struc type: none
    waveform summary for BP45.FIR     stol, RI. Single-owner lic. #1.
                                         FIR FILTER CONSTRUCTION
       TIME DOMAIN
                                                  FILENAME:   BP45
     ▪ waveform display/edit          DESIGN (KW/PM/DIF/HIL):  KW
     ▪ FFT generation                 SAMPLING FREQUENCY (Hz): 10000
     ▪ FIR FILTER CONSTRUCTIO         CENTER FREQUENCY (Hz):   1000
     ▪ convolution/correlatio                   BANDWIDTH (Hz): 200
     ▪ LPC autocorrelation            TRANSITION BANDWIDTH1 (Hz): 300
     ▪ IIR filter constructio         TRANSITION BANDWIDTH2 (Hz): 300
     ▪ recursive filtering            STOPBAND ATTENUATION (dB):  26
     ▪ difference equations           FILTER LENGTH (# OF TAPS):  45
     ▪ digital oscilloscope           PASSBAND RIPPLE (dB):       6
     ▪ userdef:sinewavegen
     ▪ user-defined
              Copyright (C) 1985 through 1990 by Hyperception

     help: F1    elete: Backspc    delete: Del    move: ↕→    menu: Esc
```

FIGURE C.10. FIR filter construction characteristics

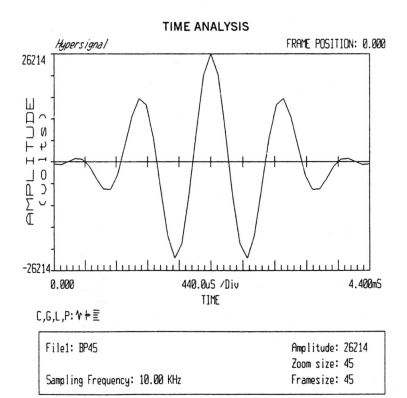

FIGURE C.11. Time response of FIR filter

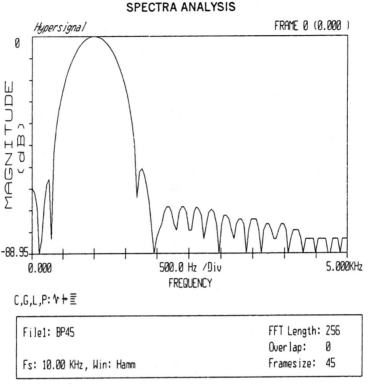

FIGURE C.12. Magnitude response of FIR filter

4. Select the FFT Generation option (as done previously in Example C.2, Figure C.4) and choose a 256-point FFT (order 8) and a Hamming window function. The magnitude of the filter is shown in Figure C.12.

To Generate C and TMS320C30 Code

The same bandpass filter characteristics as in Example C.3 will be used to illustrate the code generator utility available with the Hypersignal package:

1. Select the system config option from the utilities main menu in order to choose the desired DSP DEVELOPMENT SYSTEM option, the relevant option for this example, or

DEVELOPMENT DSP : TI30-C

C.3 FIR Filter Design and Code Generation

```
                    HYPERSIGNAL-PLUS DSP SOFTWARE, V3.01
       Text file type: filter spec
         Filter type: BP (FIR)             DRIVE: C:
   Design method used: KW (class)       SUBDIRECTORY: \HSP\
   Filter order/length: 45              TIME FILENAME: BP45
   Impulse resp length: 45              FREQ FILENAME: BP45
   Cascade struc type: none
   waveform summary for BP45.FIR     sto1, RI. Single-owner lic. #1.
       DIGITAL SIGNAL PROCESSOR CODE GENERATOR
           [TEXAS INSTRUMENTS TMS320C30]              UTILITIES

                          FILENAME: BP45        • file acquisition
    IIR STRUCTURE(S1/S2/S3-see help):            • analog conversion
                  LANGUAGE (ASM/C): c           • binary-ASCII-binary
                QUANTIZATION (Q8-Q31):           • user setup
   COEFF BASE ADDRESS (HEX) / MODE:              • directory
          ASM. CODE ROUNDING (Y/N): Y           • ext module config
    SIMULATE QUANTIZED H[n] (Y/N,INP): Y,0.25   • graphics/mouse config
                                                • serial port config
                                                • system config
                                                • CODE GENERATE/QUANTIZE
                                                • user-defined
         Copyright (C) 1985 through 1990 by Hyperception

   help: F1    elete: Backspc    delete: Del    move: ↕→    menu: Esc
```

FIGURE C.13. Code generate / quantize option

Several other digital signal processors are also supported by the Hypersignal package.

2. Select the CODE GENERATE / QUANTIZE option as in Figure C.13. This will create a file BP45.C that contains the source filter program in C, including the coefficients. Rename this file BP45SC.C, which is shown in Figure C.14.
3. Replace C with ASM in Figure C.13 to generate the source file in TMS320C30 code. Rename this file BP45S.ASM. Press E to save and Exit.

A brief description of these two files follows.

Source File BP45SC.C
The source file BP45SC.C includes the filter code and the coefficients. However, no I/O code is incorporated, because the C code generator is not specific to the Texas Instruments processor. Hence, input and output code need to be added. This program uses the modulo operator %, resulting in a relatively short filter routine. However, the compiled code can be slow to execute, compared to an implementation without the use of the modulo operator (as discussed in Chapter 4). In applications where time is not

```
/*BP45SC.C-SUBROUTINE FROM HYPERSIGNAL
*****************************************************************
* Texas Inst. TMS320C30 C Lang. Filter Realization Copyright (C) by HYPERCEPTION, INC.
*****************************************************************
*
*       Filter Generated from File: bp45.FIR
*
*       Filter Order=45
*
*       Direct Form Realization of the following convolution sum:
*
*                          N-1
*                          ___
*                          \
*                  y(n) =  /___  h(k)x(n-k)
*                          k=0
*
*
*           -1       -1                  -1
*            z        z                   z
*   o--->---o--->---o--->---o--->  -  -o--->----o
*   x(n)    |       |       |          |        |
*           |       |       |          |        |
*           v h(0)  v h(1)  v h(2)     v h(N-2) v h(N-1)
*           |       |       |          |        |
*           |       |       |          |        |
*           o--->-- + --->-- + ---> - - + --->--- + --->---o
*                                                         y(n)
*
*
*/
#define N 45            /* length of FIR filter impulse response */

int start_index = { 0 };        /* circular buffer starting position */

double H[N] = { /* filter coefficients */
                -1.83989092791E-0003, -2.65748927587E-0003, -4.31240164415E-0010,
                 3.15489863431E-0003,  2.59504949565E-0003, -4.15992854430E-0003,
                -1.54063990002E-0002, -2.50783676799E-0002, -2.54745141379E-0002,
                -1.17986263157E-0002,  1.39224582794E-0002,  4.20645113719E-0002,
                 5.88840954593E-0002,  5.30715079282E-0002,  2.22550923780E-0002,
                -2.41095421284E-0002, -6.754302919241E-0002, -8.83147433676E-0002,
                -7.47571889853E-0002, -2.95655400338E-0002,  3.03027241599E-0002,
                 8.05076375048E-0002,  1.00000000600E-0001,  8.05076375048E-0002,
                 3.03027241599E-0002, -2.95655400338E-0002, -7.47571889853E-0002,
```

FIGURE C.14. FIR bandpass filter program in C generated by Hypersignal (BP45SC.C)

critical, this code using the modulo operator can be very useful and can serve as a base to improve on.

Subroutine BP45S.ASM and Main Program BP45M.ASM in TMS320C30 Code

A partial program, BP45SPT (listed in Figure C.15), is obtained from the subroutine BP45S.ASM generated by Hypersignal, by deleting the COMMENT fields and the coefficients. The next-to-last line (with .WORD DELAY) should be deleted because it is not used here.

A main program, BP45M.ASM (which calls the subroutine BP45S.ASM), is listed in Figure C.16. Figure C.17 shows the command file that can be used to link the main program BP45M.OBJ with the subroutine BP45S.OBJ. Al-

```
                    -8.83147433676E-0002, -6.75430291924E-0002, -2.41095421284E-0002,
                     2.22550923780E-0002,  5.30715079282E-0002,  5.88840954593E-0002,
                     4.20645113719E-0002,  1.39224582794E-0002, -1.17986263157E-0002,
                    -2.54745141379E-0002, -2.50783676799E-0002, -1.54063990002E-0002,
                    -4.15992854430E-0003,  2.59504949565E-0003,  3.15489863431E-0003,
                    -4.31240164415E-0010, -2.65748927587E-0003, -1.83989092791E-0003
                    };

double DLY[N];      /* delay taps (z-shifts) */

double filt(stage_input)   /* actual filter routine */
double stage_input;
{
    double acc;
    int    i, j;

    DLY[start_index] = stage_input;
    j = --start_index;
    acc = 0.0;

    for (i=0; i<N; i++)
    {
        j = ++j % N;             /* circular buf requires modulo inc */
        acc += H[i] * DLY[j];
    }
    start_index = j;

    return acc;
}

main ()
{
#define IMPULSE_LENGTH 45   /* length of FIR filter impulse response */

double out_val[IMPULSE_LENGTH];
int    n;

    out_val[0]=filt(10000.0);   /* the "impulse" input */

    for (n=1; n<IMPULSE_LENGTH; n++)
    {
        out_val[n]=filt(0.0); /* zero input for rest of impulse response length */
    }
}
```

FIGURE C.14. (*Continued*)

though the input and output port addresses are specified in the command file, they could have been initialized instead in the main program.

Running the FIR Filter

Assemble and link the filter subroutine from Hypersignal BP45S.OBJ and the main program BP45M.OBJ, to create an output file BP45MS.OUT. Download this output file into the simulator, and configure the ports such that the simulator output file BP45MS.DAT is at port address 804001h. The pseudorandom noise sequence NOISEINT.DAT, generated in Chapter 3, or the impulse sequence IMPULSE.DAT (on the accompanying disk) can be

```
;BP45SPT
            .DEF    FILT, OUTDATA, IMPULSE, COEFF
ORDER:      .SET    45
FILT:       STF     R3, *AR1++%
            LDF     0, R0
            LDF     0, R2
            RPTS    ORDER-1
            MPYF    *AR0++, *AR1++%, R0
||          ADDF    R0, R2
            ADDF    R0, R2
            LDF     R2, R3
            RETS
.DATA
IMPULSE  .FLOAT  10000.0
OUTDATA  .FLOAT  0.0
DELAY_ADR .WORD  DELAY
COEFF_ADR .WORD  COEFF
```

FIGURE C.15. Partial subroutine program in TMS320C30 code (BP45SPT)

used as the input to the filter at port address 804000h. Run this filter, and produce a simulator output file.

Using the file acquisition and FFT utilities as in the previous example, the frequency response of this filter is plotted in Figure C.18.

C.4 IIR FILTER DESIGN AND CODE GENERATION

This section describes the use of the Hypersignal package for the design of IIR filters. From the time-domain menu, there is an option for IIR filter construction, which we will illustrate using the following example.

Example C.4 Sixth-Order IIR Bandpass Filter Design. This example illustrates the design of a sixth-order Butterworth bandpass filter. Other designs such as elliptic, Bessel, and Chebychev are also available.

1. The characteristics of this filter (from the IIR filter construction screen) are shown in Figure C.19. It has a center frequency of 1,250 Hz and a bandwidth of 200 Hz, with a sampling frequency of 10 kHz. The transition bandwidth represents how sharp or selective the filter is. For example, decreasing the value of the transition bandwidth will require a higher-order filter. A higher value for the stopband attenuation will also require a higher order filter. Construct this filter.

```
;BP45M.ASM-MAIN PROGRAM TO CALL FIR SUBROUTINE FROM HYPERSIGNAL
              .TITLE   "BP45M.ASM"         ;MAIN FILE FOR FIR
              .GLOBAL  MAIN,BEGIN,FILT,COEFF ;REF/DEF SYMBOLS
              .DATA                        ;ASSEMBLE INTO DATA SECT
IN_ADDR       .WORD    IN                  ;INIT VALUE IN
OUT_ADDR      .WORD    OUT                 ;INIT VALUE OUT
XN_ADDR       .WORD    XN+LENGTH-1         ;NEWEST(LAST) IN SAMPLE
HN_ADDR       .WORD    COEFF               ;COEFF TABLE ADDR
LENGTH        .SET     45                  ;FILTER LENGTH
XN            .USECT   "XN_BUFF",LENGTH    ;SAMPLES BUFFER SIZE
              .SECT    "VECS"              ;ASSEMBLE INTO VECT SECT
MAIN          .WORD    BEGIN
IN            .USECT   "IN_PORT",1         ;1 SPACE FOR IN
OUT           .USECT   "OUT_PORT",1        ;1 SPACE FOR OUT
              .TEXT                        ;ASSEMBLE INTO TEXT SECT
BEGIN         LDP      IN_ADDR             ;INIT DATA PAGE
              LDI      @IN_ADDR,AR5        ;INPUT ADDR -> AR5
              LDI      @OUT_ADDR,AR6       ;OUTPUT ADDR -> AR6
              LDI      @XN_ADDR,AR1        ;"LAST"SAMPLE ADDR
              LDI      LENGTH,BK           ;BK = FILTER LENGTH
              LDF      0.0,R0              ;INIT R0=0
              RPTS     LENGTH-1            ;INIT INPUT SAMPLES TO 0
              STF      R0,*AR1--(1)%       ;START/END AT LAST SAMPLE
              LDI      LENGTH,R4           ;COUNTER TO CALL FILT SUB
LOOP          FLOAT    *AR5,R3             ;INPUT NEW SAMPLE -> R3
              LDI      @HN_ADDR,AR0        ;"FIRST" COEFF H(N-1)->AR0
              CALL     FILT                ;GO TO SUBROUTINE FILT
              FIX      R3,R1               ;R1=INTEGER(R3)
              STI      R1,*AR6             ;OUTPUT INTEGER VALUE
              SUBI     1,R4                ;DECREMENT R4
              BNZ      LOOP                ;BRANCH UNTIL R4 <> 0
WAIT          BR       WAIT                ;WAIT
              .END
```

FIGURE C.16. FIR bandpass filter main program in TMS320C30 code (BP45M.ASM)

2. Generate the code in C for this IIR filter design. Figure C.20 shows the code generate / quantize screen. Note that both C code and TMS320C30 code can be generated. Replace C by ASM to generate TMS320C30 code for this filter. The IIR structure S2 implements a filter with three second-order direct form II sections in cascade (use S3 for a direct form II transpose structure). The generated C-code file IIR6BP.C, on the accompanying disk, contains the coefficients and the C code for this filter, but without I/O. The code generator (in C) is not specific to the TMS320C30. The program IIR6BP.C can be modified to incorporate I/O. The modified program is discussed in Chapter 5.

```
/*BP45MS.CMD      COMMAND FILE                    */
BP45S.OBJ         /*FIR SUB FROM HYPERSIGNAL      */
BP45M.OBJ         /*FIR MAIN PROGRAM              */
-E BEGIN          /*SPECIFIES ENTRY POINT FOR OUTPUT*/
-O BP45MS.OUT /*LINKED COFF OUTPUT FILE           */
MEMORY
{
  VECS: org = 0          len = 0x40   /*VECTOR LOCATIONS*/
  SRAM: org = 0x40       len = 0x3FC0 /*16K EXTERNAL RAM*/
  RAM : org = 0x809800   len = 0x800  /*2K INTERNAL RAM */
  IO  : org = 0x804000   len = 0x2000 /*-IOSTRB I/O     */
}
SECTIONS
{
  .data:  {} > RAM       /*DATA SECTION IN INTERNAL RAM*/
  .text:  {} > SRAM      /*CODE                        */
  .cinit: {} > SRAM      /*INITIALIZATION TABLES       */
  .stack: {} > RAM       /*SYSTEM STACK                */
  .bss:   {} > RAM       /*BSS SECTION IN RAM          */
  vecs:   {} > VECS      /*RESET & INTERRUPT VECTORS   */
  XN_BUFF  ALIGN(64): {} > RAM  /*CIRCULAR BUFFER      */
  IN_PORT  804000H  : {} > IO   /*INPUT PORT ADDRESS   */
  OUT_PORT 804001H  : {} > IO   /*OUTPUT PORT ADDRESS  */
}
```

FIGURE C.17. Command file for FIR bandpass filter program (BP45MS.CMD)

Example C.5 Sixth-Order IIR Direct Form II Transpose Filter Using TMS320C30 Code. The Hypersignal package is used to generate the TMS320C30 subroutine for the sixth-order IIR bandpass filter, centered at 1,250 Hz. Use S3 and ASM within the code generate / quantize option (see Figure C.20) and generate TMS320C30 code for a direct form II transpose filter structure. Save this subroutine as IIR6BP.ASM, which is on the accompanying disk. A main program, IIRMAIN.ASM, also on the accompanying disk, is provided such that the impulse response using the file IMPULSE.DAT (or the response due to some other input, such as random noise) can be obtained without changing the program. As a result, the suggested main program, commented in the Hypersignal subroutine, is not used. This main program, IIRMAIN.ASM, can be used to call the subroutine IIR6BP.ASM. Create a command file IIR6BP.CMD to link the main program and the Hypersignal subroutine.

Run this program example through the simulator. Use the file IMPULSE.DAT (on the accompanying disk) as input. The frequency response of the IIR filter is shown in Figure C.21. Note that a different input file

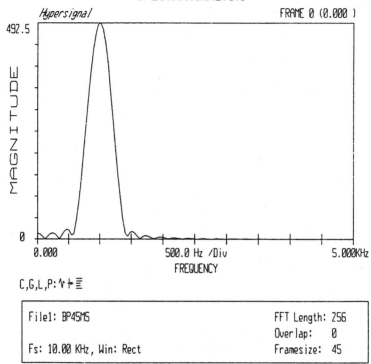

FIGURE C.18. Frequency response of FIR bandpass filter

```
                HYPERSIGNAL-PLUS DSP SOFTWARE, V3.01
       Text file type: filter spec
           Filter type: BP (IIR)              DRIVE: C:
  Design method used: B (class)         SUBDIRECTORY: \HSP\
  Filter order/length: 6                TIME FILENAME: iir6bp
  Impulse resp length: 345              FREQ FILENAME: iir6bp
    Cascade struc type: S2
waveform summary for iir6bp.IIR    stol, RI. Single-owner lic. #1.
                                        IIR FILTER CONSTRUCTION
    TIME DOMAIN
                                    FILENAME:              iir6bp
  ▪ waveform display/edit    DESIGN (S/B/BES/C1/C2/E):     B
  ▪ FFT generation           SAMPLING FREQUENCY (Hz):      10000
  ▪ FIR filter constructio    CENTER FREQUENCY (Hz):       1250
  ▪ convolution/correlatio           BANDWIDTH (Hz):       200
  ▪ LPC autocorrelation      TRANSITION BANDWIDTH1 (Hz):   200
  ▪ IIR FILTER CONSTRUCTIO   TRANSITION BANDWIDTH2 (Hz):   200
  ▪ recursive filtering      STOPBAND ATTENUATION (dB):    30
  ▪ difference equations        ORDER (STEIGLITZ ONLY):    6
  ▪ digital oscilloscope    RIPPLE (dB)/STEIGLITZ ERR:     5
  ▪ userdef:sinewavegen
  ▪ user-defined
              Copyright (C) 1985 through 1990 by Hyperception

    help: F1    elete: Backspc    delete: Del    move: ↕→    menu: Esc
```

FIGURE C.19. Filter construction characteristics for a sixth-order IIR bandpass filter

```
                HYPERSIGNAL-PLUS DSP SOFTWARE, V3.01
       Text file type: filter spec
          Filter type: BP (IIR)                DRIVE: C:
   Design method used: B (class)         SUBDIRECTORY: \HSP\
   Filter order/length: 6                TIME FILENAME: iir6bp
   Impulse resp length: 345              FREQ FILENAME: iir6bp
   Cascade struc type: S2
waveform summary for iir6bp.IIR    stol, RI. Single-owner lic. #1.
         DIGITAL SIGNAL PROCESSOR CODE GENERATOR
              [TEXAS INSTRUMENTS TMS320C30]         UTILITIES

                     FILENAME: iir6bp          • file acquisition
   IIR STRUCTURE(S1/S2/S3-see help): s2        • analog conversion
                 LANGUAGE (ASM/C): c           • binary-ASCII-binary
               QUANTIZATION (Q8-Q31):          • user setup
   COEFF BASE ADDRESS (HEX) / MODE:            • directory
          ASM. CODE ROUNDING (Y/N): Y          • ext module config
   SIMULATE QUANTIZED H[n] (Y/N,INP): Y,0.25   • graphics/mouse config
                                               • serial port config
                                               • system config
                                               • CODE GENERATE/QUANTIZE
                                               • user-defined
              Copyright (C) 1985 through 1990 by Hyperception

   help: F1    elete: Backspc    delete: Del    move: ↕→   menu: Esc
```

FIGURE C.20. Code generator menu for sixth-order IIR bandpass filter

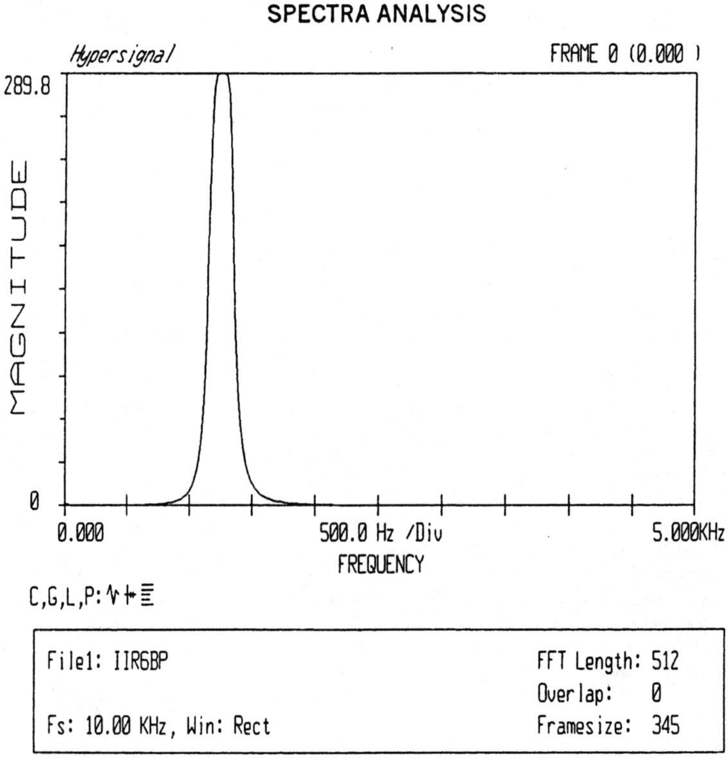

FIGURE C.21. Frequency response of sixth-order IIR bandpass filter

C.4 IIR Filter Design and Code Generation

```
;IIRMAINR.ASM-REAL-TIME MAIN PROGRAM. CALLS IIR FILTER SUBROUTINE
              .TITLE    "IIRMAINR"        ;CALLS HYPERSIGNAL SUB
              .GLOBAL   RESET,BEGIN,FILT,AICSET,AICSEC,AICIO_P
              .SECT     "VECTORS"         ;ASSEMBLE INTO VECTOR SECTION
RESET         .WORD     BEGIN             ;RESET VECTOR
              .DATA                       ;ASSEMBLE INTO DATA SECTION
STACKS        .WORD     809F00h           ;INIT STACK POINTER DATA
AICSEC        .WORD     1428h,1h,4A96h,67h ;AIC CONFIG DATA
              .TEXT                       ;ASSEMBLE INTO TEXT SECTION
BEGIN         LDP       STACKS            ;INIT DATA PAGE
              LDI       @STACKS,SP        ;SP => 809F00h
              CALL      AICSET            ;INIT AIC
LOOP          CALL      AICIO_P           ;AIC I/O INPUT=>R6,OUT=>R7
              FLOAT     R6,R3             ;INPUT NEW SAMPLE ->R3
              CALL      FILT              ;GO TO SUBROUTINE FILT
              FIX       R3,R7             ;R7=INTEGER(R3)
              BR        LOOP              ;BRANCH UNTIL R4 < 0
              .END
```

FIGURE C.22. Real-time main program to call IIR subroutine from Hypersignal (IIRMAINR.ASM)

(representing noise, for example) could have been used without changing the program. This bandpass IIR filter is further discussed in Chapter 5.

Example C.6 *Real-Time IIR Bandpass Filter Using TMS320C30 Code.* The real-time version of the previous IIR bandpass filter example can be obtained. The subroutine from Hypersignal remains the same. The real-time version of the previous main program, IIRMAINR.ASM, is shown in Figure C.22. Link this main program with the Hypersignal subroutine and the AIC communication routines in AICCOMA.ASM. Create an output file IIR6BPR.OUT to be downloaded into the EVM. With random noise as input, the output can be obtained (see Chapter 5).

Example C.7 *Second-Order IIR Lowpass Filter.* This example illustrates the use of the IIR filter construction option for the design of a lowpass filter (see also Figure C.19). Use the following:

1. a SAMPLING FREQUENCY of 100000
2. a CENTER FREQUENCY of 0
3. a BANDWIDTH of 24000
4. a TRANSITION BANDWIDTH1 of 0
5. a TRANSITION BANDWIDTH2 of 12000
6. a STOPBAND ATTENUATION of 18
7. an ORDER of 2
8. a RIPPLE (dB) of 6

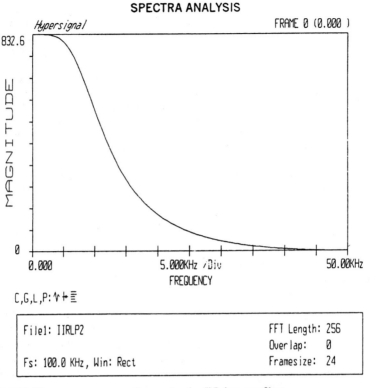

FIGURE C.23. Frequency response of second-order IIR lowpass filter

This set of characteristics will yield a second-order IIR lowpass filter with a cutoff frequency of 12,000 Hz (not 24,000 Hz), using a Butterworth design. The frequency response of this filter is plotted in Figure C.23.

REFERENCES

[1] *Hypersignal-Plus DSP Software*, Hyperception, Inc., Dallas, Tex., 1991.
[2] *Digital Filter Design Package (DFDP)*, Atlanta Signal Processors, Inc., Atlanta, Ga., 1991.
[3] The MathWorks, Inc., South Natick, Mass.

D
Burr–Brown Analog Evaluation Fixture

A reasonably priced analog evaluation fixture with dual channels A/D and D/A is available from Burr–Brown. The interfacing diagram of the Burr–Brown analog evaluation fixture to the serial port 1 on the **TMS320C30**, *through a 10-pin connector on the EVM, is shown in Figure D.1. The functional block diagram of the Burr–Brown analog evaluation fixture, shown in Figure 3.27 (Chapter 3), is based on the Burr–Brown* **DSP102** **18-bit ADC** *and the* **DSP202 18-bit DAC**.

The following header jumpers on the Burr–Brown fixture should be set.

1. For single I/O 18-bit and double I/O 16-bit A/D and D/A conversions:
   (a) HDR5/A through HDR8/A and HDR5/B through HDR8/B should be in position 1 to insert the I/O filters; otherwise, they should be in position 2
   (b) HDR9 across "DSP CC"
   (c) HDR10 across "DSP CC" and "/ENABLE"
   (d) HDR13 across "INT"
   (e) HDR15 across 1
   (f) Header jumpers on each 20-kHz filter board in position closest to the ICs
2. For single I/O 18-bit A/D and D/A conversions only: change HDR9 and HDR10 to be across "CASC"
3. For double I/O 16-bit A/D and D/A conversions only: place HDR11 across 3

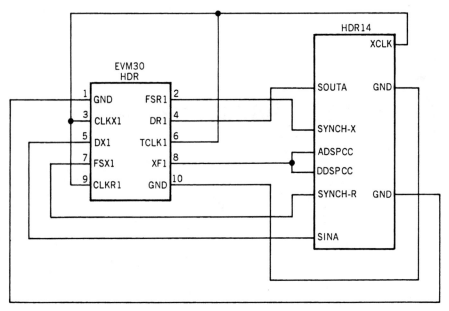

FIGURE D.1. Burr–Brown/EVM interface diagram

```
;BBLOOP.ASM-TMS320C30/BURR-BROWN LOOP FOR CHANNEL A
        .title   "BBLOOP.ASM"  ;LOOP FOR CHANNEL A
        .global  RESET, BEGIN, S_RATE, BB_SET, BB_IO_P
        .sect    "vectors"     ;VECTOR SECTION
RESET   .word    BEGIN         ;RESET VECTOR
        .data                  ;ASSEMBLE INTO DATA SECTION
STACKS  .word    809F00h       ;INIT STACK POINTER
S_RATE  .word    150           ;Fs=7.5 MHz/(2*S_RATE)
        .text                  ;ASSEMBLE INTO TEXT SECTION
BEGIN   LDP      STACKS        ;INIT DATA PAGE
        LDI      @STACKS,SP    ;SP -> 809F00h
        CALL     BB_SET        ;INIT TCLK1 AND SERIAL PORT REG
        LDI      0,R6          ;CLEAR INPUT REGISTER
LOOP    LDI      R6,R7         ;R7 = NEW INPUT SAMPLE (OUTPUT)
        CALL     BB_IO_P       ;OUTPUT R7 AND INPUT INTO R6
        BR       LOOP          ;BRANCH BACK AGAIN
        .END                   ;END
```

FIGURE D.2. Loop program to test channel A (BBLOOP.ASM)

D.1 TESTING CHANNEL A USING POLLING

Figure D.2 shows a loop program BBLOOP.ASM to test channel A only. The sampling rate is

$$F_s = \frac{7.5 \text{ MHz}}{2 \times \text{S\_RATE}} = 25 \text{ kHz}$$

This rate is handled through polling.

D.2 TESTING BOTH CHANNELS USING POLLING

To test both channels A and B, make the following changes to BBLOOP.ASM:

1. Before the instruction LDI 0,R6, add

    LDI   0,R4

    to clear the input register.
2. After the instruction LDI R6,R7, add the instruction

    LDI   R4,R5

    to input a new sample.
3. Change the CALL statement to

    CALL   BB_IO2P

    to output R5 and R7 and input R4 and R6. The resulting file is on the accompanying disk as BBLOOP2.ASM.

D.3 TESTING CHANNEL A USING INTERRUPT

Figure D.3 shows a loop program BBLOOPI.ASM to test channel A, using interrupt. The interrupt service vector ISR is at memory location 9h. This vector contains the address of the interrupt service routine, where BB_IO_I from the communication program is called to output R7 and input R6. Note the initialization of the status and interrupt enable registers for interrupt.

376 Burr-Brown Analog Evaluation Fixture

```
;BBLOOPI.ASM-TMS320C30/BURR-BROWN LOOP FOR CHANNEL A WITH INTERRUPT
           .title  "BBLOOPI.ASM"  ;INTERRUPT-DRIVEN LOOP FOR CHANNEL A
           .global RESET, BEGIN, S_RATE, BB_SET, BB_IO_I
           .sect   "vectors"     ;VECTOR SECTION
RESET      .word   BEGIN         ;RESET VECTOR
           .space  8             ;SKIP 8 WORDS
           .word   ISR           ;TIMER 0 INTERRUPT SERVICE ROUTINE
           .SPACE  54            ;REMAINDER OF VECTOR SECTION
           .data                 ;ASSEMBLE INTO DATA SECTION
STACKS     .word   809F00h       ;INIT STACK POINTER
S_RATE     .word   150           ;Fs=7.5 MHz/(2*S_RATE)
IE_REG     .word   100h          ;ENABLE TIMER 0 INTERRUPT
ST_REG     .word   2000h         ;SET STATUS REGISTER
           .text                 ;ASSEMBLE INTO TEXT SECTION
BEGIN      LDP     STACKS        ;INIT DATA PAGE
           LDI     @STACKS,SP    ;SP -> 809F00h
           CALL    BB_SET        ;INIT TCLK1 AND SERIAL PORT REG
           LDI     @IE_REG,IE    ;ENABLR XINT1
           OR      @ST_REG,ST    ;ENABLE GIE BIT IN STATUS REG
           LDI     0,R6          ;CLEAR INPUT REGISTER
LOOP       IDLE                  ;WAIT FOR RECEIVE INTERRUPT
           LDI     R6,R7         ;R7 = NEW INPUT SAMPLE (OUTPUT)
           BR      LOOP          ;BRANCH BACK AGAIN
ISR        CALL    BB_IO_I       ;OUTPUT R7 AND INPUT INTO R6
           RETI                  ;RETURN FROM INTERRUPT
           .END                  ;END
```

FIGURE D.3. Loop program to test channel A using interrupt (BBLOOPI.ASM)

D.4 TESTING CHANNELS A AND B USING INTERRUPT

Make the following changes to BBLOOPI.ASM:

1. Replace the instruction LDI 0,R6 by

   ```
   LDI     0,R5
   LDI     0,R7
   ```

   to clear the output A and B registers.
2. After LDI R6,R7, add the instruction

   ```
   LDI     R4,R5
   ```

   to input a new sample into R5.
3. Change the CALL statement to

   ```
   CALL    BB_IO2I
   ```

   The resulting file is on the accompanying disk as BBLOOPI2.ASM.

D.4 Testing Channels A and B Using Interrupt

```
;BBCOM.ASM-TMS320C30/BURR-BROWN COMM ROUTINES.POLLING OR INTERRUPT
        .title  "BBCOM.ASM"      ;COMM ROUTINES FOR CHANNELS A&B
        .global S_RATE,BB_SET,INITSP,BB_IO_I,BB_IO_P,BB_IO2I,BB_IO2P
        .data                    ;ASSEMBLE INTO DATA SECTION
P_CNTL  .word   808064h          ;PRIMARY BUS CONTROL REGISTER
PBASE   .word   808000h          ;PERIPHERAL BASE ADDR
SETSP   .word   2BC0000h         ;SERIAL PORT SET-UP DATA
GOSP    .word   0C000000h        ;SERIAL PORT START DATA
        .text
BB_SET  PUSH    AR0              ;SAVE AR0
        PUSH    R0               ;SAVE R0
        PUSH    R1               ;SAVE R1
        LDI     -249,R0          ;DATA FOR 0 WAIT STATE
        LDI     @P_CNTL,AR0      ;AR0->PRIMARY BUS CONTROL REG
        AND     *AR0,R0          ;AND CONTROL REG WITH 0 WAIT DATA
        STI     R0,*AR0          ;INITIATE SRAM TO 0 WAIT STATE
        LDI     @PBASE,AR0       ;AR0 -> 808000h
        LDI     020h,IOF         ;SET FX1 LOW
INITT1  XOR     R1,R1            ;CLEAR R1
        LDI     01h, R0          ;R0 = DATA FOR TIMER 1 PERIOD REG
        STI     R1,*+AR0(38h)    ;INIT TIMER 1 PERIOD REG(3.75 MHz)
        STI     R1,*+AR0(34h)    ;START TIMER 1 COUNTER REG @ 0
        LDI     03C1h,R0         ;INIT TIMER 1 GLOBAL REG
        STI     R0,*+AR0(30h)    ;RESET TIMER 1
        STI     R1,*+AR0(58h)    ;CLEAR DATA TRANSMIT REG (DTR)
INITSP  LDI     @SETSP,R0        ;RESET->SP:32 BITS,EXT CLKS,STD MODE
        STI     R0,*+AR0(50h)    ;FSX=OUTPUT,& INT ENABL(SP GLOBAL REG)
        LDI     131h,R0          ;X & R PORT CONTROL REGS DATA
        STI     R0,*+AR0(52h)    ;FSX/DX/CLKX = SP OPERATIONAL PINS
        STI     R0,*+AR0(53h)    ;FSR/DR/CLKR = SP OPERATIONAL PINS
        LDI     @GOSP,R0         ;R0 = VALUE TO START THE SP
        OR      *+AR0(50h),R0    ;R0 = DATA TO END RESET OF SP
        STI     R0,*+AR0(50h)    ;START C30 SERIAL COM
INITT0  LDI     @S_RATE,R0       ;TIMER CLK 0 = 7.5 MHZ / (2 * S_RATE)
        STI     R0,*+AR0(28h)    ;INIT TIMER 0 PERIOD REG (TCLK0)
        STI     R1,*+AR0(24h)    ;START TIMER 0 COUNTER REG @ 0
        LDI     3C1h,R0          ;INIT TIMER 0 GLOBAL REG
        STI     R0,*+AR0(20h)    ;RESET TIMER 0
        POP     R1               ;RESTORE R1
        POP     R0               ;RESTORE R0
        POP     AR0              ;RESTORE AR0
        RETS                     ;RETURN TO MAIN PROGRAM
;D/A IN R7 AND A/D IN R6. POLLS TIMER 0 INTERRUPT FLAG
BB_IO_P PUSH    AR0              ;SAVE AR0
        PUSH    R0               ;SAVE R0
```

FIGURE D.4. Burr–Brown/TMS320C30 communication routines (BBCOM.ASM)

```
WAIT1    LDI    IF,R0            ;R0=INTERRUPT FLAG REG CONTENT
         AND    100h,R0          ;SEE IF TIMER 0 INTER FLAG SET
         BZ     WAIT1            ;BRANCH TO WAIT IF NOT SET
         AND    -257,IF          ;CLEAR TIMER 0 INTERRUPT FLAG
         OR     60h,IOF          ;SET FX1 HIGH
         LDI    @PBASE,AR0       ;AR0 -> 808000h
         CALL   IN_OUT           ;OUTPUT D/A AND INPUT A/D
         POP    R0               ;RESTORE R0
         POP    AR0              ;RESTORE AR0
         LDI    20h,IOF          ;SET FX1 LOW TO INIT A/D-D/A CONVERSION
         RETS                    ;RETURN TO MAIN PROGRAM
;D/A IN R7 AND A/D IN R6.CALLED FROM TIMER 0 INTERRUPT SERVICE ROUTINE
BB_IO_I  PUSH   AR0              ;SAVE AR0
         OR     60h,IOF          ;SET FX1 HIGH
         LDI    @PBASE,AR0       ;AR0 -> 808000h
         CALL   IN_OUT           ;OUTPUT D/A AND INPUT A/D
         POP    AR0              ;RESTORE AR0
         LDI    20h,IOF          ;SET FX1 LOW TO INIT A/D-D/A CONVERSION
         RETS                    ;RETURN TO MAIN PROGRAM
;I/O UTILITY ROUTINE
IN_OUT   LDI    R7,R6            ;SAVE CONTENT OF THE OUTPUT REG
         LSH    14,R6            ;MULT OUTPUT BY 2**14
         STI    R6,*+AR0(58h)    ;DTR=NEXT DATA FOR D/A
         LDI    *+AR0(5Ch),R6    ;R6=DRR DATA FROM A/D
         ASH    -14,R6           ;DIVIDE INPUT BY 2**14 KEEPING SIGN
         RETS                    ;RETURN TO MAIN PROGRAM
;TWO INPUTS AND OUTPUTS ROUTINES WITH 16-BIT ACCURACY
;INPUT:R5 AND R7 (A AND B D/A).   OUTPUT:R4 AND R6 (A AND B A/D)
BB_IO2P  PUSH   AR0              ;SAVE AR0. POLLS TIMER 0 INTER FLAG
         PUSH   R0               ;SAVE R0
WAIT2    LDI    IF,R0            ;R0=INTERRUPT FLAG REG CONTENT
         AND    100h,R0          ;SEE IF TIMER 0 INTER FLAG IS SET
         BZ     WAIT2            ;BRANCH TO WAIT IF NOT SET
         AND    -257,IF          ;CLEAR TIMER 0 INTERRUPT FLAG
         OR     60h,IOF          ;SET FX1 HIGH
         LDI    @PBASE,AR0       ;AR0 -> 808000h
         CALL   IN_OUT2          ;OUTPUT D/A AND INPUT A/D
         POP    R0               ;RESTORE R0
         POP    AR0              ;RESTORE AR0
         LDI    20h,IOF          ;SET FX1 LOW TO INIT A/D-D/A CONVERSION
         RETS                    ;RETURN TO MAIN PROGRAM
BB_IO2I  PUSH   AR0              ;SAVE AR0.CALLED FROM TIMER 0 INTER
         OR     60h,IOF          ;SET FX1 HIGH
         LDI    @PBASE,AR0       ;AR0 -> 808000h
```

FIGURE D.4. *(Continued)*

```
            CALL    IN_OUT2        ;OUTPUT D/A AND INPUT A/D
            POP     AR0            ;RESTORE AR0
            LDI     020h,IOF       ;SET FX1 LOW TO INIT A/D-D/A CONVERSION
            RETS                   ;RETURN TO MAIN PROGRAM
;I/O  2 CHANNELS UTILITY ROUTINE
IN_OUT2     LDI     R5,R4          ;SAVE CONTENT OF OUTPUT A
            LDI     R7,R6          ;SAVE CONTENT OF OUTPUT B
            LSH     16,R4          ;OUTPUT A IN UPPER 16(CLEAR LOWER 16)
            AND     0FFFFh,R6      ;CLEAR OUTPUT B UPPER 16 BITS
            OR      R4,R6          ;R6 = R4 (OUTPUTS A AND B)
            STI     R6,*+AR0(58h)  ;DTR = NEXT DATA FOR D/A
            LDI     *+AR0(5Ch),R6  ;R6 = DRR DATA FROM A/D
            LDI     R6,R4          ;R4=R6(A IN UPPER 16 AND B IN LOWER 16)
            ASH     -16,R4         ;DIVIDE INPUT A BY 2**16 KEEPING SIGN
            LSH     16,R6          ;PUT INPUT B IN UPPER 16 BITS
            ASH     -16,R6         ;DIVIDE INPUT B BY 2**16 KEEPING SIGN
            RETS                   ;RETURN TO MAIN PROGRAM
            .END                   ;END
```

FIGURE D.4. *(Continued)*

D.5 BURR – BROWN COMMUNICATION ROUTINES

The Burr–Brown communication routines are shown in the program BBCOM.ASM, listed in Figure D.4. These routines can handle one or both channels, with a sampling rate determined by either polling or interrupt. Note that the maximum sampling rate is slightly lower than 200 kHz because the master clock on the EVM is 7.5 MHz. The 200 kHz would be obtained with an 8-MHz master clock frequency.

E
Extended Development System

PETER MARTIN

E.1 THE XDS1000 EMULATOR

A target module, used as an application board, was built to interface to the XDS1000 emulator. Figure E.1 shows a diagram of the various devices used. A parts list follows:

  U1–U8: TMS6788-25N RAM
  U9: 32-MHz clock oscillator
  U10: TMS320C30 processor
  U11: 7400 Quad dual-input NAND gates
  R1, R2: 22-kΩ bussed resistor network (15 each)
  R3: 470-kΩ resistor
  C1: 1-μF capacitor
  J1: connector to interface target module to AIC module
  J2: connector to interface target module to XDS1000 emulator
  J3: connector to interface target module to AIB

A photo of the target module connected to the XDS1000 and to an AIC module (through J1) is shown in Figure E.2. The target module also shows (right side) another connector (J3) used to interface to the Texas Instruments, Inc., analog interface board (AIB). An interface diagram to the 16K of SRAM is shown in Figure E.3. Each of the eight SRAM chips (TMS6788) contains 16K × 4 memory locations, with an access time of 25 ns. These SRAM chips can be tested using the program SRAMTEST.ASM (on the

E.1 The XDS1000 Emulator

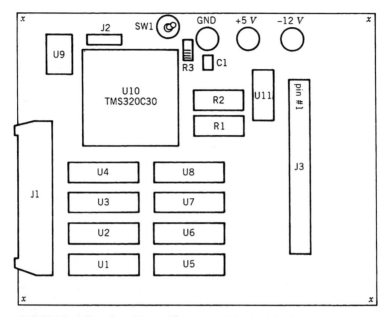

FIGURE E.1. Top-view diagram of target module that interfaces to XDS1000

FIGURE E.2. Photo of target module connected to an AIC module and the XDS1000

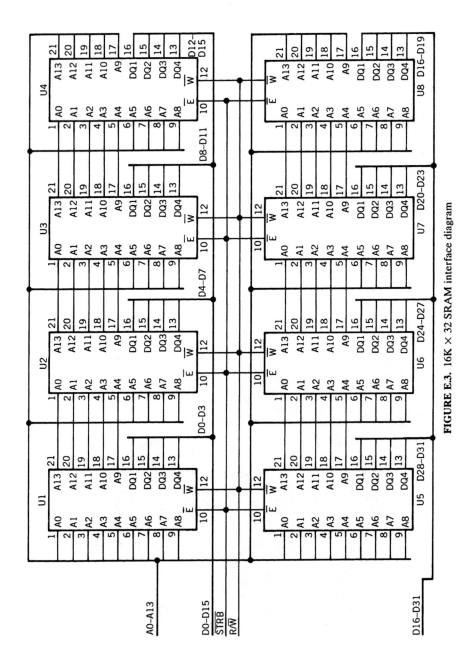

FIGURE E.3. 16K × 32 SRAM interface diagram

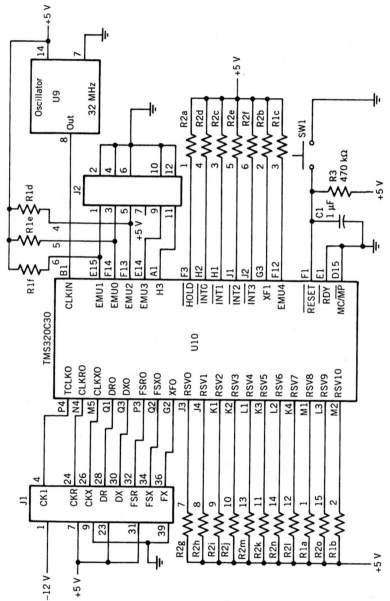

FIGURE E.4. XDS1000–AIC interface diagram

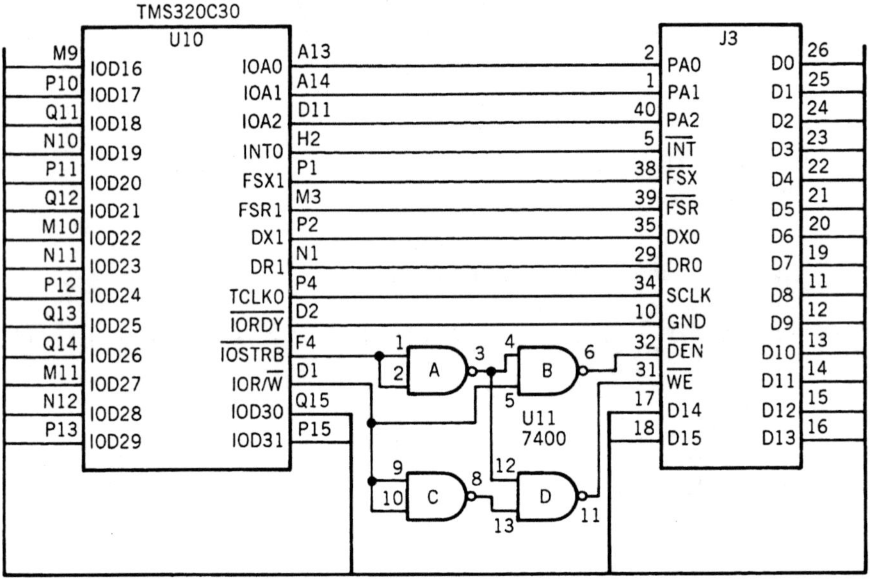

384 Extended Development System

FIGURE E.5. XDS1000–AIB interface diagram

accompanying disk). The memory locations 0h–3 FFFh (16K) are tested with "ones," then with "zeros." Figure E.4 shows the XDS1000–AIC interface diagram. It shows how the external AIC module can be interfaced to the TMS320C30 through connector J1 on the target module. It also shows the interfacing to the XDS1000 through connector J2 (on the target module). This diagram includes a 32-MHz oscillator circuit (not 30 MHz as on the EVM) and pull-up resistors. Figure E.5 shows the TMS320C30–AIB interface diagram through connector J3. All the programs, run previously with the AIC on-board the EVM (port 0), can also be run with the AIC module connected to the XDS1000, with the provision that the AIC master clock is (32 MHz)/4, or 8 MHz (not 7.5 as with the EVM).

A brief description of the schematic connections follows.

Internal Board Connections

1. RSV0–RSV10, EMU0–EMU2, and EMU4 pins have 22-kΩ pull-up resistors to 5 V.
2. $\overline{INT0}$–$\overline{INT3}$ pins have 22-kΩ pull-up resistors to 5 V, to place an inactive level on the interrupt pins.
3. $\overline{HOLD}$ pin has a 22-kΩ pull-up resistor to 5 V, allowing the TMS320C30 to continue controlling the primary bus.
4. XF1 pin has a 22-kΩ pull-up resistor to 5 V, specified as an unused I/O pin at this time.

5. $\overline{\text{RDY}}$ pin is tied to ground, to supply an active signal to the TMS320C30 internal RDY signal to allow a zero wait state when using the external ready capability for the primary bus.
6. MC/$\overline{\text{MP}}$ pin is tied to ground, to allow the TMS320C30 to run in the microprocessor mode.
7. $\overline{\text{RESET}}$ pin is connected to the reset circuitry, to allow a 167-ms active signal upon power-up.

External AIC Module Interface Through J1

1. CLKR0, CLKX0, DR0, DX0, FSR0, and FSX0 are AIC serial port pins.
2. XF0 pin is used to allow a software reset of the AIC.
3. TCLK0 is used for the AIC master clock.

External XDS1000 Interface Through Connector J2

EMU0–EMU3 and H3 are emulator pins.

External AIB Interface Through J3

1. TCLK0, DR1, DX1, FSR1, and FSX1 are AIB (AIC) serial port pins.
2. IOA0–IOA2 are used to address AIB ports including the A/D and D/A.
3. IOD16–IOD31 are used for data access to and from the AIB, including D/A and A/D conversion data.
4. $\overline{\text{IORDY}}$ is tied to ground and can be ignored or used to AND with the software internal ready signal for the expansion bus.
5. $\overline{\text{INT0}}$ is used to allow the AIB to interrupt the TMS320C30.
4. $\overline{\text{IOSTRB}}$ and IOR/$\overline{\text{W}}$ are connected to U11 to allow the compatibility between the TMS320C30 signals and the data enable (DEN) and write enable (WE) TMS3201X signals. (See the following table.)

$\overline{\text{IOSTRB}}$	IOR/$\overline{\text{W}}$	$\overline{\text{DEN}}$	$\overline{\text{WE}}$
0	0	1	0
1	0	1	1
0	1	0	1
1	1	1	1

Example E.1 *Generation of a Square Wave with the XDS1000/AIB, Using TMS320C30 Code.* This program example tests the XDS1000 emulator interfaced to the AIB. This example can also be very useful to measure the execution time of a main processing routine. For example, one can output a positive value (such as 7FFFh) before executing a time-critical routine. At the end of the time-critical routine, this positive value is negated, then sent for output. The duration of the positive output value as seen on an oscilloscope (before it becomes negative) is measured as the real time (more by a couple of instructions) it takes to execute the time-critical routine.

The emulator is accessed with the command EMU30. Support software is included with the XDS1000 package. Press X then RESET to connect to the

```
;SQUARE.ASM-SQUARE WAVE GENERATOR WITH XDS1000/AIB
          .TITLE    "SQUARE"        ;GENERATE SQUARE WAVE WITH INTER
          .SECT     "VECTORS"       ;ASSEMBLE INTO VECTOR SECTION
RESET     .WORD     BEGIN           ;RESET VECTOR
          .SPACE    8               ;SKIP 8 WORDS
TIMER0    .WORD     TIM_INT         ;TINT0 VECTOR LOCATION @ 9h
          .SPACE    54              ;REMAINDER OF VECTOR SECTION
          .DATA                     ;ASSEMBLE INTO DATA SECTION
STACKS    .WORD     809F00H         ;INIT STACK POINTER DATA
PERIOD    .WORD     190H            ;INTER RATE=8MHz/(2*PERIOD)=10 kHz
IE_REG    .WORD     100H            ;ENABLE (TINT0) TIMER 0 INTERRUPT
PER_ADDR  .WORD     808028H         ;(TINT0) TIMER 0 PERIOD REG LOCATION
TCNTL     .WORD     3C3H            ;CONTROL REGISTER VALUE
ST_REG    .WORD     2000H           ;SET STATUS REG
IO_ADDR   .WORD     804000H         ;AIB I/O ADDRESS
          .TEXT                     ;ASSEMBLE INTO TEXT SECTION
BEGIN     LDP       STACKS          ;INIT DATA PAGE
          LDI       @STACKS,SP      ;SP-> 809F00H
          LDI       @PER_ADDR,AR0   ;TINT0 PERIOD REG =>AR0
          LDI       @PERIOD,R0      ;INTERRUPT RATE =>R0
          STI       R0,*AR0--(8)    ;SET TINT0 PERIOD
          LDI       @TCNTL,R0       ;CONTROL REGISTER VALUE =>R0
          STI       R0,*AR0         ;SET TINT0 GLOBAL CNTRL REG
          LDI       @IE_REG,IE      ;ENABLE TINT0 INTERRUPT
          OR        @ST_REG,ST      ;SET STATUS REG (ENABLE GIE BIT)
          LDI       @IO_ADDR,AR0    ;1st ADDRESS OF EXP BUS -> AR0
          LDI       7FFFH,R5        ;POSITIVE VALUE -> R5
WAIT      IDLE                      ;WAIT FOR INTERRUPT
          BR        WAIT            ;BRANCH BACK TO WAIT
;    INTERRUPT ROUTINE
TIM_INT   NEGI      R5,R7           ;R7=-R5, NEGATIVE PEAK OF WAVE
          LDI       0,R5            ;REINITIALIZE R5
          ADDI      R7,R5           ;NEW VALUE OF R5
          LSH       16,R7           ;SHIFT OUTPUT INTO UPPER 16 BITS
          STI       R7,*+AR0(2)     ;OUTPUT TO AIB
          RETI                      ;RETURN FROM SUBROUTINE
          .END                      ;END
```

FIGURE E.6. Program for square-wave generation with XDS1000/AIB, using TMS320C30 code (SQUARE.ASM)

```
;LOOPAICB.ASM-LOOP PROGRAM USING AIC ON THE AIB. CALLS AICBCOM
        .TITLE   "LOOPAICB"      ;LOOP USING AIC ON THE AIB
        .OPTION  X               ;FOR SYMBOL XREF
        .GLOBAL  RESET,BEGIN,AICSET,AICSEC,AICIO  ;REF/DEF SYMBOLS
        .SECT    "VECTORS"       ;ASSEMBLE INTO VECT SECTION
RESET   .WORD    BEGIN           ;RESET VECTOR
        .DATA                    ;ASSEMBLE INTO DATA SECTION
STACKS  .WORD    809F00H         ;INIT STACK POINTER DATA
AICSEC  .WORD    03H,1224H,03H,3A76H,03H,67H,0H ;    Fs=10 kHz
        .TEXT                    ;ASSEMBLE INTO TEXT SECTION
BEGIN   LDP      STACKS          ;INIT DATA PAGE
        LDI      @STACKS,SP      ;SP => 809F00H
        CALL     AICSET          ;INIT AIC/TMS320C30
        LDI      0,R7            ;INIT R7 = OUTPUT TO 0
LOOP    CALL     AICIO           ;AIC I/O ROUTINE,INPUT=>R6,OUT=>R7
        LDI      R6,R7           ;OUTPUT R7=NEW INPUT SAMPLE
        BR       LOOP            ;LOOP CONTINUOUSLY
        .END                     ;END
```

FIGURE E.7. Loop program with the XDS1000 to test the AIC on the AIB, Using TMS320C30 code (LOOPAICB.ASM)

target module. The emulator windows are identical to the windows in the simulator. Figure E.6 shows the listing of a program, SQUARE.ASM, which generates a square wave with interrupt. The interrupt rate is (8 MHz)/(2 × PERIOD) = 10 kHz. Note that the clock generator in the target module operates at 32 MHz, not 30 MHz as with the EVM. The initialization of the various registers for interrupt is discussed in Chapter 3. The interrupt routine is specified by TIM_INT, and the output is through port address 804002h. Verify that the AIB output (through an oscilloscope) is a square wave.

Example E.2 Loop Program with the AIC on the AIB, Using TMS320C30 code. This example tests the TLC32040 analog interface chip (AIC), which is onboard the AIB. The loop program, LOOPAICB.ASM, listed in Figure E.7 calls the communication routines in the program AICBCOM.ASM (on the accompanying disk). This AIC communication program uses the TMS320C30 serial port 1 (see also the AIC communication program, AICCOMA.ASM, listed in Chapter 3). The AIC is set for a sampling rate of 10 kHz, using polling. From the AIC communication routines, the input is in R6 and the output in R7. Run the loop program with an input sinusoid (the AIC input is through the AIB input connector), and verify that a sinusoid of the same frequency is obtained at the output (through the AIB connector). Note that the AIC input/output is through the AIB.

Example E.3 Real-Time FFT with the XDS1000/AIB, Using TMS320C30 Code. This programming example performs a 256-point real-time FFT and provides an additional supporting example for the XDS1000 emulator in conjunction with the AIB. The FFT is discussed in Chapter 6. The following programs, all on the accompanying disk, are used for this example:

1. **FFT_256.ASM**: This main program, which calls the FFT subroutine FFT_MM.ASM and the AIB communication program AIBFFT.ASM, is set for a 256-point FFT. It includes a square-root routine to find the output's magnitude. This square-root routine is taken from the manual in [1]. A sampling rate of 8 kHz is set using interrupt. On interrupt, the I/O communication routines supporting the AIB are called.
2. **FFT_MM.ASM**: This generic FFT program, taken from the *TMS320C30 User's Guide* [2], performs a complex radix-2 FFT. This program is based on the Fortran version in [3], which can be used for different FFT sizes.
3. **AIBFFT.ASM**: This program initializes the AIB for interrupt and includes a routine for I/O with the AIB.
4. **SIN256.DAT**: This program contains three-fourths of a cycle of a sine function, which represents 192 twiddle constants (see Chapter 6 for a program to generate twiddle constants for different FFT sizes).

The appropriate command file links these four supporting programs. Run this FFT program, using a sinusoid input of frequency 2 kHz. The output,

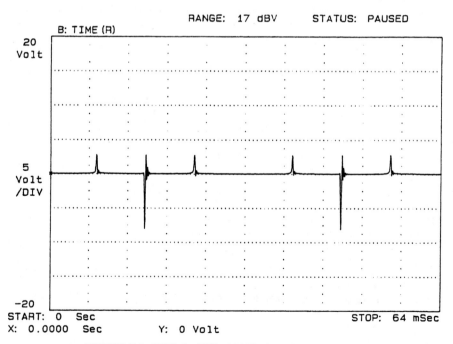

FIGURE E.8. 256-Point FFT of 2-kHz sinusoidal input signal

connected through an oscilloscope is shown in Figure E.8 (similar to the plot in Figure 6.25 for the 512-point FFT). Negative spikes (as in Figure 6.25) represent the beginning of each frame of 256 points. This technique is used in order to provide the 8-kHz (the sampling rate) range using an oscilloscope. A delta function occurs at a frequency of 2 kHz, along with a second delta function at its folding frequency. One can use this example as a crude spectrum analyzer. Observe the output spectrum using music as input.

Exercise E.1 provides a way for testing the XDS1000 emulator and the AIB. The program MR10.ASM (on the accompanying disk), which simulates the multirate filter discussed in Chapter 8, also can be used (without any changes in MR10.ASM) to test the XDS1000 with the AIB.

Exercise E.1 Testing the XDS1000 with the AIB. Test the XDS1000 emulator with the AIB using the following:

1. The sine generator program SINE4.ASM, listed in Figure 3.9. Determine the frequency of the sinusoid (use a clock value of 8 MHz in lieu of 7.5 MHz).
2. The noise generator program NOISEINT.ASM, listed in Figure 3.6. Change the output port address IO_ADDR to 804002h. Verify that the noise spectrum is flat up to the cutoff frequency set with the analog filters on the AIB.
3. An interrupt-driven loop program that can be obtained by modifying appropriately either SINE4.ASM or NOISEINT.ASM. The same port address of 804002h is used for both input and output. Use the following segment of code as the interrupt service routine, with IO_ADDR specifying the output port address:

```
TIM_INT   LDI   @IO_ADDR,AR0   ;AR0 = INPUT SAMPLE ADDRESS
          LDI   *AR0,R0        ;R0 = INPUT VALUE
          STI   R0,*AR0        ;OUTPUT THIS VALUE
```

REFERENCES

[1] P. Papamichalis (editor), *Digital Signal Processing with the TMS320 Family— Theory, Algorithms, and Implementations*, Vol. 3, Texas Instruments, Inc., Dallas, Tex., 1990.

[2] *TMS320C3X User's Guide*, Texas Instruments, Inc., Dallas, 1991.

[3] C. S. Burrus and T. W. Parks, *DFT/FFT and Convolution Algorithms: Theory and Implementation*, Wiley, New York, 1988.

F
Programs for Multirate Filter and Video Line Rate Analysis Projects

F.1 MULTIRATE FILTER PROGRAM

The program `MR10MCR.ASM`, listed in Figure F.1, implements the 10-band multirate filter discussed in Chapter 8. This interrupt-driven program includes a macro for noise generation, three sets of 41 coefficients associated with the lower, middle, and higher $\frac{1}{3}$-octave bands, and two sets of 11 coefficients associated with the interpolation filters. The program is initialized to turn on the middle $\frac{1}{3}$-octave band 10 filter, as set in `SCALE_10`.

F.2 VIDEO MODULE PARTS LIST

A parts list of the components used in the video module in Figure 8.28 follows:

Device	Function
(1) 74HC00	NAND gate
(1) 74HC04	Inverter
(1) 74HC08	AND gate
(1) 74HC107	JK flip-flop
(1) 74HC174	D flip-flop
(1) 74HC164	Serial-to-parallel shift register
(2) 74HC40103	Binary down-counter
(1) IDT72131	Parallel-to-serial FIFO
(1) TDA8708	ADC
(1) LM1881	Sync separator
(1)	9.8-MHz clock

```
;MR10MCR.ASM-10-BAND MULTIRATE FILTER USING THE AIC
;WITH 5 WARNINGS (NO ERRORS)
          .title   "MR10MCR.ASM"          ;10-BAND WITH MACROS
          .option  X                      ;FOR XREF
          .global  NOISE, FILTER, INTRPLTR, RST_WBFR
          .global  BEGIN, LOOP, SET_UP, PROCESS, SETAR7_1, SETAR7_2
          .global  PROCLOOP, F_REPEAT, I_REPEAT, EXIT, WAIT
          .global  AICSEC,AICSET_I,AICIO_I
          .sect    "VECTORS"              ;ASSEMBLE INTO VECTOR SECTION
RESET     .word    BEGIN                  ;RESET VECTOR
          .space   4                      ;SKIP 4 WORDS
TIMER0    .word    ISR                    ;TINT0 VECTOR LOCATION @ 5h
          .space   58                     ;REMAINDER OF VECTOR SECTION
;***************************************************************************
;*    MACRO DEFINITIONS BELOW                                              *
;***************************************************************************
RST_WBFR  $MACRO                          ;MACRO TO PING-PONG WORKING BUFFERS
          LDI      1, R5
          AND      R6, R5
          BNZ      SWITCH?
          LDI      1, R5
          AND      IR1, R5
          BNZ      CHNG_1?
          BD       CONT?
          LDI      @B3_ADR, AR5
          LDI      @B2_ADR, AR6
          NOP
CHNG_1?   BD       CONT?
          LDI      @B2_ADR, AR5
          LDI      @B3_ADR, AR6
          NOP
SWITCH?   LDI      1, R5
          AND      IR1, R5
          BNZ      CHNG_2?
          BD       CONT?
          LDI      @B3_ADR, AR5
          LDI      @B1_ADR, AR6
          NOP
CHNG_2?   BD       CONT?
          LDI      @B1_ADR, AR5
          LDI      @B3_ADR, AR6
          NOP
CONT?     NOP
          $ENDM
NOISE     $MACRO                          ;****RANDOM NOISE GENERATOR****
          LDI      0,R4                   ;CLEAR R4
          LDI      R0,R2                  ;PUT SEED IN R2
          LSH      -17,R2                 ;MOVE BIT 17 TO LSB
          ADDI     R2,R4                  ;ADD TO R4 (17)
          LSH      -11,R2                 ;MOVE BIT 28 TO LSB
          ADDI     R2,R4                  ;ADD TO R4 (28+17)
          LSH      -2,R2                  ;MOVE BIT 30 TO LSB
          ADDI     R2,R4                  ;ADD TO R4 (30+(28+17))
          LSH      -1,R2                  ;MOVE BIT 31 TO LSB
          ADDI     R2,R4                  ;ADD TO R4 (31+(30+(28+17)))
          AND      1,R4                   ;MASK LSB OF R4
          LDIZ     @MINUS,R2              ;IF R4 = 0    R2 = MINUS VALUE
          LDINZ    @PLUS,R2               ;IF R4 = 1    R2 = PLUS VALUE
          LSH      1,R0                   ;SHIFT SEED LEFT 1
          OR       R4,R0                  ;PUT R4 INTO LSB OF R0 (SEED)
          $ENDM
```

FIGURE F.1. Multirate filter program (MR10MCR.ASM)

```
;*********************************************************************
;*      THIS SECTION DEFINES MEMORY FOR SCALE VALUES, COEFFS, ETC.    *
;*********************************************************************
            .data                   ;ASSEMBLE INTO DATA SECTION
SCALE_1     .float  0.0, 0.0, 0.0   ;FILTER SCALE VALUES   (LOW, MID, UPPER)
SCALE_2     .float  0.0, 0.0, 0.0
SCALE_3     .float  0.0, 0.0, 0.0
SCALE_4     .float  0.0, 0.0, 0.0
SCALE_5     .float  0.0, 0.0, 0.0
SCALE_6     .float  0.0, 0.0, 0.0
SCALE_7     .float  0.0, 0.0, 0.0
SCALE_8     .float  0.0, 0.0, 0.0
SCALE_9     .float  0.0, 0.0, 0.0
SCALE_10    .float  0.0, 1.0, 0.0
                                    ;LOW 1/3 OCTAVE COEFFICIENTS #40 TO #0
LFILT_40    .float  0.000000000,  0.000000000,  0.001861572,  0.009765625
            .float  0.004577636, -0.007446289, -0.005493164, -0.000274658
            .float -0.015899658, -0.019592285,  0.037536621,  0.089355468
            .float  0.005004882, -0.159637451, -0.143707275,  0.124420166
            .float  0.299743652,  0.065185546, -0.322631835, -0.308929443
            .float  0.140655517,  0.421447753,  0.140655517, -0.308929443
            .float -0.322631835,  0.065185546,  0.299743652,  0.124420166
            .float -0.143707275, -0.159637451,  0.005004882,  0.089355468
            .float  0.037536621, -0.019592285, -0.015899658, -0.000274658
            .float -0.005493164, -0.007446289,  0.004577636,  0.009765625
LFILT_0     .float  0.001861572
                                    ;MID 1/3 OCTAVE COEFF 40
MFILT_40    .float -0.001342773, -0.000488281, -0.003692626,  0.003509521
            .float  0.009277343, -0.004486083, -0.007781982,  0.000396728
            .float -0.011413574,  0.009765625,  0.057342529, -0.023529052
            .float -0.131866455,  0.035247802,  0.225738525, -0.038635253
            .float -0.319488525,  0.030212402,  0.389099121, -0.011444091
            .float -0.414855957, -0.011444091,  0.389099121,  0.030212402
            .float -0.319488525, -0.038635253,  0.225738525,  0.035247802
            .float -0.131866455, -0.023529052,  0.057342529,  0.009765625
            .float -0.011413574,  0.000396728, -0.007781982, -0.004486083
            .float  0.009277343,  0.003509521, -0.003692626, -0.000488281
MFILT_0     .float -0.001342773
                                    ;UPPER 1/3 OCTAVE COEFFICIENTS #40 TO #0
UFILT_40    .float  0.001953125, -0.003295898, -0.000396728,  0.001861572
            .float  0.001007080,  0.003356933, -0.010589599,  0.002441406
            .float  0.006927490,  0.003692626,  0.006896972, -0.064880371
            .float  0.055999755,  0.110687255, -0.229766845,  0.020507812
            .float  0.334716796, -0.306884764, -0.166229248,  0.463836669
            .float -0.166229248, -0.306884764,  0.334716796,  0.020507812
            .float -0.229766845,  0.110687255,  0.055999755, -0.064880371
            .float  0.006896972,  0.003692626,  0.006927490,  0.002441406
            .float -0.010589599,  0.003356933,  0.001007080,  0.001861572
            .float -0.000396728, -0.003295898,  0.001953125,  0.000000000
UFILT_0     .float  0.000000000
                                    ;INTRPLTR COEFFICIENTS 1,3,...,21
INTR_01     .float  0.012542724, -0.034423828,  0.079803466, -0.180267334
            .float  0.625427246,  0.625427246, -0.180267334,  0.079803466
            .float -0.034423828,  0.012542724,  0.000000000
                                    ;INTRPLTR COEFFICIENTS 0,2,...,20
INTR_00     .float  0.007263183, -0.012023925,  0.019012451, -0.025939941
            .float  0.030700683,  0.967346191,  0.030700683, -0.025939941
            .float  0.019012451, -0.012023925,  0.007263183
```

FIGURE F.1. Multirate filter program (MR10MCR.ASM) *(Continued)*

```
FLT1_IN    .usect   "F_DATA", 64        ;START OF FILTER'S INPUT SAMPLES
FLT2_IN    .usect   "F_DATA", 64
FLT3_IN    .usect   "F_DATA", 64
FLT4_IN    .usect   "F_DATA", 64
FLT5_IN    .usect   "F_DATA", 64
FLT6_IN    .usect   "F_DATA", 64
FLT7_IN    .usect   "F_DATA", 64
FLT8_IN    .usect   "F_DATA", 64
FLT9_IN    .usect   "F_DATA", 64
FLT10_IN   .usect   "F_DATA", 64
INT1_IN    .usect   "I_DATA", 16        ;START OF INTRPLTR'S INPUT SAMPLES
INT2_IN    .usect   "I_DATA", 16
INT3_IN    .usect   "I_DATA", 16
INT4_IN    .usect   "I_DATA", 16
INT5_IN    .usect   "I_DATA", 16
INT6_IN    .usect   "I_DATA", 16
INT7_IN    .usect   "I_DATA", 16
INT8_IN    .usect   "I_DATA", 16
INT9_IN    .usect   "I_DATA", 16
           .bss     F1_COEFF, 41
           .bss     F2_COEFF, 41
           .bss     F3_COEFF, 41
           .bss     F4_COEFF, 41
           .bss     F5_COEFF, 41
           .bss     F6_COEFF, 41
           .bss     F7_COEFF, 41
           .bss     F8_COEFF, 41
           .bss     F9_COEFF, 41
           .bss     F10_COEFF, 41
           .bss     I_COEFF, 22
           .bss     BUFFER_3, 256
           .bss     FCP_STRG, 10
           .bss     FIP_STRG, 10
           .bss     ICP_STRG, 1
           .bss     IIP_STRG, 9
BUFFER_1   .float   0.0
           .space   511
BUFFER_2   .float   0.0
           .space   511
I_LENGTH   .set     11                  ;# OF INTRPLTR COEFFS
F_LENGTH   .set     41                  ;# OF FILTER COEFFS
STACKS     .word    809F00H             ;INIT STACK POINTER DATA
PLUS       .word    000001000H          ;POSITIVE LEVEL (+2^12)
MINUS      .word    0FFFFF000H          ;NEGATIVE LEVEL (-2^12)
SEED       .word    7E521603H           ;INITIAL SEED VALUE
AICSEC     .word    0E1CH,1h,4286H,67H  ;Fs=16234 Hz
SRAM_CNTL  .word    808064H             ;PRIMARY BUS CONTROL REGISTER
FLTC_ADR   .word    F1_COEFF            ;POINTS TO START OF FILTER COEFFS
FLTI_ADR   .word    FLT1_IN             ;POINTS TO START OF FILTER INPUTS
INTC_ADR   .word    I_COEFF             ;POINTS TO START OF INTRPLTR COEFFS
INTI_ADR   .word    INT1_IN             ;POINTS TO START OF INTRPLTR INPUTS
B1_ADR     .word    BUFFER_1            ;POINTS TO START OF BUFFER #1
B2_ADR     .word    BUFFER_2            ;POINTS TO START OF BUFFER #2
B3_ADR     .word    BUFFER_3            ;POINTS TO START OF BUFFER #3
FC_PNTR    .word    FCP_STRG            ;POINTS TO START OF FLTR COEFF PNTRS
FI_PNTR    .word    FIP_STRG            ;POINTS TO START OF FLTR INPUT PNTRS
IC_PNTR    .word    ICP_STRG            ;POINTS TO START OF INTRPLTR COEFF PNTRS
```

FIGURE F.1. *(Continued)*

```
II_PNTR     .word   IIP_STRG            ;POINTS TO START OF INTRPLTR INPUT PNTRS
SCALER      .float  0.25                ;OUTPUT SCALE FACTOR
;************************************************************************
;*                          INITIALIZATION                              *
;************************************************************************
            .text                       ;ASSEMBLE INTO TEXT SECTION
BEGIN       LDP     STACKS              ;INIT DATA PAGE
            LDI     @STACKS, SP         ;SP-> 809F00H
            LDI     -249, R0            ;0 WAIT STATE DATA
            LDI     @SRAM_CNTL, AR0     ;PRIMARY BUS CNTL REG => AR0
            AND     *AR0, R0            ;SET DESIRED BITS
            STI     R0, *AR0            ;REPLACE CONTENTS PRIMARY BUS CNTL REG
;************************************************************************
;*                      CREATE ADJUSTED COEFFS                          *
;*   AR0 = SCALE VALUES ADDRESS         AR3 = UPR 1/3 FLTR COEFF ADDR   *
;*   AR1 = LOW 1/3 FLTR COEFF ADDR      AR4 = ADJUSTED COEFF ADDR (10 SETS) *
;*   AR2 = MID 1/3 FLTR COEFF ADDR      R7  = LOOP COUNTER 1 TO 10(10 FILTERS) *
;*   REGISTERS USED FOR CALCULATIONS: R0, R1, R2, R3, R4, R5, R6        *
;************************************************************************
            LDI     SCALE_1, AR0        ;1ST SCALE VALUE ADDR -> AR0
            LDI     @FLTC_ADR, AR4      ;ADJUSTED FLTR COEFFS STARTING ADDR -> AR4
            LDI     10, R7              ;SET LOOP COUNTER TO 10 (10 -> R7)
LOOP        LDI     LFILT_40, AR1       ;LOW 1/3 FLTR COEFFS ADDR -> AR1
            LDI     MFILT_40, AR2       ;MID 1/3 FLTR COEFFS ADDR -> AR2
            LDI     UFILT_40, AR3       ;UPR 1/3 FLTR COEFFS ADDR -> AR3
            LDF     *AR0++(1), R0       ;LOW 1/3 SCALE VALUE -> R0
            LDF     *AR0++(1), R1       ;MID 1/3 SCALE VALUE -> R1
            LDF     *AR0++(1), R2       ;UPR 1/3 SCALE VALUE -> R2
            LDI     F_LENGTH-1, RC      ;LENGTH-1 -> RC
            RPTB    RPTB1_E             ;REPEAT BLOCK ("LENGTH" TIMES)
              LDF   0.0, R6             ;CLEAR R6
              LDF   *AR1++(1), R3       ;LOW 1/3 FLTR COEFF -> R3
              MPYF  R0, R3              ;COEFF * SCALE VALUE -> R3
              LDF   *AR2++(1), R4       ;MID 1/3 FLTR COEFF -> R4
              MPYF  R1, R4              ;COEFF * SCALE VALUE -> R4
              LDF   *AR3++(1), R5       ;UPR 1/3 FLTR COEFF -> R5
              MPYF  R2, R5              ;COEFF * SCALE VALUE -> R5
              ADDF3 R3, R4, R6          ;ADD R3 + R4 -> R6
              ADDF  R5, R6              ;ADD R5 + R6 -> R6
RPTB1_E       STF   R6, *AR4++(1)       ;STORE RESULTS
            SUBI    1, R7               ;DECREMENT LOOP COUNTER (R7-1)
            BNZ     LOOP                ;BRANCH ON NOT ZERO TO LOOP

                                        ;MOVE INTRPLTR COEFFS FROM
                                        ;SRAM TO INTERNAL RAM
            LDI     INTR_01, AR0        ;AR0=ADDR OF INTR COEFF IN SRAM
            LDI     @INTC_ADR, AR1      ;AR1=ADDR OF INTR COEFF IN RAM
            LDF     *AR0, R0
            RPTS    I_LENGTH+I_LENGTH-2
            LDF     *++AR0, R0
||          STF     R0, *AR1++
            STF     R0, *AR1
```

FIGURE F.1. Multirate filter program (MR10MCR.ASM) *(Continued)*

```
;******************************************************************
;*                CLEAR INPUTS FOR ALL FILTERS AND INTRPLTRS      *
;*    AR0 = FILTER & INTRPLTR INPUT ADDRESS          R0 = 0       *
;*    NOTE: CLEAR ALL INPUT DATA LOCATIONS FOR EACH FILTER/INTRPLTR *
;*          (10 FLTRS x 64 LOCATIONS) + (9 INTPLTRS x 16 LOCATIONS) *
;******************************************************************
            LDF     0.0, R0                 ;CLEAR R0
            LDI     @FLTI_ADR, AR0          ;ADDR OF FLTRS INPUT DATA -> AR0
            RPTS    10*64+9*16              ;REPEAT CNTR = SIZE OF ALL INPUT BUFFERS
            STF     R0, *AR0++(1)           ;STORE R0 INTO CURRENT ADDRESS
;******************************************************************
;*    STORE STARTING DATA ADDRS INTO POINTER ADDR FOR ALL FLTRS/INTPLTRS *
;*    AR0 = FILTER AND INTRPLTR INPUT DATA AND COEFFS ADDRESSES   *
;*    AR1 = FILTER AND INTRPLTR POINTER STORAGE ADDRESSES         *
;******************************************************************
            LDI     @FC_PNTR, AR1           ;PNTR STORAGE ADDRESS
            LDI     @FLTC_ADR, AR0          ;FILTER COEFF ADDRESS
            LDI     9, RC
            RPTB    RPTB2_E
            STI     AR0, *AR1++
RPTB2_E     ADDI    41, AR0
            LDI     @FLTI_ADR, AR0          ;FILTER INPUT ADDRESS
            LDI     9, RC
            RPTB    RPTB3_E
            STI     AR0, *AR1++
RPTB3_E     ADDI    64, AR0
            LDI     @INTC_ADR, AR0          ;INTRPLTR COEFF ADDRESS
            STI     AR0, *AR1++
            LDI     @INTI_ADR, AR0          ;INTRPLTR INPUT ADDRESS
            LDI     8, RC
            RPTB    RPTB3
            STI     AR0, *AR1++
RPTB3       ADDI    16, AR0
;******************************************************************
;*                   MAIN PROCESSING INITIALIZATION               *
;******************************************************************
SET_UP      LDI     @SEED, R0               ;NOISE SEED VALUE -> R0
            LDI     0, R6                   ;CLEAR BUFFER
            CALL    AICSET_I                ;INIT AIC
                                            ;NOTE: 1st time thru output buffer => 0
;******************************************************************
;*                        MAIN PROCESSING                         *
;*    AR0 = D/A OUTPUT ADDRESS          R0 = SEED FOR RND NOISE GENERATOR  *
;*    AR1 = FLTR & INTRP INPUT DATA PNTR R1 = WORKING REG (USED IN FLTR SUB) *
;*    AR2 = FLTR & INTRP COEFF PNTR     R2 = WORKING REG (USED IN FLTR SUB) *
;*    AR3 = STORAGE PNTR                R3 = OUTPUT SMPL FROM FILT/INTRPLTR *
;*    AR4 = USED FOR REPEAT LOOPS       R4 = NOT USED                       *
;*    AR5 = WORKING BUFFER POINTER      R5 = WORKING REG                    *
;*    AR6 = WORKING BUFFER POINTER      R6 = BUFFER STATUS REG              *
;*    AR7 = OUTPUT BUFFER POINTER       R7 = OUTPUT SAMPLE                  *
;*                                      IR0 = OUTPUT CNTR (0-512)           *
;*                                      IR1 = INDICATES LAST BAND# COMPLETED *
;******************************************************************
```

FIGURE F.1. *(Continued)*

```
PROCESS    LDI      0, IR0                ;****BEGIN PROCESSING****
           LDI      0, IR1                ;SET REGS TO DEFAULT VALUES
           LDI      1, AR4
           LDI      0, RC
           RST_WBFR
           LDF      0.0, R5
           STF      R5, *AR5
           LDI      1, R5
           AND      R6, R5
           BNZ      SETAR7_2
SETAR7_1   LDI      @B1_ADR, AR7
           BR       PROCLOOP
SETAR7_2   LDI      @B2_ADR, AR7

PROCLOOP   RST_WBFR                       ;BEGIN PROCESSING LOOP
           LDI      @FI_PNTR, AR3
           LDI      *+AR3(IR1), AR1
           LDI      F_LENGTH, BK
           LDI      1, R4
           SUBI3    R4, AR4, RC
F_REPEAT   RPTB     RPTB_E1
           LDI      @FC_PNTR, AR3
           LDI      *+AR3(IR1), AR2
           NOISE                          ;RETURNS NOISE SAMPLE IN R2
           FLOAT    R2, R1
           STF      R1, *AR1++%           ;STORE NOISE SAMPLE IN FILTER INPUT
           CALL     FILTER                ;RETURNS FILTERED DATA IN R3
           ADDF     *AR5, R3              ;SUM PREVIOUS BAND'S SAMPLE
RPTB_E1    STF      R3, *AR5++            ;STORE RESULTS
           LDI      @FI_PNTR, AR3
           STI      AR1, *+AR3(IR1)
           RST_WBFR
           CMPI     9, IR1                ;IF THIS IS BAND 10 THEN BRANCH TO EXIT
           BZ       EXIT                  ;     (INTERPOLATION NOT NEEDED)
           LDI      @II_PNTR, AR3
           LDI      *+AR3(IR1), AR1
           LDI      I_LENGTH, BK
           LDI      1, R4
           SUBI3    R4, AR4, RC
I_REPEAT   RPTB     RPTB_E2
           LDI      @IC_PNTR, AR3
           LDI      *AR3, AR2
           LDF      *AR5++, R3
           STF      R3, *AR1++%
           CALL     INTRPLTR
           STF      R3, *AR6++
           CALL     INTRPLTR
RPTB_E2    STF      R3, *AR6++
           LDI      @II_PNTR, AR3
           BD       PROCLOOP
           STI      AR1, *+AR3(IR1)
           ADDI     1, IR1
           LSH      1, AR4
```

FIGURE F.1. Multirate filter program (MR10MCR.ASM) *(Continued)*

```
EXIT        ADDI    1, IR1
            LDI     512, R5
                                        ;finished filling up next output buffer
                                        ;wait for current output buffer to empty
WAIT        IDLE
            CMPI    R5, IR0
            BNZ     WAIT
            ADDI    1, R6
            B       PROCESS
;***************************************************************************
;*                      SUBROUTINES DEFINED BELOW                          *
;***************************************************************************
FILTER      PUSH    ST
            PUSH    RC
            PUSH    RS
            PUSH    RE
            LDF     0.0,R1
            LDF     0.0,R3
            LDI     F_LENGTH-1, RC
            RPTB    RPT_FLT
RPT_FLT     MPYF3   *AR1++%,*AR2++,R1
    ||      ADDF3   R1,R3,R3
            ADDF3   R1,R3,R3
            POP     RE
            POP     RS
            POP     RC
            POP     ST
            OR      2000h, ST
            RETS


INTRPLTR    PUSH    ST
            PUSH    RC
            PUSH    RS
            PUSH    RE
            LDF     0.0, R1
            LDF     0.0, R3
            LDI     I_LENGTH-1, RC
            RPTB    RPTB_I?
RPTB_I?     MPYF3   *AR1++%, *AR2++, R1
    ||      ADDF3   R1, R3, R3
            ADDF3   R1, R3, R3
            POP     RE
            POP     RS
            POP     RC
            POP     ST
            OR      2000h, ST
            RETS
;***************************************************************************
;*      ISR             INTERRUPT SERVICE ROUTINE - OUTPUTS SAMPLE         *
;***************************************************************************
```

FIGURE F.1. *(Continued)*

```
ISR     PUSH    ST                      ;PUSH ST REG TO STACK
        PUSH    IE                      ;PUSH IE REG TO STACK
        PUSH    R4                      ;PUSH R4 REG TO STACK
        LDF     *AR7++, R4              ;GET SAMPLE FROM OUTPUT BUFFER
        MPYF    @SCALER,R4              ;SCALE OUTPUT SAMPLE
        FIX     R4, R7                  ;R7=INTEGER(R4)
        PUSH    R6                      ;SAVE R6
        CALL    AICIO_I                 ;CALL AIC I/O ROUTINE
        POP     R6                      ;RESTORE R6
        ADDI    1, IR0                  ;INCREMENT OUTPUT COUNTER
        POP     R4                      ;POP LAST R4 REG FROM STACK
        POP     IE                      ;POP LAST IE REG FROM STACK
        POP     ST                      ;POP LAST ST REG FROM STACK
        OR      2000h, ST               ;MAKE SURE GIE BIT REMAINS SET
        RETI                            ;RETURN FROM INTERRUPT
        .end
```

FIGURE F.1. Multirate filter program (MR10MCR.ASM) *(Continued)*

```
/*VIDEO.C-PROGRAM FOR IMAGE PROCESSING (VIDEO LINE RATE) ANALYSIS*/
#include <c:\c30hll\math.h>
#include <c:\c30hll\stdlib.h>
#include "c30_1.h"
#include "projcmd.h"
#include "serial1.c"
#define buffer 512
#define STAGES 3            /* number of 2nd-order stages in filter */
int buffer1 = buffer/2;
int dma_int2 = 0x00040000;
volatile int *bus = (volatile int *) 0x808060;
volatile int *dma = (volatile int *) 0x808000;
volatile int *host = (volatile int *) 0x804000;
int index = 0, loop, ramp = 0, arrow = 0;
int *input, *intermediate, *average;

float A0[STAGES][3] = { /* numerator coefficients */
                { 2.186845E-02, 2.186845E-02, 0.000000E+00 },
                { 2.186845E-02, 4.373690E-02, 2.186845E-02 },
                { 2.186845E-02, 4.373690E-02, 2.186845E-02 }};
float B0[STAGES][2] = { /* denominator coefficients */
                { -8.047146E-01, 0.000000E+00 },
                { -1.665480E+00, 7.049447E-01 },
                { -1.832568E+00, 8.759922E-01 }};

float A1[STAGES][3] = { /* numerator coefficients */
                { 6.308639E-02, 6.308639E-02, 0.000000E+00 },
                { 6.308639E-02, 1.261727E-01, 6.308639E-02 },
                { 6.308639E-02, 1.261727E-01, 6.308639E-02 }};
float B1[STAGES][2] = { /* denominator coefficients */
                { -6.401314E-01, 0.000000E+00 },
                { -1.356731E+00, 4.939717E-01 },
                { -1.608209E+00, 7.708878E-01 }};
```

FIGURE F.2. Program for image processing algorithms (VIDEO.C)

```c
float A2[STAGES][3] = { /* numerator coefficients */
                  { 2.989367E-01, 2.989367E-01, 0.000000E+00 },
                  { 2.989367E-01, 5.978735E-01, 2.989367E-01 },
                  { 2.989367E-01, 5.978735E-01, 2.989367E-01 }};
float B2[STAGES][2] = { /* denominator coefficients */ .
                  { -1.256041E-01, 0.000000E+00 },
                  { -2.772673E-01, 1.211474E-01 },
                  { -3.806422E-01, 5.391505E-01 }};
float A[STAGES][3];
float B[STAGES][2];
float DLY[STAGES][2] = {0};   /* delay storage elements */
/* 6th order IIR filter */
void IIR(int *IO_in, int *IO_out, int n, int len)
{
  int i, loop = 0, loop1, int_val;
  float dly, yn, input;
  while (loop < len)
  {
    ++loop;
    input = IO_in[loop] & 0xFF;
    for (i = 0; i < n; i++)
    {
      dly = input - B[i][0] * DLY[i][0] - B[i][1] * DLY[i][1];
      yn = A[i][2] * DLY[i][1] + A[i][1] * DLY[i][0] + A[i][0] * dly;
      DLY[i][1] = DLY[i][0];
      DLY[i][0] = dly;
      input = yn;
    }
    if (yn >= 0) int_val = (int)yn;
    else int_val = 0;                       /* clip any negative values */
    IO_out[loop] = int_val;
  }
  DLY[0][0] = 0;
  DLY[0][1] = 0;
  DLY[1][0] = 0;
  DLY[1][1] = 0;
  DLY[2][0] = 0;
  DLY[2][1] = 0;
}
/* function to perform averaging */
void AVG(int *IO_in, int *IO_out, int loop, int rate)
{
  int index, temp;
  if (loop == 0)
    for (index = 0; index < buffer; index++)
      IO_out[index] = IO_in[index];
  else
    for (index = 0; index < buffer; index++)
      IO_out[index] += IO_in[index];
  if((rate-loop) == 1)
    for (index = 0; index < buffer; index++)
      IO_out[index] = IO_out[index] / rate;
}

main()
```

FIGURE F.2. *(Continued)*

```
{
  int i;
  int *temp;
  int avg = 0;
  int lpa = 0;
  int lpb = 0;
  int lpc = 0;
  int edge1 = 0;
  int edge2 = 0;
  int RATE = 10;
  int line = 128;
  int sel_rate = 1;
  int status;
  int tmp;
  input = (int *) calloc(buffer, sizeof(int));
  average = (int *) calloc(buffer, sizeof(int));
  intermediate = (int *) calloc(buffer1, sizeof(int));
  init_evm();
  asm(" LDI    400h,IE");
  asm(" OR     2000h,ST");
  SERIAL1SET();
  PBASE[0x58] = line;                    /* select line number to be sampled */
  for (;;)
  {
    for(loop = 0; loop < RATE; loop++) /* input data from SP1 */
    {
      for (index = 0; index < buffer1; index++)
      {
        RWAIT();
        input[index] = PBASE[0x5C];
      }
      for (index = buffer1-1; index >= 0; index--)  /* amplify and shift data*/
      {
        tmp = index*2;
        input[tmp+1] = ((input[index] & 0xFF) - 64) * 1.25;
        input[tmp] = (((input[index] >> 8) & 0xFF) - 64) * 1.25;
      }
      if (avg)                          /* call averaging function */
        AVG((int *)input, (int *)average, loop, RATE);
      else if (lpa)                     /* select 500 KHz low pass filter */
      {
        for (index = 0; index < 3; index++)
        {
          A[index][0] = A0[index][0];
          B[index][0] = B0[index][0];
        }
        for (index = 0; index < 3; index++)
        {
          A[index][1] = A0[index][1];
          B[index][1] = B0[index][1];
        }
        for (index = 0; index < 3; index++)
        {
          A[index][2] = A0[index][2];
        }
        IIR((int *)input, (int *)average, STAGES, buffer);
```

FIGURE F.2. Program for image processing algorithms (VIDEO.C) *(Continued)*

```
      }
    else if (lpb)                     /* select 1 MHz low pass filter */
    {
      for (index = 0; index < 3; index++)
      {
        A[index][0] = A1[index][0];
        B[index][0] = B1[index][0];
      }
      for (index = 0; index < 3; index++)
      {
        A[index][1] = A1[index][1];
        B[index][1] = B1[index][1];
      }
      for (index = 0; index < 3; index++)
      {
        A[index][2] = A1[index][2];
      }
      IIR((int *)input, (int *)average, STAGES, buffer);
    }
    else if (lpc)                     /* select 3 MHz low pass filter */
    {
      for (index = 0; index < 3; index++)
      {
        A[index][0] = A2[index][0];
        B[index][0] = B2[index][0];
      }
      for (index = 0; index < 3; index++)
      {
        A[index][1] = A2[index][1];
        B[index][1] = B2[index][1];
      }
      for (index = 0; index < 3; index++)
      {
        A[index][2] = A2[index][2];
      }
      IIR((int *)input, (int *)average, STAGES, buffer);
    }
    else if (edge1)                   /* first edge enhancement algorithm */
      for (index = 1; index < buffer-1; index++)
      {
        average[index] = 3*input[index]-input[index-1]-input[index+1];
        if (average[index] > 255) average[index] = 255;
        if (average[index] < 0) average[index] = 0;
      }
    else if (edge2)                   /* second edge enhancement algorithm */
      for (index = 1; index < buffer-1; index++)
      {
        average[index] = (4*input[index]-input[index-1]-input[index+1])/2;
        if (average[index] > 255) average[index] = 255;
        if (average[index] < 0) average[index] = 0;
      }
}
/* clip data to 8 bits */
for (index = 0; index < buffer; index++)
```

FIGURE F.2. *(Continued)*

```c
{
  if (input[index] < 0) input[index] = 0;
  if (input[index] > 255) input[index] = 255;
  if (average[index] < 0) average[index] = 0;
  if (average[index] > 255) average[index] = 255;
}
/* prepare to send data to PC HOST with 16 bit transfer */
for (index = 0; index < buffer1; index++)
{
  tmp = index*2;
  input[index] = input[tmp] << 8;
  average[index] = average[tmp] << 8;
  input[index] |= input[tmp+1] & 0xFF;
  average[index] |= average[tmp+1] & 0xFF;
}
PBASE[0x58] = line;    /* select next line number to be sampled */
status = 0;            /* status of EVM to be sent to PC HOST*/
if (lpa) status += 1;  /* 500 KHz low pass filter status */
if (lpb) status += 2;  /* 1 MHz low pass filter status */
if (lpc) status += 4;  /* 3 MHz low pass filter status */
if (avg) status += 8;  /* averaging status */
if (edge1) status += 16;  /* first edge enhancement status */
if (edge2) status += 32;  /* second edge ehnancement status */
input[buffer1-2] = status << 8;
input[buffer1-2] |= line & 0xFF;  /* number of the line sampled */
input[buffer1-1] = RATE << 8;     /* current rate of the display */
input[buffer1-1] |= sel_rate & 0xFF;  /* # lines to skip for line select*/
average[buffer1-2] = input[buffer1-2];
average[buffer1-1] = input[buffer1-1];
while(dma[TRANSFER]);
if (!avg && !lpa && !lpb && !lpc && !edge1 && !edge2)
  temp = input;
else temp = average;
intermediate = temp;
i = *host & 0xFF;
if (i && i < SEND)
{
  *host = i;
  switch(i)
  {
    case RESET: {avg = 0; lpa = 0; RATE = 10; arrow = 0; lpb = 0;
                 lpc = 0; sel_rate = 1; edge1 = 0; edge2 = 0; break;}
    case INCREASE: if (RATE < 120) {RATE = RATE << 1; break;}
                   else break;
    case DECREASE: if (RATE > 1) {RATE = RATE >> 1; break;}
                   else break;
    case LPA: if (lpa) {lpa=0;avg=0;lpb=0;lpc=0;edge1=0;edge2=0;break;}
              else {lpa=1;avg=0;lpb=0;lpc=0;edge1=0;edge2=0;break;}
    case LPB: if (lpb) {lpb=0;avg=0;lpa=0;lpc=0;edge1=0;edge2=0;break;}
              else {lpb=1;avg=0;lpa=0;lpc=0;edge1=0;edge2=0;break;}
    case LPC: if (lpc) {lpc=0;avg=0;lpa=0;lpb=0;edge1=0;edge2=0;break;}
              else {lpc=1;avg=0;lpa=0;lpb=0;edge1=0;edge2=0;break;}
    case AVER: if (avg) {avg=0;lpa=0;lpb=0;lpc=0;edge1=0;edge2=0;break;}
               else {avg=1;lpa=0;lpb=0;lpc=0;edge1=0;edge2=0;break;}
    case EDGE1: if (edge1) {lpc=0;avg=0;lpa=0;lpb=0;edge1=0;break;}
                else {edge2=0;edge1=1;lpc=0;avg=0;lpa=0;lpb=0;break;}
    case EDGE2: if (edge2) {lpc=0;avg=0;lpa=0;lpb=0;edge2=0;break;}
                else {edge1=0;edge2=1;lpc=0;avg=0;lpa=0;lpb=0;break;}
    case L_MINUS: if (line > (0 + sel_rate)) {line -= sel_rate; break;}
                  else break;
```

FIGURE F.2. Program for image processing algorithms (VIDEO.C) *(Continued)*

```
          case L_PLUS: if (line < (255 - sel_rate)) {line += sel_rate; break;}
                      else break;
          case SEL_RATE: if (sel_rate < 16) {sel_rate = sel_rate << 1; break;}
                         else {sel_rate = 1; break;}
        }
        if (RATE == 4) RATE = 5;
        if (RATE == 8) RATE = 10;
        if (RATE == 15) RATE = 20;
        if (RATE == 40) RATE = 30;
        while (*host & 0xFF);
        *host = NONE;
      }
      else if (i == SEND)
      {
        for (loop = 0;loop < buffer;loop++) /*discard field for synchronization*/
        {
          RWAIT();
          temp[0] = PBASE[0x5C];
        }
        recording((int *)intermediate, SEND); /* transfer results to PC HOST */
      }
    }
  }
}
void init_evm()
{
  bus[EXPANSION] = 0x0;
  bus[PRIMARY] = 0x1000;
  *host = NONE;
  asm(" OR        800h,ST");
}

void c_int11()
{
  while(*host & 0xFF);
  *host = NONE;
  dma[GLOBAL] = 0;
  asm(" AND       0FFFFh,IE");
}

void recording(int *source, int rec_cmd)
{
  asm(" AND       0FFF8h,IF");
  asm(" OR        @_dma_int2,IE");
  dma[SOURCE] = (int) source;
  dma[DEST] = (int) host;
  dma[TRANSFER] = buffer1;
  dma[GLOBAL] = 0x0E13;
  *host = rec_cmd;
}
void c_int05()
{
  for (;;);
}

void c_int99()
{
  for (;;);
}
```

FIGURE F.2. (*Continued*)

```c
/*VIDEOPC.C-PC HOST PROGRAM FOR IMAGE PROCESSING PROJECT*/
#include <stdlib.h>
#include <stdio.h>
#include <conio.h>
#include <graphics.h>
#include "pc.c"
#include "pc_1.h"
#include "projcmd.h"
void send_cmd();
void init_graphic();
void transfer();
int pause = 0;
int update = 1;

void main()
{
  init_evm();
  init_graphic();
  buffer = 256;
  send_cmd();
  for(;;)
  {
    if (!pause) transfer();
    if (kbhit()) update = 1;
    if (kbhit()) send_cmd();
  }
}

void send_cmd()
{
  int command;
  command = get_command();
  switch(command)
  {
    case RESET:
    case LPA:
    case LPB:
    case LPC:
    case SEL_RATE:
    case INCREASE:
    case DECREASE:
    case AVER:
    case EDGE1:
    case EDGE2:
    case L_PLUS:
    case L_MINUS:
        break;
    case PAUSE: if (pause) {pause = 0; break;}
                else {pause = 1; break;}
    case QUIT:
        closegraph();
        exit(EXIT_SUCCESS);
        break;
    default:
        command = NONE;
        break;
  }
  if (command) send_command(command);
```

FIGURE F.3. PC host program for image processing (VIDEOPC.C)

```c
}

void init_graphic()
{
  int m, loop;
  int X0 = 65, Y0 = 50;
  int X1 = X0 + 508, Y1 = Y0 + 265;
  int gdriver = DETECT, gmode, errorcode;
  initgraph(&gdriver, &gmode, "");
  errorcode = graphresult();
  if (errorcode != grOk)
  {
     printf("Graphics error: %s\n", grapherrormsg(errorcode));
     printf("Press any key to halt:");
     getch();
     exit(1);
  }
  setbkcolor(BLUE);
  rectangle(X0-1, Y0-1, X1+1, Y1+1);
  settextjustify (CENTER_TEXT, CENTER_TEXT);
  for (loop = 0; loop < 11; loop++)
  {
    int vert;
    char string[4] = {0};
    vert = Y0+5+loop*25;
    itoa(100-(loop*10), string, 10);
    line (X0-5, vert, X0, vert);
    outtextxy (X0-20, vert, string);
  }
  settextstyle(DEFAULT_FONT, HORIZ_DIR, 2);
  outtextxy (getmaxx()/2, Y0-30, "VIDEO LINE RATE PROCESSING");
  settextstyle(DEFAULT_FONT, VERT_DIR, 1);
  outtextxy (X0-50, Y0+128, "IRE units");
  settextstyle(DEFAULT_FONT, HORIZ_DIR, 0);
  settextjustify(LEFT_TEXT, CENTER_TEXT);
  outtextxy (94, 360, " F1   RESET    ");
  outtextxy (94, 375, " F2   500K LPF ");
  outtextxy (94, 390, " F3   1Meg LPF ");
  outtextxy (94, 405, " F4   3Meg LPF ");
  outtextxy (94, 420, " F5   AVERAGING");
  outtextxy (94, 442, "      LINE SEL ");
  outtextxy (360,360, " F6   # LINES  ");
  outtextxy (360,375, " F7   EDGE1    ");
  outtextxy (360,390, " F8   EDGE2    ");
  outtextxy (360,405, " F9   PAUSE    ");
  outtextxy (360,420, "F10   QUIT     ");
  outtextxy (360,442, "      RATE     ");
  line (108,432,111,432);              // start of up arrow
  line (107,433,112,433);
  line (106,434,113,434);
  line (109,431,109,439);
  line (110,431,110,439);
  line (109,446,109,454);              // start of down arrow
  line (110,446,110,454);
  line (106,451,113,451);
  line (107,452,112,452);
```

FIGURE F.3. *(Continued)*

```c
    line (108,453,111,453);
    line (372,435,380,435);                         // start of left arrow
    line (372,436,380,436);
    line (375,439,375,432);
    line (374,438,374,433);
    line (373,437,373,434);
    line (372,450,380,450);                         // start of right arrow
    line (372,451,380,451);
    line (377,454,377,447);
    line (378,453,378,448);
    line (379,452,379,449);
    rectangle (210,350,285,460);
    rectangle (470,350,545,460);
    setfillstyle(SOLID_FILL,LIGHTGRAY);
    setviewport(X0, Y0, X1, Y1, 1);
    outtextxy(150,128, "PRESS SPACE BAR TO CONTINUE");
}

void transfer()
{
  int i, value, index, status, line, sel_rate, rate;
  char string[4] = {0};
  int temp[720];
  while(READ_CMD);
  WRITE_CMD(128);
  while(READ_CMD != 128);
  CLR_READ_ACK;
  READ_DATA;
  index = 0;
  for(i = 0; i < buffer; i++)
  {
    do
    {
      UPDATE_STATUS0;
    }
    while(!IS_READ_ACK);
    CLR_READ_ACK;
    value = (int)READ_DATA;
    temp[index++] = (value >> 8) & 255;
    temp[index++] = value & 255;
  }
  WRITE_CMD(NONE);
  bar(0,0,508,265);
  setcolor(LIGHTGRAY);
  lineto (0, (256- temp[0]));
  setcolor(WHITE);
  for (i = 0; i < 508; i++)
    lineto (i, (256 - temp[i]));
  if (update)
```

FIGURE F.3. PC host program for image processing (VIDEOPC.C) *(Continued)*

```
{
  update = 0;
  status = temp[508];
  line = temp[509];
  rate = temp[510];
  sel_rate = temp[511];
  setviewport(211,351,284,459,1);
  bar(0,0,73,108);
  if (status & 1) outtextxy(25, 24,"ON");
  else outtextxy(25,24, "OFF");
  status = status >> 1;
  if (status & 1) outtextxy(25, 39,"ON");
  else outtextxy(25,39, "OFF");
  status = status >> 1;
  if (status & 1) outtextxy(25, 54,"ON");
  else outtextxy(25,54, "OFF");
  status = status >> 1;
  if (status & 1) outtextxy(25, 69,"ON");
  else outtextxy(25,69, "OFF");
  itoa(line, string, 10);
  outtextxy (25,91, string);
  setviewport(471,351,544,459,1);
  bar (0,0,73,108);
  itoa(sel_rate, string, 10);
  outtextxy (33, 9, string);
  status = status >> 1;
  if (status & 1) outtextxy(25, 24,"ON");
  else outtextxy(25, 24, "OFF");
  status = status >> 1;
  if (status & 1) outtextxy(25, 39,"ON");
  else outtextxy(25, 39, "OFF");
  rate = 60/rate;
  if (rate > 1)
  {
    itoa(rate, string, 10);
    outtextxy (25, 92, "1/  ");
    outtextxy (45, 92, string);
  }
  else
  {
    rate = temp[510];
    rate = rate/60;
    if (!rate) rate = 1;
    itoa(rate, string, 10);
    outtextxy (33, 92, string);
  }
  setviewport(65,50,573,315,1);
 }
}
```

FIGURE F.3. *(Continued)*

```
/*PROJCMD.H-INTERACTIVE PC COMMANDS DEFINITION */
/*----------------------------------------------*/
/*     PC EXTENDED KEYBOARD VARIABLES           */
/*----------------------------------------------*/
#define F1              59
#define F2              60
#define F3              61
#define F4              62
#define F5              63
#define F6              64
#define F7              65
#define F8              66
#define F9              67
#define F10             68
#define L_ARROW         75
#define R_ARROW         77
#define UP              72
#define DOWN            80
/*----------------------------------------------*/
/*     COMMANDS FROM THE PC TO THE EVM          */
/*----------------------------------------------*/
#define RESET           F1
#define LPA             F2
#define LPB             F3
#define LPC             F4
#define AVER            F5
#define SEL_RATE        F6
#define EDGE1           F7
#define EDGE2           F8
#define PAUSE           F9
#define QUIT            F10
#define L_MINUS         UP
#define L_PLUS          DOWN
#define SEND            128
#define INCREASE        R_ARROW
#define DECREASE        L_ARROW
```

FIGURE F.4. Interactive PC commands definition program for image processing (PROJCMD.H)

F.3 SUPPORTING PROGRAMS FOR VIDEO LINE RATE ANALYSIS

Figure F.2 shows a listing of the main program VIDEO.C for the image processing (video line rate analysis) algorithms. This program includes the functions for IIR filtering, averaging, and edge enhancement. Figure F.3 is a listing of the program VIDEOPC.C, which contains the PC host communication functions. The program PROJCMD.H, in Figure F.4, defines the PC commands. Additional support programs are on the accompanying disk.

Index

Adaptive filters:
  adaptation using C code, 242–245
  algorithms for restoring signal properties, 242
  for noise cancellation using C code, 244–250, 279–287
    adaptation with shifted signal, 282–286
    difference equations option, 279
    execution time, 249–250
    shifting the input signal, 282–286
    using arcsine and arccosine, 279
    using table lookup, 279
  for noise cancellation using TMS320C30 code, 260–262
    real-time, 262–267
  linear combiner, 237, 239
  mean squared error function, 239
  notch filter using TMS320C30 code, 250–256, 274–279
    50-coefficient adaptive filter, 276, 277
    60-Hz artifact, 275
    baseline artifacts, 275
    difference equations utility option, 250
    electrocardiographic (ECG) monitoring, 275
    frequency-selective 60-Hz notch filter, 276, 277
    two-weight adaptive notch filter, 278
  performance measure, 239–242
  predictor using TMS320C30 code, 254, 257–260
  rate of convergence, 237
  recursive least squares (RLS) algorithm, 242
  structures, 238, 239
    for noise cancellation, 238, 239, 244
    for system identification, 238
Addition of five values using TMS320C30 code, 30–33
Addition with C and C-called function using TMS320C30 code, 45, 46
Addressing modes, 22, 23
  bit-reversed, 23
    reversed-carry propagation, 23
  circular, 22, 23
  direct, 22
  indirect, 22
    bit-reversed, 22
    circular, 22
    using displacement and indexing, 22
  register, 22
  short- and long-immediate, 23
ALIGN command, 38
Analog interface board (AIB), 88

Index

Analog interface chip (AIC):
  communication program `AICCOM.C`, 68
  communication program `AICCOM01.C`, 75
  communication program `AICCOMA.ASM`, 78–80
  configuration data, 55, 75
  control functions, 52
  data receive register, 52
  data transmit register, 52
  for obtaining different sampling rate, 148, 149
  functional block diagram, 52
  input and output filters, 52
  internal timing configuration, 52
  master clock frequency, 55
  primary input (IN), 71, 81
  secondary communication, 52, 55
  serial port 0 transmit interrupt, 71
  serial port registers, 52, 55
  switched capacitor filter (SCF), 55, 69
  through serial port 1, 73–75
  TLC32046 AIC, 52
  transmit and receive registers TA, RA, 54
  transmit and receive registers TB, RB, 54
Architecture and memory organization, 18–21
Assembler format, 9, 38
Assembler directive, 9, 38
  `.ALIGN` for circular buffering, 38
  `.USECT`, 38
Assembling to create an object file, 4

Bilinear transformation (BLT), 166–174
  analog and digital frequencies, 166–168
  calculation of $H(z)$ from $H(s)$, 170
  design procedures, 168–174
  for transforming analog filter into discrete, 166
  frequency warping, 167
  utility programs, 170–174
bss-block started by symbol, 5
Burr–Brown analog evaluation fixture, 87, 88, 373–379
  analog-to-digital converter DSP102, 87
  block diagram, 87, 88
Burr–Brown/EVM interface diagram, 373, 374
Burr–Brown/TMS320C30
  communication routines, 377–379
  conversion rate of 200 kHz, 87
  digital-to-analog converter DSP202, 87
  double I/O 16-bit A/D and D/A, 373
  single I/O 18-bit A/D and D/A, 373
  testing channels A and B, 376–379
  to test channel A, 374, 375
  using interrupt, 375, 376

C-identifiers, 45
Circular buffer and data transfer:
  using TMS320C30 code, 35–38
Circular buffering, 28
CLK from CPU status window, 32
Compiling:
  invoke the code generator, 3
  with optimization, 3, 6
  g option, 12
  k option, 1
  o option, 5
  q option, 5
  s option, 5
  z option, 6
Compiling, assembling, linking, 5
  executable file, 6
C-Source debugger, 12–15
  for debugging C code, 13

Data and floating point formats, 28–30
  extended precision, 28, 29
  floating-point, 28
  IEEE format, 28
  implied bit, 29
  short, 28
  single precision, 28, 29
  unsigned integer, 28
Debugger screen, 14
Decimation-in-frequency, 199–205
  8-point FFT, 203, 204
  8-point FFT flow graph, 203
  16-point FFT, 204–207
  16-point radix-4, 212
  flow graph for a 16-point radix-4, 214
Decimation-in-time, 205–209
  8-point FFT flow graph, 209

Dedicated registers, 46
Digital filtering background
    programming, 38–42
  using TMS320C30 code, 38
Digital signal processing applications:
  real-time with C and the TMS320C30:
    student projects, 269–327
Digital signal processors, 1
  applications, 2
  first-generation TMS32010, 1, 17
    instruction cycle time, 17
  second-generation TMS320C5, 1, 17
  third-generation TMS320C30, 2
Digital signal processing tools, 2, 353–372
  AIC module, 2
  analog interface board (AIB), 2
  assembler, linker, and C compiler, 2
  evaluation module (EVM), 2
  filter design packages, 2
  generation of a sinusoid, 353
  Hypersignal-Plus DSP Software, 353
  MATLAB, 353
  simulator, 2
  XDS1000 emulator, 2
Discrete signals, 96, 97
  shift invariance, 96
  superposition property, 96
DMA communication between PC host
    and TMS230C30, 87
Download and run executable file, 13

Edge enhancement, 306, 309, 358–360
  using C code, 358–360
EVMLOAD, 13
EMVTEST, 12
Extended development system, 88, 89,
    380–389
  16K × 32 SRAM interface diagram,
    382
  diagram of target module, 381
  interface diagram to 16K SRAM, 380
  interface target module to AIB, 380,
    385
  interface target module to AIC, 380,
    385
  interface target module to XDS1000,
    380, 385
  generation of square wave with
    XDS1000/AIB, 385–387

loop program with AIC on the AIB,
    386, 387
  real-time FFT with the XDS1000/AIB,
    387–389
  XDS1000-AIC interface diagram, 383
  XDS1000 emulator, 380–389

Fast Fourier Transform (FFT):
  bit reversal for unscrambling, 209–211
  decimation-in-frequency, 199–205
  decimation-in-time, 205–209
  decomposition or decimation process,
    201
  development of the FFT algorithm,
    198, 199
  digit reversal, 213, 214
  discrete Fourier transform, 197, 198
  8-point FFT flow graph, 203, 209
  8-point using C code, 222–227
  8-point using mixed C and TMS320C30
    code, 227–229
  FFT algorithm: radix-4, 211–214
  inverse, 215, 232
  program to generate twiddle constants,
    226
  for real-valued input, 231
  real-time 512-point using mixed C and
    TMS320C30 code, 229–232
  twiddle factors, 198
    for 16-point FFT radix-4, 214
    periodicity and symmetry, 198
  two-point FFT butterfly, 202
Fast Hartley transform (FHT), 215–222
  discrete Hartley transform, 215
  8-point, 219–220
  8-point FHT flow graph, 218
  16-point, 220–222
Filter design packages, 109, 110, 139, 140,
    145, 146
Finite impulse response (FIR) filters,
    90–152
  approximated transfer function, 103
  bandpass using C and C-called
    assembly function, 120–123
  bandpass using C code with the modulo
    operator, 123
  bandpass using TMS320C30 code,
    116–120
  bandstop, 145, 147–150
  convolution in time, 98

Index

Finite impulse response (*Continued*)
design of a filter, 98
differentiator, 145, 146
filter design packages, 109–110
frequency-selective filters, 105
impulse response coefficients, 98, 104
impulse response duration, 144
lattice structure, 99–103
  image polynomials, 101
  k-parameters, 101
  reflection coefficients, 99
  reversal recursion, 101
linear phase, 99
lowpass filter example, 106
lowpass using TMS320C30 code, 111–114
lowpass with two circular buffers, using TMS320C30 code, 114–116
memory organization, 110
multiband, 145, 146
nonrecursive, 99
Nyquist frequency, 103
Parks-McClellan approximation, 109
real-time using C code, 130
real-time using mixed C and TMS320C30, 130–132
real-time using noncommercial filter development package, 140–144
real-time using TMS320C30 code, 132–136
real-time using TMS320C30 code with macro, 135, 136
tapped delay, 99
transversal, 99
using Fourier series, 103–106
  Fourier series coefficients, 103
  normalized frequency variable, 103
with data move using C code, 123
with noise generation using TMS320C30 code, 137, 138
with sample rate control, 148–150
with samples shifted using C code, 125–130
without modulo, using C code, 123–125
Float instruction, 40
Four-channel multiplexer for fast data acquisition, 321–323
  8-bit flash analog-to-digital converter, 321
  circuit diagram, 321–322

PC screen display of FFT of signal, 321, 323
Frame pointer FP, 45, 46
Frequency shift using modulation:
  carrier frequency signal, 320
  implementation, 320, 321

Graphics driver file `EGAVGA.BGI`, 15

Harvard, architecture, 18
Hypersignal package:
  code generation, 359, 362–366
  difference equations utility option, 250
  FFT generation, 355–357
  file acquisition option, 354
  FIR bandpass filter, 360–366
    magnitude response, 362
  FIR filter construction, 360
  FIR filter design, 359–366
  FIR frequency response, 369
  frequency response of IIR lowpass filter, 372
  IIR filter design, 366–372
    code generation, 366–368
    frequency response, 370
  magnitude display option, 355
  running the FIR filter, 365, 366
  time and frequency domain utilities, 354–357
    plot of a sine sequence, 354–357
  waveform display option, 355
Hypersignal-Plus DSP Software Package, 109, 110, 354–372

IEEE format, 28
Introduction to C programming, 331–339
  arrays and loops, 334, 335
  compiler directive, 332
  file pointer, 338
  function called main ( ), 332
  interrupt function, 336
  modulo (remainder) operation, 333
  recursive IIR filter, 336–339
  variables and arithmetic operations, 332–334
  volatile variable, 335
Introduction to image processing:
  Video line rate analysis, 301–309
  interactive algorithms, 301
    digital filtering, 301

edge enhancement, 301
image averaging, 301
Infinite impulse response (IIR) filters:
  alarm generator, 191–195
  bandpass filter using C code, 182–184
    real-time, 184–186
  bandpass filter using TMS320C30 code, 184–189
    real-time, 189–191
  bandstop filter using BLT.BAS, 170–174
  bilinear transformation, 166–174
  Butterworth bandstop, 169
  Butterworth highpass, 168, 169
  cosine generation using TMS320C30 code, 181, 182
  digital oscillator, 155
  general input–output equation, 153
  lattice structures, 160–166
    all-pole, 161, 162
    k-parameters, 165
    ladder portion, 163, 164
    with poles and zeros, 163–166
  magnitude and phase responses programs, 171–174
    AMPLIT.CPP, 171–174
    MAGPHSE.BAS, 171–174
  marginally stable, 154
  sine generation using C code, 174–177
  sinusoid generation, 154
  stable and unstable system, 154
  structures, 155–166
    cascade, 158, 159
    direct form I, 155, 156
    direct form II, 156–161
    direct form II transpose, 157, 158
    lattice, 160–166
    parallel, 159, 160
Input/Output, 50
  analog interface board, 50
  analog interface chip, 52–57
Input and Output:
  signals, 51
  using C code, 43–45
Instruction set, 23–28
Instruction types:
  branch, 25
  circular buffering, 28
  input and output, 25
  load and store, 24

logical, 25
math, 24, 25
parallel, 27
repeat, 26
Interrupts, 56, 57
  enable register, 56, 58
  global interrupt enable bit, (GIE), 56
  handlers in C, 56
  rate, 58
  vectors, 57
Inverse fast Fourier transform, 215, 232

Lattice structures, 99–103, 160–166
Least mean square (LMS) algorithm, 237, 239–242
  sign-data, 241
  sign-error, 241
  sign-sign, 241
Linking:
  common object file format (COFF), 4
  convention, 5
  E BEGIN option, 32
  E option, 10
Loading an executable file into the simulator, 6
  COFF executable output file, 6
  execution modes, 8
  memory configuration file, 6
  port configuration file, 6
  trace execution, 8
Loop program:
  amplitude control, 15
  for either port 0 or 1, 75–77
  to test both AIC inputs, 81, 82
  to use serial port 1, 80, 81
  using C code, 10–13, 72, 73
  using TMS320C30 code, 77–80
  with amplitude control using C code, 83–85
  with the EVM and AIC, 10–13

Matrix multiplication:
  using C code, 3–8, 45, 46
  using TMS320C30 code, 8–10, 42, 43
  with C and C-called function, 46–48
Multiplication of two arrays using TMS320C30 code, 33–35
Multirate filter:
  1/3 octave filter, 292, 294, 296–300
  32-bit noise generator, 292, 294

Multirate filter (*Continued*)
  Burr–Brown analog evaluation fixture, 300
  decimation, 291
  functional block diagram, 291, 293
  implementation, 296–300
  interpolation, 291, 295
  programs for, 390–398
  sampling rate
    increase, 292
    reduction, 292
  single rate and multirate coefficients, 295
  with the XDS1000 emulator, 300

Neural network for signal recognition:
  back-error propagation, 324
  design and implementation, 324–327
  hidden layer, 324
  processing element, 325
  three-layer neural network, 324, 325
  weighted sum of input, 324

Output rate controlled by interrupt using TMS320C30 code, 57–59

Parametric equalizer, 269–274
  bass range and treble range, 270
  design considerations, 269, 270
  graphic equalizer, 270
  multiband filter, 269, 270
  PC host communication, 273
PID controller, 310–318
  antiwindup algorithm, 318
  control algorithm:
    proportional, integral, and derivative, 310
  control voltage, 314–318
  digital speed control system, 310, 312
    block diagram, 310
  driver circuit, 310, 312, 313
  effects of PID gain constants on motor speed, 314–318
  implementation, 310–318
  PI controller, 318
  PID constants, 312
  servomotor, 310
  tachgenerator, 310
Pipelining, 6

PC-host TMS320C30 communication, 82–87
  between PC and EVM, 82
  DMA data transfer, 82, 83
Pointers to memory addresses, 44
Programming example with the simulator, 30–48
Pseudorandom noise generator, 59–62
  using TMS320C30 code, 59–62

Recursive least squares (RLS) algorithm, 242
Registers:
  auxiliary, 19
  BK to specify circular buffer, 19
  data page, 19
  dedicated, 121
  extended precision, 19
  for indexing, 19
  frame pointer, 122
  interrupt enable and flag, 19
  I/O flag, 19
  program counter PC, 19
  repeat count RC, 19
  start and end, 19
  system stack pointer, 19
Register formats:
  direct memory access (DMA), 351, 352
  external interface control, 349
  interrupt enable (IE), 340, 347
  interrupt flag (IF), 340, 348
  I/O flag (IOF), 340, 348
  peripheral bus, 349, 350
  peripheral timers, 349, 350
  serial port peripheral addresses, 350, 351
  status (ST), 340, 346
Register formats and memory maps, 346–352
  reset and interrupt vector locations, 349
Reset interrupt vector, 68
Resetting and running the TMS320C30:
  `evmctrl go`, 86
  `evmctrl stop`, 86
  outport ( ) function, 85
  with PC-host, 85–87

Serial ports, 57
  memory maps registers, 57

Simulator screen, 6, 32
Sine generation:
  real-time using C code, 177–179
  using C code, 174–177
  using TMS320C30 code, 177–181
Sine generation with 4 points:
  using C code, 67, 68
    real-time, 68–71
  using TMS320C30 code, 62–67
Single-step, 6
Swept frequency response:
  design considerations and
    implementations, 287–291
  frequency response of bandpass filter
    as on PC screen, 291
Symbolic debugging, 12

Testing the hardware tools:
  alternative I/O, 10
  analog interface chip (AIC), 10
  Burr–Brown analog fixture, 10
  evaluation module (EVM), 10
Testing the software tools, 3
  develop executable program, 3
TI-tagged format, 5
Time critical applications, 8
Timers, 57
  global control register, 57
  timer counter register, 57
  timer period register, 57
Timer interrupt 0 (TINT0), 58
TMS320C30:
  expansion bus, 21
  functional block diagram, 19
  instruction cycle time, 2, 19
  instruction set, 340–345
  instructions in parallel, 19
  memory map, 21
  memory organization for convolution, 110
  parallel instructions, 344, 345
  pipelining levels, 18
TMS320C30-AIC communication
    routines, 68–72
  AICSET function, 72
  data receive register, 72
  data transmit register, 72
  interfacing diagram, 74
  timer period register, 72
TMS320C31, 18

Utility package from Hyperception, 65

VECSIR.ASM, 68
Video line rate analysis:
  bandwidth of a broadcast signal, 301
  circuit diagram, 301, 302
  hardware considerations, 304–306
  image processing algorithms, 306–309
    digital filtering, 306, 308
    edge enhancement, 306, 309
    image averaging, 306, 308, 309
  programs for, 390, 398–408
  television picture (frame), 302
  timing diagram, 304
  transfer rate, 305
  video information, 303
  video module parts list, 390
Volatile declaration, 44

Window function:
  Blackman, 108
  Hamming, 108
  Hanning, 107, 108
  Kaiser, 108, 109
  rectangular, 104, 106
    high sidelobes or oscillations, 107
  trade-offs, 107
Wireguided submersible:
  design and implementation, 318–320
  PC screen display of submersible
    course, 319

XDS1000 emulator, 88, 89
  AIC or AIB, 89
  interfacing between XDS1000, 89
  target system, 89

Z-transform, 91–96
  analysis of discrete-time signals, 91
  of exponential sequence, 92
  of sinusoidal sequence, 92
  mapping from s-plane to z-plane, 93, 94
    stable system, 94
    unstable system, 94
  to solve difference equation, 91, 95